Laboratory Investigations in

Molecular Biology

Steven A. Williams
Smith College and the University of Massachusetts

Barton E. Slatko
New England Biolabs, Inc.

John R. McCarrey
University of Texas at San Antonio

JONES AND BARTLETT PUBLISHERS
Sudbury, Massachusetts
BOSTON   TORONTO   LONDON   SINGAPORE

World Headquarters

Jones and Bartlett Publishers	Jones and Bartlett Publishers	Jones and Bartlett Publishers
40 Tall Pine Drive	Canada	International
Sudbury, MA 01776	6339 Ormindale Way	Barb House, Barb Mews
978-443-5000	Mississauga, Ontario L5V 1J2	London W6 7PA
info@jbpub.com	CANADA	UK
www.jbpub.com		

Jones and Bartlett's books and products are available through most bookstores and online booksellers. To contact Jones and Bartlett Publishers directly, call 800-832-0034, fax 978-443-8000, or visit our website www.jbpub.com.

Substantial discounts on bulk quantities of Jones and Bartlett's publications are available to corporations, professional associations, and other qualified organizations. For details and specific discount information, contact the special sales department at Jones and Bartlett via the above contact information or send an email to specialsales@jbpub.com.

Production Credits

Chief Executive Officer: Clayton Jones
Chief Operating Officer: Don W. Jones, Jr.
President, Higher Education and Professional Publishing: Robert W. Holland, Jr.
V.P., Design and Production: Anne Spencer
V.P., Sales and Marketing: William Kane
V.P., Manufacturing and Inventory Control: Therese Connell
Acquisitions Editor, Science: Cathleen Sether
Managing Editor, Science: Dean W. DeChambeau
Editorial Assitant: Molly Steinbach
Senior Production Editor: Louis C. Bruno, Jr.
Marketing Manager: Andrea DeFronzo
Text Design: Anne Spencer
Cover Design: Kristin Ohlin
Illustrations: Elizabeth Morales
Composition: Group 360, Inc./NK Graphics
Printing and Binding: Courier Kendallville
Cover Printing: Courier Kendallville
Cover Photo: © Scott Bowlin/ShutterStock, Inc.

Library of Congress Cataloging-in-Publication Data
Williams, Steven A., 1951–
  Laboratory investigations in molecular biology / Steven A. Williams, Barton E. Slatko, John R. McCarrey.
     p. ; cm.
  Includes bibliographical references.
  ISBN-13: 978-0-7637-3329-2 (alk. paper)
  1. Diagnosis, Laboratory. 2. Molecular biology—Technique. 3. Biomolecules—Analysis. 4. DNA probes.
5. Gene expression. I. McCarrey, John R. II. Slatko, Barton E. III. Title.
  [DNLM: 1. DNA Probes—Laboratory Manuals. 2. Laboratory Techniques and Procedures—Laboratory Manuals. 3. Gene Expression Profiling—methods—Laboratory Manuals. 4. Genomic Library—Laboratory Manuals. 5. Mice—genetics—Laboratory Manuals. 6. Sequence Analysis, DNA—methods—Laboratory Manuals. QU 25 W727L 2006]
  RB37.W552 2006
  616.07'5—dc22
                                                                                2006010786

Printed in the United States of America
10 09 08 07 06   10 9 8 7 6 5 4 3 2 1

DEDICATION

We dedicate this book to our families who have endured our long absences during the 20 years of our teaching the NEB/MB summer courses, upon which this endeavor is based. In particular, we acknowledge Janet and Kalani Williams, Betty and Rebeckah Slatko, Lynne and Elissa Dorsky, and Scott and Sean McCarrey. We also dedicate this book to the inspiration provided by our Ph.D. advisors Richard Snow, Yuichiro Hiraizumi, and Ursula Abbott. We thank our parents, James and Gail Williams, Flora and Leon Slatko, and Stanley and Mary McCarrey for the unwaivering support and encouragement they provided throughout our lives and careers.

CONTENTS

The projects described in this manual are the outgrowth of over 20 years of successful instruction of more than 3000 participants in the New England Biolabs Molecular Biology Summer Workshops held annually at Smith College, Northampton, MA. The experiments described represent protocols that the authors have repeatedly tested and refined since 1986 to reflect some of the most efficient and reliable methods in molecular biology. We feel that these experiments provide a unique opportunity for a realistic "hands on" laboratory experience that greatly enhances the students' learning, regardless of their level of beginning expertise. Participants in the New England Biolabs Molecular Biology Summer Workshops ranged from experienced molecular biologists to individuals with no molecular biology experience (and many with relatively little experience in laboratory science). Those enrolled have included undergraduate students, graduate students, technicians, postdoctoral fellows, high school teachers, M.D. and Ph.D. principal investigators, scientific illustrators, venture capitalists, and, yes, even a few lawyers.

Regardless of their education and research background, participants have found our program to be a unique learning experience that is valuable both for those that do bench research and those that just want to know how molecular biology experiments are done. We believe that this book reflects the philosophy of the course: that learning science is best done hands-on doing real experiments. All of the laboratory periods described in this book, therefore, are parts of larger projects studying two different mouse genes (organized as Projects I–IV). It is this overriding philosophy that has guided us in writing and organizing this book.

The unique organization of this volume provides students and instructors with significant flexibility. The four projects are designed to be modular and, yet, are provided as an integrated set of experiments to achieve a longer-term research goal. This enables a student to initiate and work on a long-term project through many related lab periods or to focus on short-term experiments to learn specific techniques. An instructor, therefore, can pick and choose which modules he or she wishes to utilize and organize them into a semester or year-long set of projects encompassing many molecular biology techniques. The availability of reagent kits that correspond to each module makes this flexible organization especially easy for instructors to implement. We strongly recommend using these kits in conjunction with the laboratory experiments described in this volume. The reagents supplied with these kits are optimized for use with the protocols described in this volume and are pretested and quality controlled. This will maximize the likelihood of successful completion of the experiments and minimize the preparation effort required for each laboratory instructor. For more information on these reagent kits, visit www.bioscience.jbpub.com or www.americanbio.com.

In addition to kits that correspond to each module, the instructors and the publisher intend to maintain a message board for scientists who use this book in the research laboratory or in the classroom and to communicate with one another and with the authors (www.bioscience.jbpub.com). Suggestions for revisions to protocols, new protocols, and improved explanations will be contributed by the authors and (hopefully) by the users of this laboratory manual. In this way, we intend to make this book a dynamic document that proves useful for many years to come.

Acknowledgments

We wish to acknowledge Dr. Don Comb of New England Biolabs, Inc. for his support of science education in general and for his support of the New England Biolabs Molecular Biology Summer Workshops that formed the foundation for this book in particular. We also acknowledge the efforts of numerous laboratory instructors and guest lecturers, as well as the more than 3000 students who participated in the summer workshops during the past 20 years. These laboratory instructors and students aided in developing, testing, and refining the protocols described in this manual. We acknowledge Dr. Wen Li and Wayne Gagnon for their technical expertise with the reagent materials and methods needed for the kits. A special thanks to Michelle Lizotte-Waniewski, Lori Saunders, and Sandra Laney for their many years of ideas and dedication to the summer workshops. We thank Courtney Smock, Dick Fish, and Eric Parham for the preparation of the photographs that were used in this text. Finally, we appreciate the patience shown by the editors and associates of Jones and Bartlett during the production of this project.

S. Williams
B. Slatko
J. McCarrey

The Three Tenureds. Left to right: John R. McCarrey, Steven A. Williams, and Barton E. Slatko.

Steven A. Williams, Ph.D.

Gates Professor of Biological Sciences, Smith College, Northampton, MA; Professor of Molecular and Cellular Biology, University of Massachusetts; Director, New England Biolabs Molecular Biology Summer Workshops

Dr. Williams received his Ph.D. degree in Genetics from the University of California, Davis in 1982. His doctoral dissertation research focused on gene expression in the yeast *Saccharomyces cerevisiae*. He later did research in the Laboratory of Parasitic Diseases at the National Institutes of Health, where he worked on the molecular biology of parasitic nematodes that cause elephantiasis and river blindness. Dr. Williams has been at Smith College and the University of Massachusetts since 1983. In 1986, he founded (with Dr. Barton Slatko) the New England Biolabs Molecular Biology Summer Workshops. These two-week workshops are the longest running and largest molecular biology courses for professionals in the world. In 1994, Dr. Williams was named the director of the Filarial Genome Project by the World Health Organization. In 1995, the Clark Foundation named Dr. Williams the director of the River Blindness Genome Project. In 2005, Dr. Williams was named Director of the new Center for Molecular Biosciences at Smith College. Dr. Williams' research accomplishments and interests include the discovery and characterization of the first repetitive DNA sequences in several species of parasitic nematodes, the development of sensitive PCR diagnostic tests for these parasites, and the sequencing of over 5000 filarial parasite genes. He has also conducted research to apply the genome sequence data of filarial parasites to the development of new drug targets and vaccines for filariaisis and river blindness. His current research interests include the aforementioned and efforts to develop RNA interference for functional studies of parasitic nematode genes and to elucidate mechanisms of gene regulation in these important parasites that infect 300 million people worldwide.

Barton E. Slatko, Ph.D.

Director of Applications and Product Development, and Director, DNA Sequencing Group, Molecular Parasitology Division, New England Biolabs, Inc., Ipswich, MA

Dr. Slatko received his Ph.D. degree in Zoology Genetics from the University of Texas, Austin in 1977. His doctoral dissertation research was conducted in the laboratory of Dr. Yuichiro Hiraizumi and was focused on the genetics and cytogenetics of the male recombination/mutator strains of *Drosphila melanogaster*, now known to be the result of transposable P elements and their associated cytotypes. He then did a postdoctoral fellowship in the laboratory of Dr. Mel Green at the University of California, Davis. Dr. Slatko was an Assistant Professor at Williams College (Williamstown, MA) before joining New England Biolabs, Inc. (NEB) in 1983. Dr. Slatko was part of the restriction endonuclease group and molecularly cloned the first thermophilic restriction endonuclease and cognate methylase (*Taq*I), before forming the DNA sequencing core facility at NEB. Since that time, he has been part of the Parasitology Division involved in the Filarial Genome Project, the goal of which is eradicating filarial diseases from human populations worldwide. Under his direction, his laboratory recently completed and published the DNA sequence and annotation of an obligate endosymbiont (*Wolbachia*) from a filarial parasite, a potential therapeutic and preventative drug target. He has served on both NSF and NIH grant review panels and is a science editor for *BioTechniques*. While maintaining his research laboratory, he now directs the new product application and development group for NEB. He is actively involved in science education and has been involved with the NEB Molecular Biology Courses, with the other co-authors, since 1986.

John R. McCarrey, Ph.D.

Professor of Cell and Molecular Biology, University of Texas at San Antonio; Professor, Departments of Cellular and Structural Biology and Obstetrics and Gynecology, University of Texas Health Science Center at San Antonio; Scientist, Department of Comparative Medicine, Southwest Foundation for Biomedical Research, San Antonio; Affiliate Scientist, Southwest National Primate Research Center, San Antonio; Director, San Antonio Institute for Cellular and Molecular Primatology

Dr. McCarrey received his Ph.D. degree in Genetics from the University of California, Davis in 1981. His doctoral dissertation research was conducted in the laboratory of Dr. Ursula Abbott and was focused on sex determination and germ cell development in the chick embryo. He then did a postdoctoral fellowship in the laboratory of Dr. Susumu Ohno at the City of Hope Research Institute in Duarte, California, where he worked on the molecular biology of the mammalian sex determination process. Dr. McCarrey's first faculty position was in the Division of Reproductive Biology in the Johns Hopkins University School of Hygiene and Public Health. He then moved to the Southwest Foundation for Biomedical Research in San Antonio, Texas. In 2001, he assumed his current position as Professor of Cell and Molecular Biology at the University of Texas at San Antonio. Dr. McCarrey's research accomplishments and interests include the discovery of the first functional retroposon in the human genome (the *Pgk2* gene), elucidation of mechanisms that regulate tissue-specific transcription in mammalian spermatogenic cells, studying the phenomenon of meiotic sex chromosome inactivation during spermatogenesis in mammals, describing the kinetics of reprogramming of epigenetic mechanisms during mammalian gametogenesis, and current studies to investigate the status of epigenetic and genetic mechanisms in cloned mice and in mouse embryos produced by assisted reproductive technologies, as well as in embryonic stem cells derived from mice and from nonhuman primates.

INTRODUCTION

Technical advances in the manipulation and analysis of DNA and RNA have initiated a revolutionary change in our understanding of genomes, gene structure, and gene function. Recombinant DNA technology and the latest advances in genomics and transcriptomics have enabled rapid advances in molecular biology, genetics, biomedical research, agriculture, and many other areas. These advances have had a dramatic influence on fields as diverse as medicine, genetics, neurobiology, evolutionary biology, ecology, plant biology, parasitology, computational biology, and agricultural science.

Recombinant DNA cloning involves the formation of new, heritable genetic material by insertion of foreign DNA into an appropriate cloning vector. Cloning vectors (e.g., plasmids and bacteriophage) facilitate the incorporation and replication of genetic material into a host organism in which it does not naturally occur. The ability to replicate plasmid or bacteriophage DNA in host bacterial cells allows large quantities of specific segments of DNA to be isolated in a pure form. Subsequently, this DNA can be subjected to a variety of analyses, including restriction site mapping, DNA sequence analysis, microarray analysis, *in vitro* mutagenesis, and labeling for use as a DNA probe, etc. The polymerase chain reaction (PCR) often provides a shortcut through many of the previously mentioned steps and represents the most important advance in DNA technology in the past 20 years. Cloned DNA can also be used to achieve high-level expression of a given gene product in a specific cell type. This technology can be used to produce large quantities of a specific protein for use in research or for use as a drug for treating disease. Other examples include the production of transgenic animals and plants, gene therapy, "designer" drugs, and so on.

To conduct recombinant DNA cloning experiments, molecular biologists require methods to

1. Cut DNA molecules at specific sites (using restriction enzymes)

2. Join DNA molecules (using DNA ligase enzymes)

3. Transfer recombinant DNA into intact host cells (using transformation, transfection, infection, or electroporation)

4. Propagate the recombinant DNA in foreign host cells (by replication of the cloning vector in the host cells)

5. Detect and analyze the recombinant clones (using a variety of methods such as plaque lifts, Southern blots, DNA sequencing, etc.)

In this laboratory manual, we have provided a detailed procedure and schedule for each of four major projects. Each project involves a related collection of multiple experiments that recreate real projects commonly conducted in modern molecular biology research labs. These include isolating and analyzing genomic DNA (Project I), cloning and sequencing a specific gene (Project II), recovering RNA and analyzing the expression of a specific gene (Project III), and preparing and characterizing a cDNA library (Project IV). Each project is divided into modules that represent a related set of procedures. Individual reagent kits are available for each module to ensure that these experiments can be easily organized for use in college and university laboratory courses; professional courses; research laboratories; and, where appropriate, advanced high school courses. We strongly recommend the use of these kits to ensure properly prepared, quality-controlled reagents for each experiment. Required materials and reagents used in each module can be found at the end of each module. Complete recipes are found in Appendix II. Appendix IV contains several additional protocols to supplement the procedures described in the four main projects.

OVERVIEW AND OBJECTIVES

The projects described in this manual are the outgrowth of 20 years of successful instruction of over 3000 participants in 60 sessions of the New England Biolabs Molecular Biology Summer Workshops held annually at Smith College, Northampton, MA. The authors have been involved in the design and implementation of these experimental methods since the beginning of these courses in 1986, and the experiments described in this volume represent protocols that have been repeatedly tested and refined to reflect state-of-the-art methods in molecular biology. It is our feeling that these experiments provide a unique opportunity for a realistic "hands-on" laboratory experience that greatly enhances the student's learning experience, regardless of the level of expertise. Participants in the New England Biolabs Workshops ranged from experienced molecular biologists to individuals with no molecular biology experience, including undergraduate students, graduate students, technicians, postdoctoral fellows, high school teachers, MD and PhD researchers, artists, and yes, even lawyers.

This volume is organized in a different fashion from most other laboratory books in that the four main projects are designed to be modular and yet are provided as a continual integrated set of experiments to achieve a longer term research goal. This enables a student to initiate and work on a long-term project through many related lab periods (or an entire semester) or to focus on short-term experiments to learn specific techniques. Thus, an instructor can pick and choose which modules he or she wishes to use and organize them into a semester or year-long set of projects encompassing many molecular biology techniques. The availability of kits that correspond to each module makes this variable organization especially easy for instructors to implement.

Four major projects are described in this book, with each project composed of a series of modules linked together in a logical fashion to create a coherent laboratory project. **Project I** involves the isolation and characterization of genomic DNA from mouse liver. Included in this project are three modules that enable the student to experience DNA isolation, PCR amplification, determination of DNA quantity and quality, agarose gel electrophoresis, Southern blot hybridization, two different nonradioactive methods for labeling and detecting DNA, and analysis of methylation patterns in genomic DNA.

PROJECT I. GENOMIC DNA ISOLATION AND ANALYSIS

Isolate and purify genomic mouse DNA from liver tissue.

Measure the quantity and quality of the isolated mouse DNA.

Amplify the mouse *Ttr* gene using the PCR from genomic DNA.

Nonradioactive 3' end labeling of a *Ttr* DNA oligonucleotide probe.

Southern blot analysis of the *Ttr* gene from PCR-amplified mouse DNA.

Nonradioactive labeling of a probe for the mouse reverse transcriptase (*Rvt*) gene using PCR.

Southern blot analysis of the *Rvt* gene from genomic mouse DNA and analysis DNA methylation patterns in the mouse genome.

Project II includes five modules describing the construction of a mouse genomic library in an *E. coli* bacteriophage vector, the identification and isolation of a particular gene from the library, subcloning the gene from the bacteriophage DNA into a plasmid vector, purification of the plasmid DNA, PCR amplification of the gene for confirmation of its size, automated fluorescent DNA sequencing of the gene, and bioinformatic analysis of the gene sequence.

PROJECT II. GENOMIC CLONING, DNA SEQUENCING, AND BIOINFORMATICS

Construction of a mouse genomic library in a lambda bacteriophage vector.

Nonradioactive labeling of a mouse reverse transcriptase (*Rvt*) gene using PCR.

Screening of the mouse genomic library using the *Rvt* nonradioactive DNA probe.

PCR subcloning the *Rvt* gene from the lambda vector to a plasmid vector.

Isolation of plasmid DNA from colonies containing recombinant plasmids with the *Rvt* gene.

DNA sequencing of PCR products containing the cloned *Rvt* gene.

Bioinformatic analysis of the DNA sequence data.

Project III includes three modules describing the isolation of total and poly A+ messenger RNA from mouse liver, the analysis of gene expression using Northern blots, and the reverse transcriptase-polymerase chain reaction (RT-PCR) to analyze gene expression.

PROJECT III. GENE EXPRESSION ANALYSIS: RNA ISOLATION, NORTHERN BLOT, AND RT-PCR

Preparation of total and poly A+ (messenger) RNA from mouse liver tissue.

Northern blot analysis to analyze *Ttr* gene expression in mouse liver and spleen tissues.

Amplification of *Ttr* mRNA by RT-PCR to study gene expression.

Project IV includes two modules describing the construction and characterization of a directional cDNA library from mouse liver poly A+ RNA (mRNA) and screening of this library for the presence of a particular expressed gene (the *Ttr* gene).

PROJECT IV. cDNA CLONING AND cDNA LIBRARY ANALYSIS

Synthesis of double-stranded cDNA from mouse liver poly A+ mRNA.

Construction of a unidirectional cDNA library in a lambda expression vector.

Analysis of library quality and average insert size in the cDNA library by PCR.

Detection of the *Ttr* cDNA in the library by PCR.

The objective of this book and the corresponding reagent kits is to provide students, scientists, and even nonscientists with a well-designed introduction to basic and advanced experiments in gene manipulation and gene expression. This hands-on laboratory experience will promote an understanding of fundamental concepts and methods in molecular biology and an appreciation of how these concepts and methods are applied to real research problems and/or are used in teaching and business. Completion of these projects will also provide an enhanced background and knowledge base that will allow students to read and better understand the vast research literature that is based on methods of molecular biology, including the unique terminology and concepts that pertain to this field.

A number of pathways can be used to organize a laboratory course using the previously mentioned projects and modules. Instructors can refer to Figures I-1 to I-4 for further advice on organizing a course using these various modules.

PRACTICAL MATTERS AND ADVICE

Laboratory Organization

Each module in each project is further partitioned into individual, 3-hour lab periods that can be combined in different ways to prepare a laboratory course. How this is done will depend on (1) the interests of the instructor, (2) the amount of time available each week for laboratory work, and (3) the length of the semester or number of laboratory periods that are available to the instructor. To assist in beginning to think about how to organize such a course, examples are provided of several tracks that could be followed for part of a semester, an entire semester, or for a year-long course (refer to the Project Flowcharts on the next 6 pages). Reagent kits (Appendix I) can be purchased for all of the modules in each project. Any reagent not defined in the text of a module is defined in the Materials and Reagents list at the end of the module. We strongly recommend the use of these reagent kits to avoid the significant labor required to prepare the individual reagents and to insure that quality-controlled reagents are used for each experiment, which will, in turn, insure the likely success of each experiment.

Notes on Terminology Used in This Book

Scientists utilize a great deal of jargon in their experiments and in their daily discussions in the laboratory. We have been careful in this manual to define our terms and minimize jargon as much as possible; however, to describe a "nanofuge" as a "small, plastic, bench-top centrifuge capable of low speed spins" would be extremely awkward. Some jargon, therefore, is necessary and useful. There are also terms that scientists use interchangeably to describe certain procedures, reagents, and equipment in the laboratory. For example, both *microcentrifuge* and *microfuge* refer to the same type of bench-top centrifuge capable of spinning 1.5- to 2.0-ml microfuge tubes at up to 19,000 rpm. A list of some of the terms follows. This is not meant to be complete but simply is a list of terms we feel the reader needs to know before reading the protocols.

Microcentrifuge = microfuge (see above)

Nanofuge = picofuge = TOMY centrifuge (see above)

Speed vacuum concentrator = speed vac = spin vac

5x agarose gel loading dye = 5x Blue Juice = 5x BJ

Ethanol = EtOH

Ethidium bromide = EtBr

Double-distilled, sterilized water = ddH20

Chisam = 24:1 (v/v) chloroform/isoamylalcohol

Membranes used for Northern and Southern blots and plaque lifts are often referred to a "filters" (i.e., a plaque lift filter is the same thing as a plaque lift membrane)

Micropipettors = pipettes = pipets = Pipetman (Gilson)

In our courses we use Gilson Pipetman, and these are referred to as P10 (for measuring up to 10 microliters), P20 (for measuring up to 20 microliters), P100 (for measuring up to 100 microliters), etc.

The commonly used DNA cloning vector, bacteriophage lambda, is often referred to as just "lambda" or with the Greek letter λ.

Working in Pairs

We strongly recommend that beginners in molecular biology work in pairs during this laboratory course. Not only does this save on the cost of reagents, but more importantly, we find that students working in pairs make far fewer mistakes. Students should be instructed to work together on all aspects of an experiment and not to divide up tasks so as to finish the laboratory period more quickly. Speed should not be the primary goal!

We also recommend that each student or pair be given a "student number" so that they can label and track their microcentrifuge tubes and reagents as the course progresses. In the text of the experiments, when participants are instructed to write their number on a tube, this refers to their individual student number or their group number (for group numbers, the two individual student numbers are combined [e.g., 5/6 or 15/16]). Numbers are useful because it is often difficult to write names on the tubes and initials may not always be unique. We have also indicated in the text how to label each tube; this ensures that the students can always find the appropriate stored tube later in the course.

Storing Reagents

Finally, we recommend that each student or pair be provided individual plastic boxes for storing their reagents in microcentrifuge tubes at −20°C and at 4°C. When storage at −80°C is noted in the text, this can be done in an ultracold freezer holding a temperature anywhere within the range of −70°C to −80°C.

Laboratory Safety

It is strongly recommended that students wear lab coats and closed-toed shoes at all times. When dealing with hazardous reagents, gloves made of disposable latex (or an appropriate alternative) and safety glasses should also be worn whenever noted in the text (the laboratory instructor should require these). There is no use of flame in any of these projects, although a few protocols require the use of flame-drawn Pasteur pipettes. We recommend that the instructor or teaching staff prepare these flame-drawn Pasteur pipettes ahead of time. All of the safety rules of the institution at which the course is taught should be strictly adhered to, as should the institution's rules for the proper disposal of biohazardous materials and chemicals. It is the instructor's responsibility to familiarize himself or herself with all relevant institutional rules, regulations, and policies and to be sure that these rules are followed by students and staff alike. Protocols in which hazardous materials are used are clearly identified in the text (e.g., any protocol with the use of phenol, chloroform, or ethidium bromide). As with any protocol, it is always advisable to err on the side of safety. Good luck and happy cloning!

GENERAL LAB SETUP

Appendix II describes general materials (and reagents) required for each project and also provides photographs of selected specialized equipment.

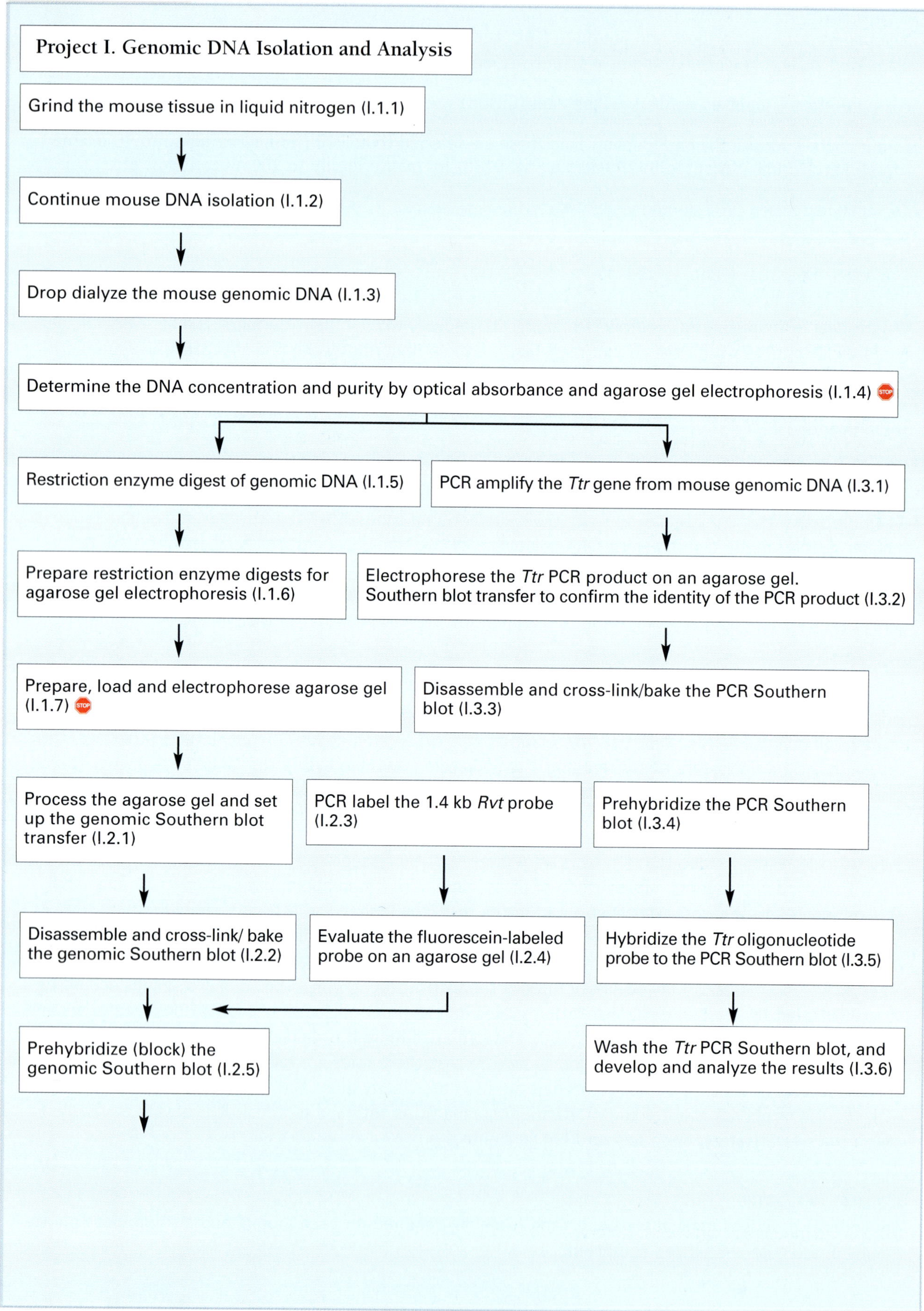

Project I. Genomic DNA Isolation and Analysis
Grind the mouse tissue in liquid nitrogen (I.1.1)
Continue mouse DNA isolation (I.1.2)
Drop dialyze the mouse genomic DNA (I.1.3)
Determine the DNA concentration and purity by optical absorbance and agarose gel electrophoresis (I.1.4)
Restriction enzyme digest of genomic DNA (I.1.5)
PCR amplify the Ttr gene from mouse genomic DNA (I.3.1)
Prepare restriction enzyme digests for agarose gel electrophoresis (I.1.6)
Electrophorese the Ttr PCR product on an agarose gel. Southern blot transfer to confirm the identity of the PCR product (I.3.2)
Prepare, load and electrophorese agarose gel (I.1.7)
Disassemble and cross-link/bake the PCR Southern blot (I.3.3)
Process the agarose gel and set up the genomic Southern blot transfer (I.2.1)
PCR label the 1.4 kb Rvt probe (I.2.3)
Prehybridize the PCR Southern blot (I.3.4)
Disassemble and cross-link/ bake the genomic Southern blot (I.2.2)
Evaluate the fluorescein-labeled probe on an agarose gel (I.2.4)
Hybridize the Ttr oligonucleotide probe to the PCR Southern blot (I.3.5)
Prehybridize (block) the genomic Southern blot (I.2.5)
Wash the Ttr PCR Southern blot, and develop and analyze the results (I.3.6)

Hybridize the *Rvt* repeat probe to the genomic Southern blot (I.2.6)

↓

Wash the genomic Southern blot, and detect the labeled probe (I.2.7)

Notes: 🛑 Potential stopping- and starting-point flexibility.

1. For mouse DNA preparation only, stop after I.1.4.
2. For mouse DNA preparation and restriction enzyme digestion only, stop after I.1.7.
3. For demonstration of Southern blots only, genomic DNA can be provided and the experiment can be initiated at I.1.5.
4. Both Southern blots can be performed simultaneously (on the same agarose gel). The gel can be cut in half and each half can be processed individually. Note that the postgel processing and detection methods differ!
5. To demonstrate PCR of a single-copy gene from genomic DNA, genomic DNA can be provided and the experiment can begin at I.3.1. The Southern blot can be used to demonstrate the identity of the product, although its size on an agarose gel may be used as an indication (but not proof) of its identity.

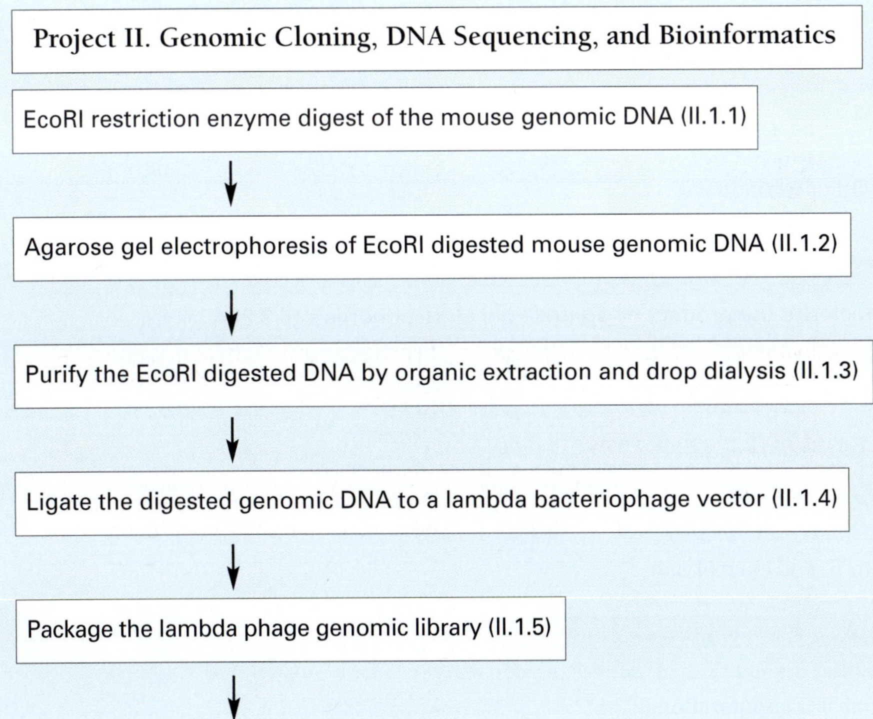

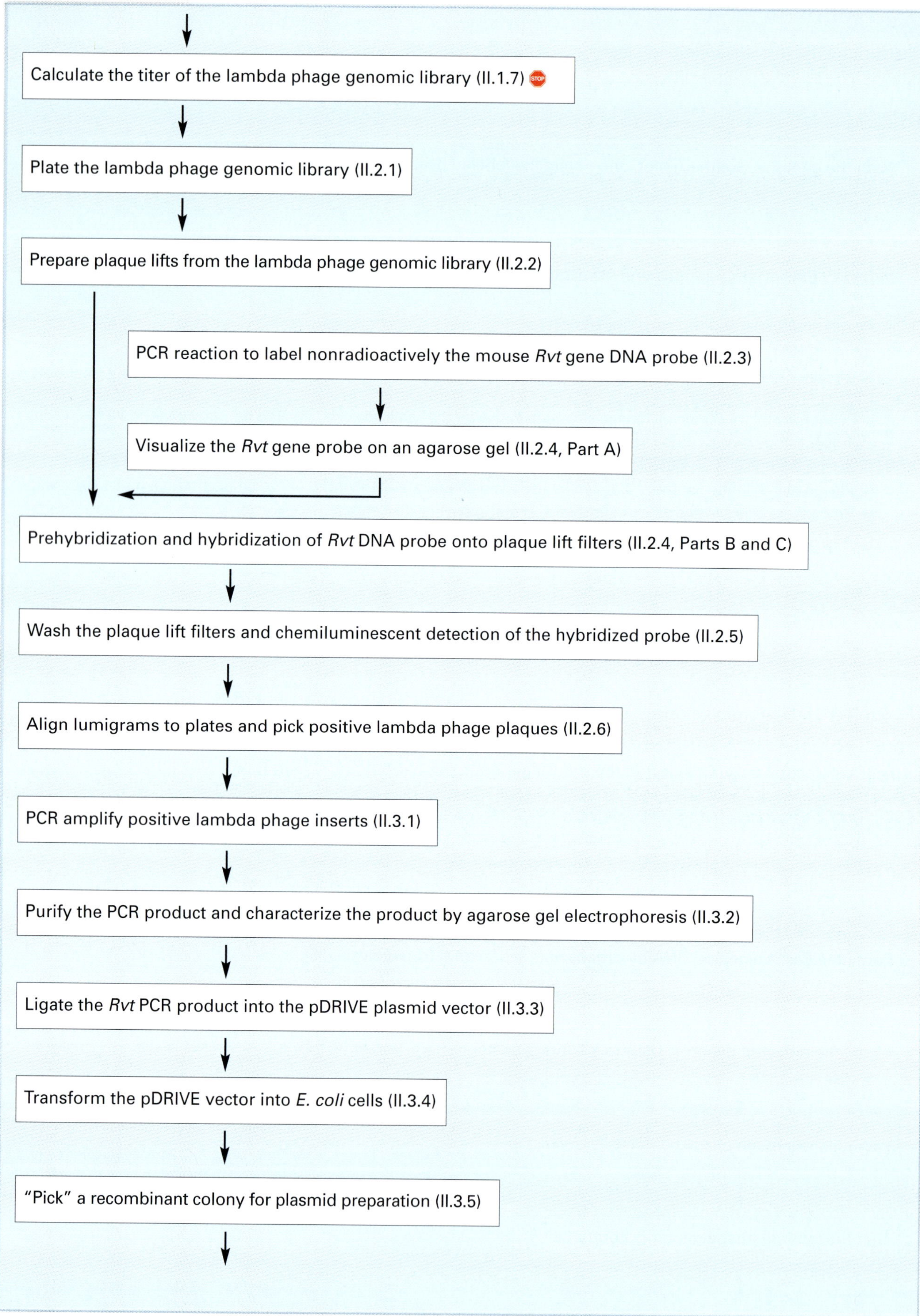

Calculate the titer of the lambda phage genomic library (II.1.7)
Plate the lambda phage genomic library (II.2.1)
Prepare plaque lifts from the lambda phage genomic library (II.2.2)
PCR reaction to label nonradioactively the mouse Rvt gene DNA probe (II.2.3)
Visualize the Rvt gene probe on an agarose gel (II.2.4, Part A)
Prehybridization and hybridization of Rvt DNA probe onto plaque lift filters (II.2.4, Parts B and C)
Wash the plaque lift filters and chemiluminescent detection of the hybridized probe (II.2.5)
Align lumigrams to plates and pick positive lambda phage plaques (II.2.6)
PCR amplify positive lambda phage inserts (II.3.1)
Purify the PCR product and characterize the product by agarose gel electrophoresis (II.3.2)
Ligate the Rvt PCR product into the pDRIVE plasmid vector (II.3.3)
Transform the pDRIVE vector into E. coli cells (II.3.4)
"Pick" a recombinant colony for plasmid preparation (II.3.5)

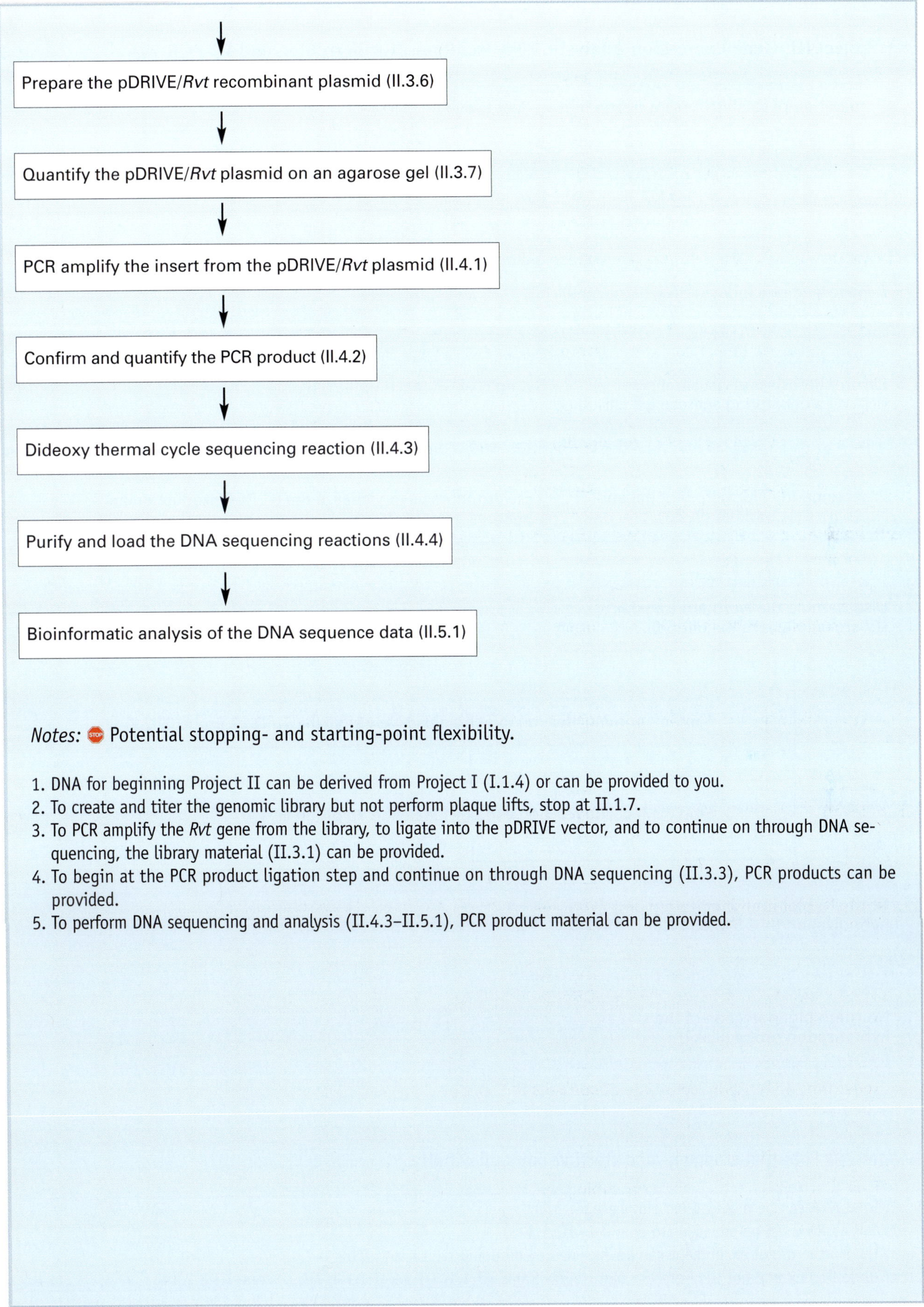

Notes: 🛑 Potential stopping- and starting-point flexibility.

1. DNA for beginning Project II can be derived from Project I (I.1.4) or can be provided to you.
2. To create and titer the genomic library but not perform plaque lifts, stop at II.1.7.
3. To PCR amplify the *Rvt* gene from the library, to ligate into the pDRIVE vector, and to continue on through DNA sequencing, the library material (II.3.1) can be provided.
4. To begin at the PCR product ligation step and continue on through DNA sequencing (II.3.3), PCR products can be provided.
5. To perform DNA sequencing and analysis (II.4.3–II.5.1), PCR product material can be provided.

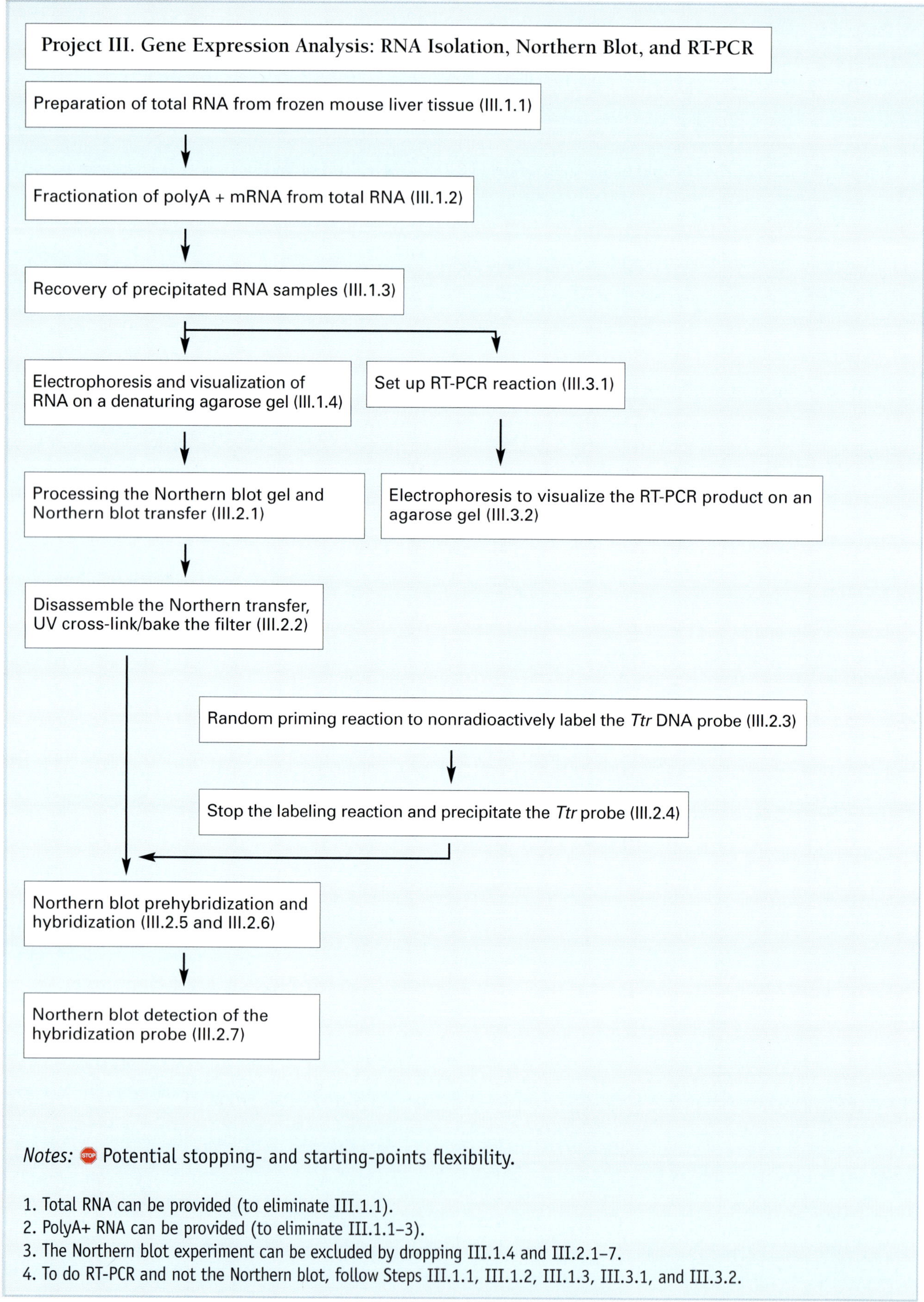

Notes: 🛑 Potential stopping- and starting-points flexibility.

1. Total RNA can be provided (to eliminate III.1.1).
2. PolyA+ RNA can be provided (to eliminate III.1.1–3).
3. The Northern blot experiment can be excluded by dropping III.1.4 and III.2.1–7.
4. To do RT-PCR and not the Northern blot, follow Steps III.1.1, III.1.2, III.1.3, III.3.1, and II.3.2.

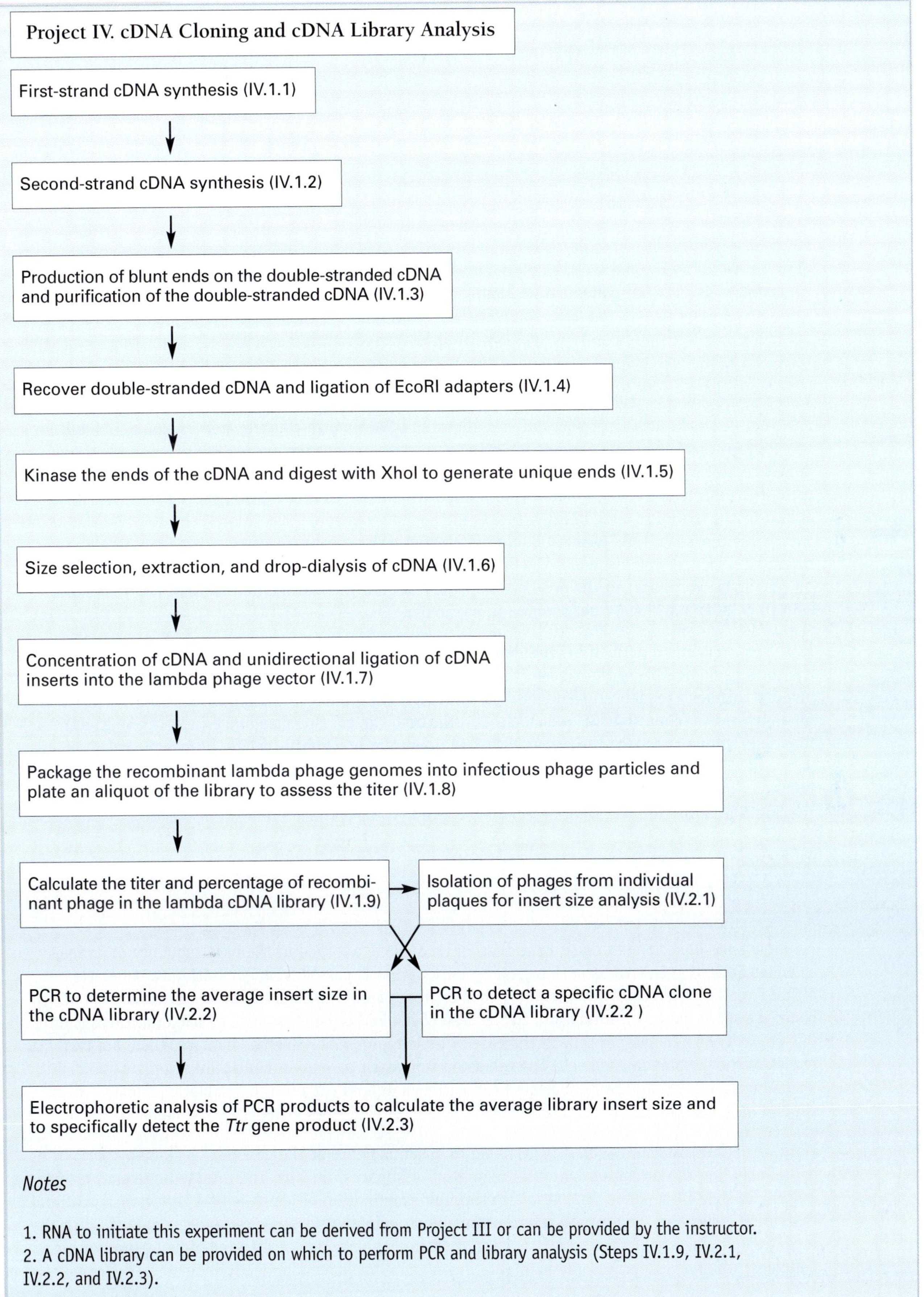

Notes

1. RNA to initiate this experiment can be derived from Project III or can be provided by the instructor.
2. A cDNA library can be provided on which to perform PCR and library analysis (Steps IV.1.9, IV.2.1, IV.2.2, and IV.2.3).

I   GENOMIC DNA ISOLATION AND ANALYSIS

PROJECT SUMMARY

In the series of experiments described in **Module I.1**, you will isolate and purify mouse genomic DNA and use two different methods to determine the concentration and purity of your sample. As an initial characterization of your genomic DNA sample, you will cleave it with the restriction endonucleases AluI, HpaII, MspI, EcoRI, and SfiI. You will then electrophorese the cleaved genomic DNA on an agarose gel to see how restriction endonucleases with different recognition specificities (four base: "four cutters," six base: "six cutters," and eight-base: "eight cutters") digest genomic DNA.

In the series of experiments described in **Module I.2**, you will use your purified mouse genomic DNA in a second genome analysis experiment. The digested DNA samples you prepared in Module I.1 will be transferred from the gel to a membrane using the Southern blot technique. You will then perform a Southern blot hybridization experiment in which you will produce a fluorescein-labeled *Rvt* gene probe using the polymerase chain reaction (PCR). The *Rvt* gene encodes the protein reverse transcriptase and is part of the highly repeated LINE-1 repeat in the mouse genome. The fluorescein-dUTP-labeled probe will be hybridized to the Southern blot and detected using a chemiluminescent reaction with an anti-fluorescein monoclonal antibody. The data obtained will be used to investigate the organization of the 1.4-kb mouse *Rvt* repeat in the genome and to investigate the methylation state of mouse genomic DNA in general and the *Rvt* gene region in particular.

In the experiments described in **Module I.3**, you will detect a different, specific gene sequence within the total genomic DNA of the mouse (the *Ttr* gene, which encodes the protein transthyretin). You will amplify this gene sequence using PCR. You will then electrophorese the products of these PCR reactions on an agarose gel and transfer the amplified DNA from the gel to a membrane (another Southern transfer). This will demonstrate the power of PCR to dramatically amplify specific DNA sequences from even very complex samples of DNA. To verify that you have amplified the correct PCR product, you will use a nonradioactively end-labeled *Ttr* oligonucleotide probe (labeled with biotin) and hybridize this probe to the Southern blot. You will then detect the *Ttr* probe using a chemiluminescent reaction.

PROJECT BACKGROUND

An organism's genetic material is contained in its genome encoded in the base sequence of double-stranded DNA. It is in this form that genetic information is passed from generation to generation, either from a single cell to two daughter cells or from parents to their offspring. A genome is one complete set of genetic information. In the case of haploid organisms, such as bacteria, there is only one copy of the genome per cell. In the case of diploid organisms, such as humans, there are two copies of the genome in each cell, one inherited from each parent. In multicellular organisms such as mammals, essentially every nucleated cell carries a diploid copy of the complete genome — even though each cell will use (or express) only a fraction of the genes encoded in the genome.

Many situations exist in which it is desirable to examine genes at the genomic level. Many parts of a gene, including the promoter and other regulatory sequences, introns, and intergenic sequences, can only be studied in genomic DNA. It is also through examination of genomic DNA that the structure of individual genes and the physical relationship among neighboring genes can be determined. With the recent emphasis on genome studies (studies of entire genomes rather than individual genes within genomes), the need to access and manipulate genomic DNA has dramatically increased. The complete sequencing of genomes has resulted in extremely valuable sequence databases from hundreds of species. This, in turn, has facilitated comparative analyses of genomes that have revealed a tremendous amount of information, including data on gene organization and information on how different species have evolved.

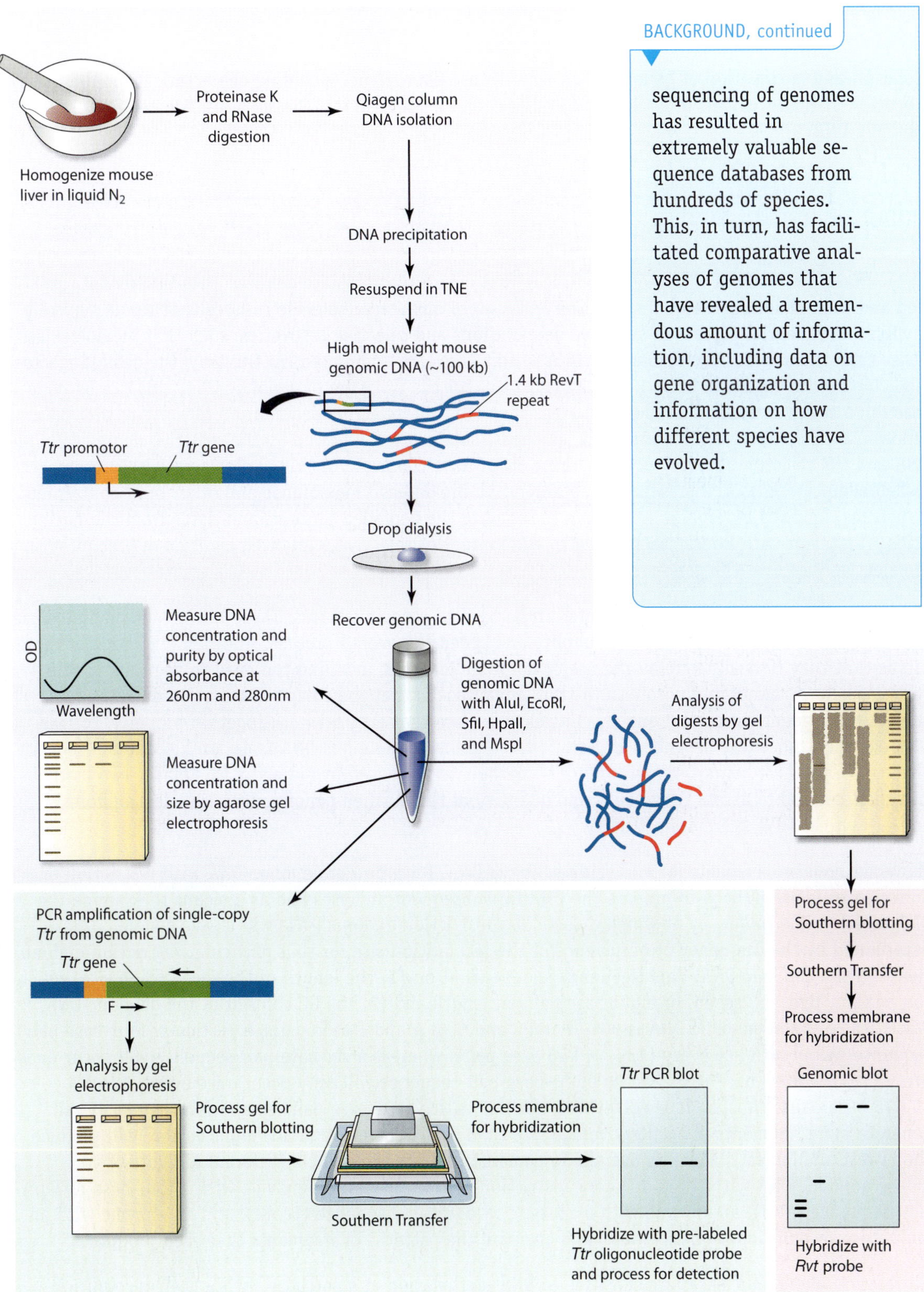

Project I flow diagram.

MODULE I.1   Mouse Genomic DNA Isolation and Analysis

MODULE SUMMARY

Isolation and purification of genomic DNA from mouse liver tissue, including characterization and quantification of the recovered DNA. Restriction digestion and gel electrophoresis of undigested and digested genomic DNA.

MODULE BACKGROUND

To facilitate the analysis of genomic DNA, it is first necessary to isolate genomic DNA in a useful (purified and high molecular weight) state. The isolation of any macromolecule from cells or tissue typically involves a sequential process of selective degradations and selective recoveries, such that biomolecules other than the desired macromolecule (DNA in this case) are eliminated. In this way, the desired macromolecule is recovered in a pure, nondegraded state. For the isolation of genomic DNA, this can be done through a series of defined biochemical steps involving the selective degradation of proteins and lipids followed by the selective recovery of nucleic acids. This is often followed by the selective degradation of RNA and the selective recovery of DNA. These steps have classically included digestion with proteinases to degrade proteins in the presence of detergents to break down membranes and other lipid-containing structures, followed by organic extraction to separate the degraded proteins and lipids from the nucleic acids. Treatment with RNase can be used to eliminate RNA selectively followed by ethanol precipitation to selectively recover the genomic DNA.

Alternatively, commercially available kits can be used to accomplish this. These kits often combine some of the degradation/recovery steps to simplify and speed the overall process. Another advantage of these kits is that they typically employ the use of nontoxic reagents, avoiding the need to use caustic substances such as phenol or chloroform. A third advantage is that the companies that produce the kits subject the reagents to careful quality control to be sure that they all work together properly. For the DNA isolation procedure described in Module I.1, the Qiagen Genomic DNA Isolation kit will be used. An alternative protocol, using a classic "nonkit" method, is presented in Appendix IV. Finally, two methods will then be used to assess the quality and quantity of the purified genomic DNA–gel electrophoresis and ultraviolet (UV) spectrophotometry.

Many methods are available for analyzing genomic DNA. Digestion of genomic DNA with restriction endonucleases is useful both to prepare the DNA for subsequent analyses and as a means to examine the DNA directly. The frequency at which a particular restriction endonuclease will cleave genomic DNA is determined by the frequency of occurrence of the recognition site for that particular restriction endonuclease in the genomic DNA. This frequency is dependent on (1) the length of the recognition site, which can be four, five, six, seven, or eight base pairs in length and (2) the GC content of the genomic DNA. Statistically, if the genomic DNA is 50% G and C and 50% A and T base pairs, a particular four-base pair sequence will occur once every $(1/4)^4 = 256$ base pairs, whereas a six-base pair sequence will occur once every $(1/4)^6 = 4096$ base pairs and an eight-base pair sequence will only occur once every $(1/4)^8 = 65,536$ base pairs. Thus, a "four cutter" (an enzyme with a four-base pair recognition sequence) will generate more fragments of smaller average size than a "six cutter," and an "eight cutter" will produce the fewest and largest fragments overall. To demonstrate this, we will cut different aliquots of your mouse genomic DNA with three different restriction endonucleases: an enzyme with a four-base recognition sequence (AluI), an enzyme with a six-base recognition sequence (EcoRI), and an enzyme with an eight-base recognition sequence (SfiI) and compare the patterns of fragments that each produce.

All restriction endonucleases are sensitive to DNA methylation at some base within their recognition site. If the DNA is methylated at this site, the restriction endonuclease is unable to cleave the DNA (this is how the bacteria that produce these enzymes normally protect their own DNA from digestion). Unlike bacterial DNA, mammalian DNA is only methylated at some cytosine residues that are located within a 5'-C-p-G-3' (CpG) dinucleotide. Overall, DNA from mammalian somatic cells is generally methylated at a majority of CpG dinucleotides. To demonstrate the effect of DNA methylation on digestion by restriction endonucleases, we will compare the patterns of fragments produced by digestion with two different restriction enzymes (HpaII and MspI) that recognize the same four-base pair sequence (5' CCGG 3'). HpaII is methylation sensitive, whereas MspI is insensitive to CpG methylation.

The agarose gel can also be used to initiate a Southern blot analysis to look at the genomic organization of the *Rvt* gene, part of a multicopy "repeat" element (LINE-1) in the mouse genome. This is done as described in Module I.2.

LAB PERIOD I.1.1.   GRIND THE MOUSE TISSUE IN LIQUID NITROGEN

EACH PAIR of students will be given about 350 mg of frozen mouse liver tissue from which genomic DNA will be isolated. These tissue samples were prepared before this experiment by dissecting a liver from a freshly euthanized mouse, cutting the tissue into uniform fragments, and then snap freezing the fragments in liquid nitrogen. The frozen fragments have been stored at −80°C before use in this experiment. Because you will be grinding the mouse tissue using a mortar and pestle with liquid nitrogen, both partners will wear lab coats and safety glasses for protection from the liquid nitrogen. The materials needed for this procedure are shown here.

Step 1   Begin by having one partner add about 10 ml of liquid nitrogen to the mortar.

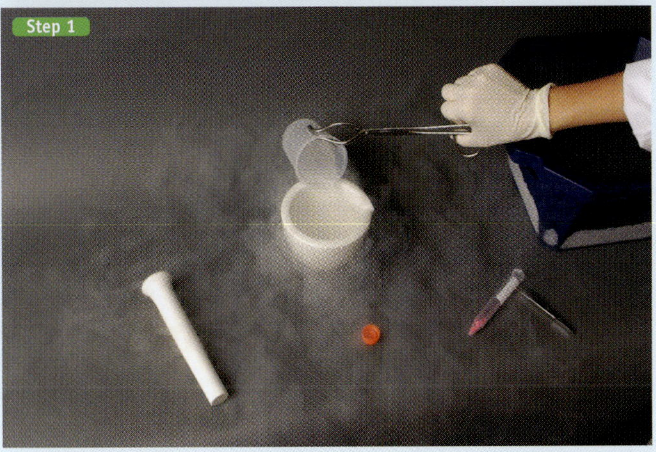

LAB PERIOD I.1.1 CONT.

Step 2   Use precooled forceps to add frozen fragments of mouse liver tissue that total about 350 mg to the mortar.

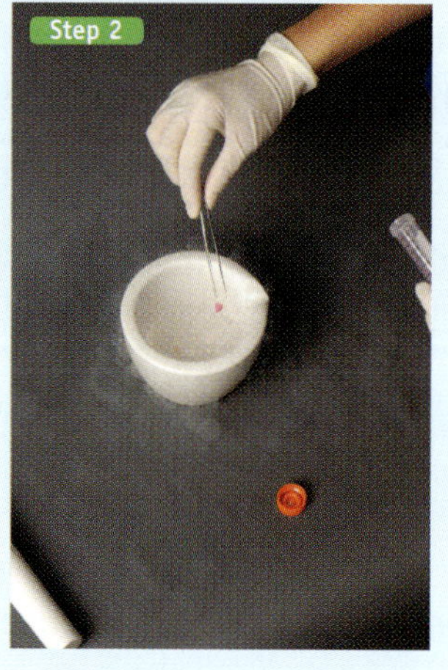

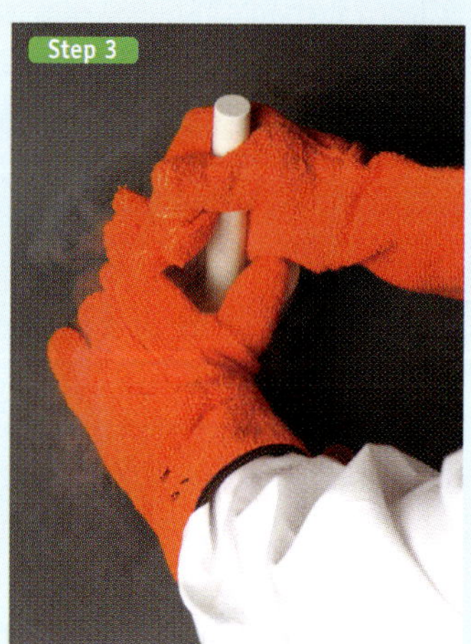

Step 3   One partner will grind the tissue into a fine powder.

Be sure to work quickly but carefully. As the liquid nitrogen evaporates, the other partner will add small aliquots of liquid nitrogen. Be sure to not let the liquid nitrogen evaporate completely. If it does, the liver tissue sample will warm and will then bond to the bottom of the mortar. It will then be very difficult to remove. Continue grinding and adding liquid nitrogen until the tissue is completely pulverized.

CAUTION!

Wear Safety Glasses!

Step 4   Quickly pour the liquid nitrogen and ground tissue from the mortar into a 50-ml conical tube. The sample will "bump" when it contacts the 50-ml polypropylene tube; thus, be careful not to look into the tube or point the tube at anyone. Set the tube aside with the cap off for 5 to 10 minutes so that the liquid nitrogen can safely evaporate.

Step 5   After the liquid nitrogen has completely evaporated and the pellet is dry, add 20 ml of Qiagen buffer G2, cap, and gently vortex the tube just long enough to dislodge the pellet from the bottom of the tube (approximately 5 seconds). The buffer G2 contains proteinase K to degrade protein, detergents to break down lipid membranes, and RNase to degrade the RNA.

Step 6   Twist the cap securely onto the 50-ml tube, and then wrap a piece of parafilm around the cap to prevent leakage. Tape your tube to a rocker platform and incubate overnight (12 to 24 hours) at 50°C with gentle rocking.

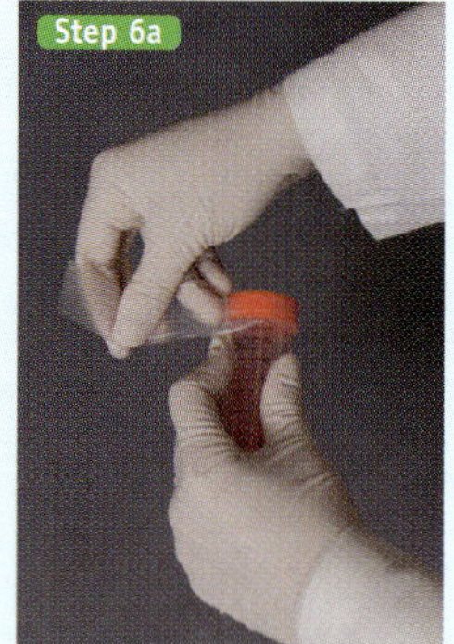

LAB PERIOD I.1.2.   CONTINUE MOUSE DNA ISOLATION

Step 1 After several hours of digestion, the tissue should be converted into a "sludge-like" state. Pour the digested mouse tissue from the 50-ml conical tube into a 50-ml round-bottom tube and label with your group number(s).

Step 2 Centrifuge the tube containing the digested mouse sample at 5000x *g* for 10 minutes to sediment any particulate matter before loading the clear sample onto the column.

Step 3 Set up the Qiagen Genomic-tip DNA column over a 50-ml plastic centrifuge tube. Equilibrate the column by adding 10 ml of Qiagen buffer QBT to the column. Allow the solution to flow through by gravity alone. The Qiagen Genomic-tip column is an anion exchange column. Here, the negatively charged DNA will bind to the positively charged diethylaminoethyl (DEAE) resin in the column under one set of conditions and will be released from the column under another set of conditions. More detailed information on these anion exchange columns is available from the manufacturer.

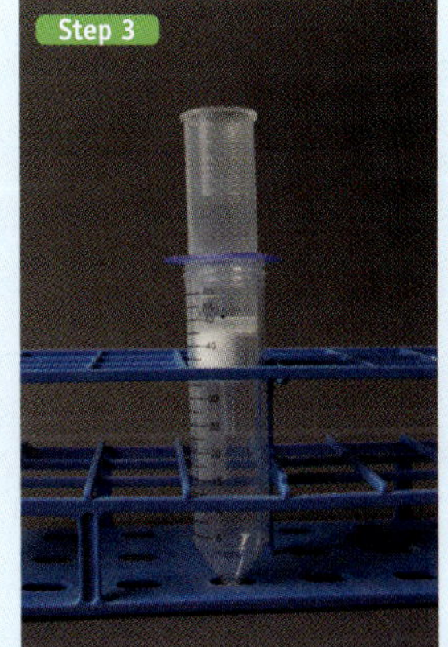

Step 4 After the centrifugation step (Step 2), carefully transfer the supernatant to a new 50-ml conical tube without disturbing the particulate matter at the bottom of the tube (use a 10-ml glass or plastic pipette). Cap the tube and vortex at maximum speed for exactly 10 seconds.

PROCEDURAL NOTE

Vortexing genomic DNA for short periods of time has minimal effect on the size of the DNA. It accelerates the Qiagen procedure by eliminating poor flow rates associated with clogging of the column. The average size of the DNA is reduced by only 10 kb when vortexed for up to 20 seconds (you will vortex for only 10 seconds).

Step 5 Immediately pour the mouse DNA sample from the new 50-ml conical tube onto the equilibrated Genomic-tip column, and allow the liquid to flow through the column and into the beaker.

PROCEDURAL NOTE

Particularly concentrated genomic DNA lysates may exhibit diminished flow rates due to increased viscosity. Flow can be assisted by the application of gentle positive pressure using a disposable syringe and an adaptor (a rubber stopper with a hole). When using positive pressure, do not allow the flow rate to exceed 20 to 40 drops per minute. It may also be helpful to dilute the lysate with an equal volume of Buffer QBT prior to loading on the column.

Step 6 Wash the Genomic-tip column twice with 15 ml Qiagen buffer QC, allowing the full volume to flow through completely into a waste beaker. The wash contains degraded macromolecules and other contaminants, whereas your DNA remains in the column. Buffer QC contains 1 M NaCl at pH 7.0; under these conditions, the DNA remains bound to the resin in the column.

Step 7 Transfer the Genomic-tip column so that it sits over a new 15-ml glass Corex tube, and elute the genomic DNA with 7 ml of 50°C Qiagen buffer QF. Buffer QF contains 1.25 M NaCl at pH 8.5; under these conditions, the DNA is released from the resin and is eluted from the column.

»»

LAB PERIOD I.1.2 CONT.

Step 8   To the glass Corex tube containing the purified mouse DNA, add 5 ml (0.7 volumes) isopropanol, and seal the top of the tube with parafilm. While holding the parafilm in place with your thumb, invert 10 to 15 times to precipitate the DNA. The DNA should immediately precipitate. This is very cool and is one of the few times you will actually see your DNA!

Step 9   Label the Corex tube with your name or group number and centrifuge at 5000x *g* for 30 minutes.

Step 10   Carefully pour off the isopropanol supernatant.

Step 11   Wash the DNA pellet by adding 2 ml ice-cold 70% ethanol. Invert the tube gently, trying not to dislodge the pellet. This step washes excess salt and other contaminants from your DNA pellet.

Step 12   Centrifuge the tubes at 5000x *g* for 5 minutes. This step insures that your pellet is stuck securely to the bottom of the tube.

Step 13   Carefully pour off the ethanol supernatant. The pellet may not stick to the tube as well after the wash, so take care not to pour away your pellet!

Step 14   Keep the tube inverted and place on a Kimwipe to air dry for 30 minutes.

Step 15   Add 1 ml TNE buffer to your DNA sample, and wrap the cap of the tube with parafilm. Do not mix or vortex the tube!

Step 16   Resuspend the samples at 40°C overnight by taping the tubes to a roller drum kept in an incubator. If no roller drum is available, tape the tubes to a rocking platform. If no appropriate incubator is available, this step can be done at room temperature.

PROCEDURAL NOTE

After the overnight incubation at 40°C to initiate the resuspension, the roller drum holding the samples should be removed from the 40°C incubator and placed at room temperature. This can be performed by the instructor if scheduling does not permit the students to be present. Continue rotating the samples at room temperature for 3 to 7 additional days. For scheduling purposes, resuspension can be continued for up to 7 days. The longer the time, the more completely the DNA is resuspended. If resuspension is allowed to proceed for longer than 7 days, it is recommended that the additional days be done at 4°C. These resuspension times are long, but in order to get high molecular weight DNA into solution, this process cannot be rushed. If high molecular weight DNA is not needed, the resuspension process can be aided by vortexing and pipetting the sample up and down using a P1000 pipette. Please note that yields of DNA from this procedure are high enough that three days of resuspension is a minimum; complete resuspension may not be achieved after 3 days. If you have some unresuspended DNA in your sample, this will be removed during the centrifugation in lab period I.1.3, Step 4 below.

LAB PERIOD I.1.3.   DROP DIALYZE THE MOUSE GENOMIC DNA

YOUR MOUSE DNA sample has been gently rotating at room temperature for several days now and should be completely resuspended in the TNE buffer. The next step will be a dialysis procedure to remove impurities from the solution of resuspended DNA.

Each pair of students will need two Petri dishes, two 250-ml beakers, and two Millipore membranes to drop dialyze their mouse DNA. Although not a commonly used procedure, drop dialysis is extremely effective at removing low molecular contaminants that can inhibit subsequent reactions (especially restriction digests and ligation reactions).

Step 1 Pour 200 ml 0.1x TE buffer into each of two clean 250-ml beakers. Place the beakers in a protected region of your bench where they are unlikely to be disturbed.

Step 2 Float a 45-mm diameter Millipore dialysis membrane filter *shiny side up* on the buffer surface in each beaker. Use two sets of forceps to hold the filter, and gently lower it onto the surface of the buffer in each beaker. Allow the filters to prewet for about 5 minutes before adding your sample.

PROCEDURAL NOTE

Use gloves and forceps while handling the membranes! Never touch these membranes with your bare fingers. You will transfer oil from your fingers to the membrane. This will prevent proper dialysis and may cause your sample to flow from the membrane into the buffer (not a desirable result!).

Step 3 While your filters are wetting, retrieve your 15-ml conical tube with your redissolved mouse genomic DNA from the roller drum.

Step 4 Centrifuge your tube for 1 minute at 1600x *g* at 4°C in the tabletop centrifuge to bring all of the contents to the bottom of the tube.

Step 5 Carefully cut off the end of a blue P1000 pipette tip using scissors (do not use a razor blade!). Cut off the bottom few millimeters of the tip.

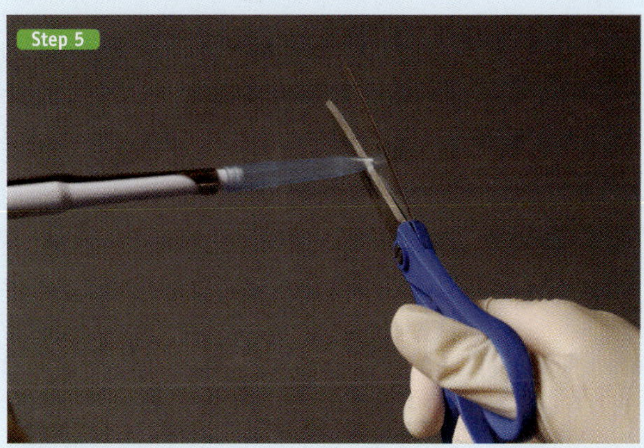

»»»

LAB PERIOD I.1.3 CONT.

Step 6 Using the cut-off blue pipette tip, *carefully* pipette half of your DNA sample (about 500 µl) onto the center of one dialysis filter. Repeat this procedure to pipette the other half of your DNA sample onto the other dialysis filter.

PROCEDURAL NOTE

Do not load more than 500 µl per filter because the volume of the sample may initially increase during dialysis.

Step 7 Cover each beaker with the lid of a Petri dish, and *do not disturb* the beakers. Do not attempt to move the beakers after the mouse DNA has been pipetted onto the membranes. Allow the dialysis to proceed overnight.

PROCEDURAL NOTE

Overnight dialysis is preferable (12–20 hours). If scheduling requires a shorter dialysis time, 6 hours is the recommended minimum time. If scheduling requires a longer dialysis time, a maximum period of 2 days of dialysis can be used.

LAB PERIOD I.1.4. **DETERMINE THE DNA CONCENTRATION AND PURITY BY OPTICAL ABSORBANCE AND AGAROSE GEL ELECTROPHORESIS**

Step 1 To recover your mouse DNA from dialysis, use a P1000 pipette to *carefully* retrieve as much of the DNA sample as possible from each filter and transfer to one new microfuge tube. Be sure the pipette tip is held at a 90° angle to the filter or you will push the filter around the surface of the buffer. *Do not* use a cut-off tip for this step (cut tips are not smooth and are not as effective for removing your sample from the filter). Although you are not using a cut-off tip, the amount of shearing induced by a single pipetting step is minimal. You will likely recover about 300 to 500 µl per filter. Cap the tube, and label it with your number and "mouse genomic DNA." Keep this tube on ice as you proceed with the following steps.

Step 2 Estimate the total volume of your sample by filling a second microcentrifuge tube with a similar amount of water and measuring this amount with a P1000 pipette.

PROCEDURAL NOTE

Record your total volume. You should not pipette your sample up and down to estimate the volume because this will cause additional shearing of your DNA sample. Remember, one goal of this experiment is to isolate very-high-molecular-weight genomic DNA.

»»

LAB PERIOD I.1.4 CONT.

Step 3 This is your "mouse genomic DNA" sample. This tube should be kept on ice or in the refrigerator at 4°C. You are now ready to determine the quantity and purity of your mouse genomic DNA sample.

Two methods can be used to assess the quality and quantity of purified genomic DNA: gel electrophoresis and UV spectrophotometry. Each of these methods provides different but overlapping information about the quality and quantity of genomic DNA. Gel electrophoresis will be used to estimate the quantity of DNA and to confirm that the isolated DNA is initially of high molecular weight and free of contaminating RNA. UV spectrophotometry will be used to provide a precise estimate of the quantity of recovered DNA and to confirm that the DNA is free of other chemical contaminants including organic solvents (Table 1-1).

Table 1-1: Characterization of Genomic DNA

Method	Quantity	Molecular Weight	Presence of RNA	Presence of Contaminants[2]
		Characteristic[1]		
Gel[3]	Approximate	Yes	Yes	No
Spec[4]	Precise[5]	No	No	Yes

[1]Characteristics of genomic DNA to be examined.
[2]Including protein, phenol, and lipids.
[3]Analysis by aragose gel electrophoresis.
[4]Analysis by UV spectrophotometry.
[5]Estimate of DNA quantity with the spectrophotometer is accurate only if the DNA sample is very pure. Contamination of the sample with protein, lipid, phenol, RNA or free nucleotides will cause an overestimation of your DNA concentration because all of these molecules contribute to absorbance at 260 nm (see Figure 1-2).

PART A | Optical Absorbance

The concentration and purity of DNA can be determined by reading the optical density (OD) of a sample at 260 and 280 nm. The reading at 260 nm allows calculation of the concentration of nucleic acid in the sample. For pure double-stranded DNA, 1 OD_{260} = 50 µg/ml (for a 1.0 cm path length cuvette). The ratio between the readings at 260 nm and 280 nm (OD_{260}/OD_{280}) provides an estimate of the purity of the nucleic acid. Pure preparations of DNA have an OD_{260}/OD_{280} value of 1.6 to 1.9. If there is contamination with significant amounts of protein, lipid, or phenol, the OD_{260}/OD_{280} will be less than 1.6. Values of greater than 1.9 indicate possible contamination with RNA.

The most informative estimate of purity of a nucleic acid preparation is obtained from a continuous scan of UV absorbance from 220 to 320 nm. In this experiment, you will obtain individual readings of the OD of your DNA sample at 260 and 280 nm and run a continuous optical scan of your sample.

Step 1 You will use an aliquot of your "mouse genomic DNA" sample that has been stored on ice or at 4°C for this procedure. To be sure that your DNA is fully resuspended, pipette it up and down several times using a P1000 and a cut-off blue pipette tip. Check with an instructor if it appears that any of the DNA is not resuspended.

PROCEDURAL NOTE

If it is imperative to have the highest molecular weight DNA possible, omit this step.

»»

LAB PERIOD I.1.4 CONT.

Step 2 Using a P200, remove 50 µl of your mouse DNA to a new microfuge tube and add 950 µl 0.1x TE. Label the tube with your group number and "Spec Mouse DNA." Mix this tube thoroughly by vortexing (you do not care whether you shear this DNA sample because you will discard it after the scan).

Step 3 Transfer the 1-ml sample into a quartz cuvette (this DNA will only be used for the spectrophotometer reading).

CAUTION!

Quartz cuvettes are expensive!

PROCEDURAL NOTE

Quartz cuvettes are necessary; plastic or regular glass cuvettes will not transmit light accurately at 260 and 280 nm.

Step 4 Read the OD at 260 nm and 280 nm as described by the instructor. Calculate the concentration and the 260/280 ratio of your diluted DNA sample. Remember that your OD was measured on a 1:20 dilution of your stock "mouse genomic DNA," and thus, the concentration of your stock DNA is 20 times that of your diluted sample. You will use this estimate of your mouse DNA concentration to set up restriction digests and other manipulations in other modules.

Step 5 Perform a continuous optical scan of your sample from 220 to 320 nm wavelength. If your spectrophotometer does not do automatics scans, take readings every 5 nm from 220 to 320 nm. If your spectrophotometer does not print out the scan, plot your results using graph paper. Plot absorbance on the Y-axis and wavelength on the X-axis. An example of a continuous scan of a pure sample of DNA is shown in **Figure 1-1**. Examples of mouse DNA samples that have various impurities are shown in **Figures 1-2a** to **1-2d**.

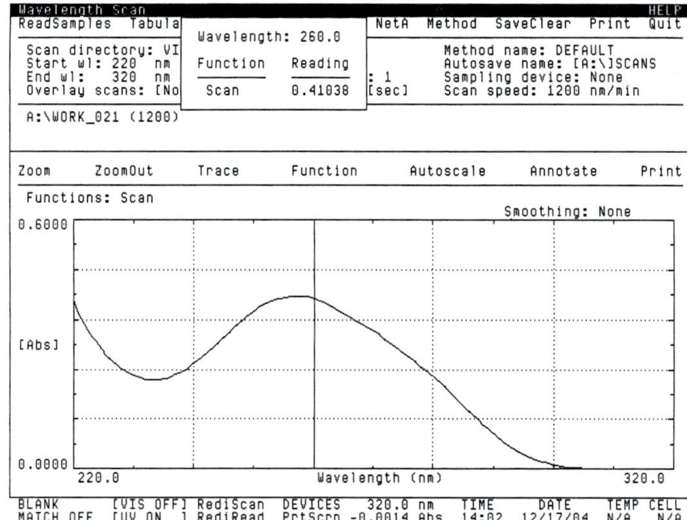

Fig 1-1 A continuous scan of pure DNA.

LAB PERIOD I.1.4 CONT.

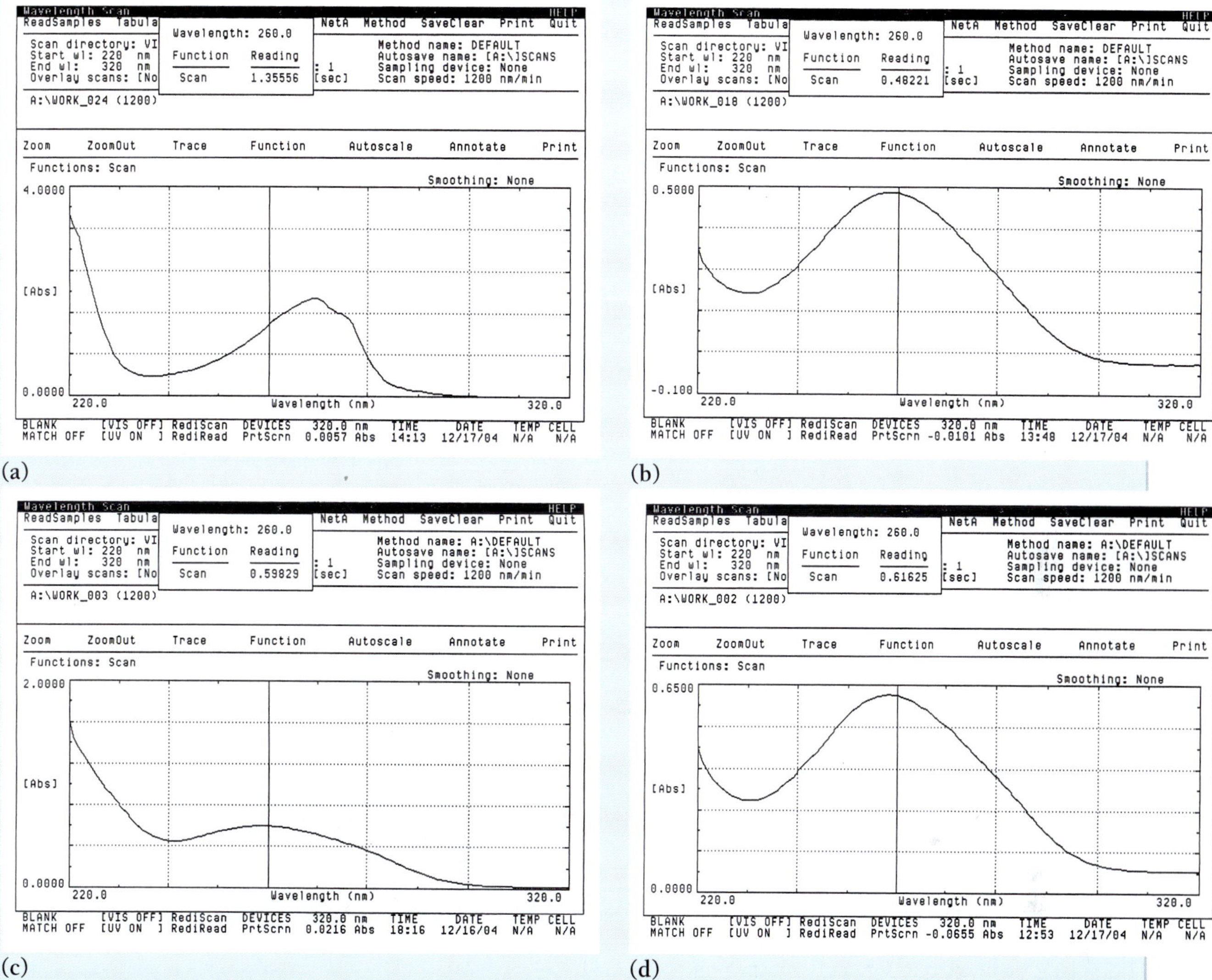

Fig 1-2 Spectrophotometer output of mouse DNA with various impurities. (a) Trace phenol. (b) 50% RNA. (c) BSA. (d) Lipid.

PART B Agarose Gel Electrophoresis

Another valuable analysis of your mouse genomic DNA sample can be obtained by electrophoresing your DNA on an agarose gel. This can provide a second assessment of DNA concentration and can also give you information regarding the molecular weight of your isolated DNA (the gel will tell you whether you have succeeded in isolating high molecular weight, nondegraded DNA). A gel can also reveal any contaminating RNA remaining in your DNA preparation. In order to assess the size and quantity of your isolated DNA samples accurately, it is very important to run "marker" DNA of known size and quantity. These markers usually consist of multiple sized molecules so that you can match the size of your DNA to the size of a particular marker band. It is often necessary to use several dilutions of the unknown DNA (in this case your purified mouse DNA) because there is a minimum and maximum amount of DNA that can be loaded on the gel for proper resolution and detection — generally between 1 ng and 1 µg for minigels. In order to aid in loading the gels and to

»»

help track the migration of the DNA during electrophoresis, a dye (called "Blue Juice") is added to the samples that migrates in the same direction as the DNA. The Blue Juice also contains Ficoll; this gives the sample greater density than the gel running buffer and enables the sample to "sink" into the wells of the gel during gel loading. Your instructor will demonstrate this in the laboratory.

Step 1 Prepare a 1% agarose gel by combining 1 g agarose and 100 ml 1x Tris-acetate buffer (TAE) in a 250-ml flask.

Step 2 Melt the agarose in a microwave oven (high power for about 1 minute) or on a hot plate using a stir bar. Cover the flask with aluminum foil (hot plate only . . . not for the microwave!), and heat until the solution just begins to boil. *Carefully* swirl the solution. Be sure that no solid agarose remains. If you can see unmelted agarose, heat the sample further and carefully swirl again.

CAUTION!

The agarose will be very hot; agarose solutions "bump" readily; thus, wear a protective glove and goggles.

Step 3 An example of the components of an agarose gel apparatus.

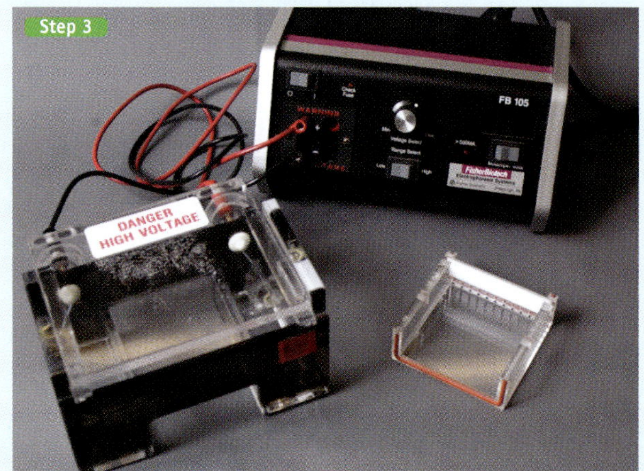

Step 4 Allow the agarose to cool to about 55°C to 65°C before pouring it. Pour approximately 35 ml (depending on your gel apparatus) of molten agarose into the gel tray. The level of the gel should not be more than half way up the teeth of the comb. Wait at least 20 minutes for the gel to cool and solidify.

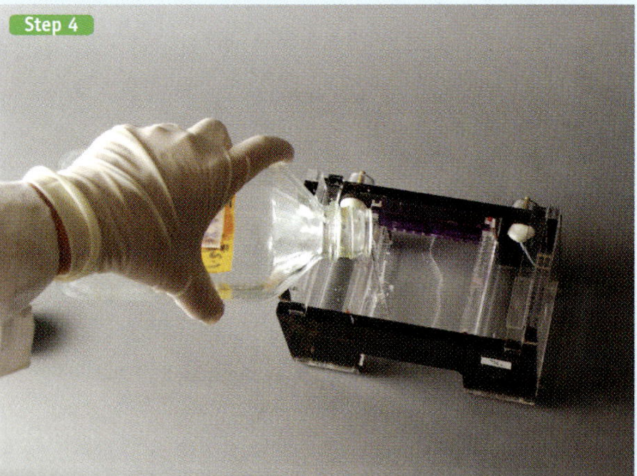

Step 5 Orient the gel tray in the gel rig with the comb closest to the negative terminal. *Carefully* remove the comb from the gel as demonstrated by the instructor.

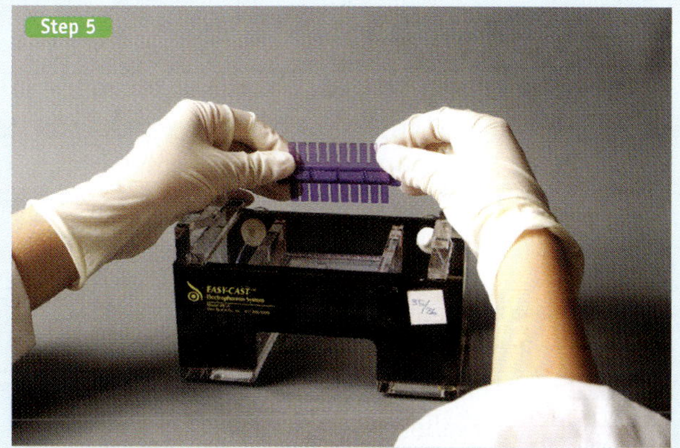

Step 6 *Carefully* pour gel running buffer (1x TAE) into both side chambers of your gel box until the gel is *completely* submerged. The thickest area of the gel is at the wells and should be covered by 1 or 2 mm of buffer.

PROCEDURAL NOTE

Gels may be stored in 1x TAE buffer overnight at room temperature or at 4°C.

Step 7 Remove 5 µl of your mouse genomic DNA sample, and place it into a new microcentrifuge tube labeled "mouse DNA gel." Remove a second 5-µl sample, and place it into a second microcentrifuge tube labeled "diluted mouse DNA." To this second tube, add 45 µl 0.1x TE, pH 8.0. Mix thoroughly by pipetting up and down. This tube now has a 1:10 dilution of your mouse genomic DNA.

Step 8 From the "diluted mouse DNA" tube, remove 5 µl of the diluted DNA to a new microfuge tube labeled "dil mouse gel." Discard the tube labeled "diluted mouse DNA" (it is no longer needed).

Step 9 Add 5 µl of Blue Juice loading dye (5x BJ) to your tubes that are labeled "mouse DNA gel" and "dil mouse gel." *Do not add 5x BJ to your total mouse DNA preparation!* Stir with your pipette tip to mix after each addition. Verify that you now have two microfuge tubes with 5x BJ in them called "mouse DNA gel" and "dil mouse gel." These are samples of (1) your undiluted mouse DNA and (2) a sample of your mouse DNA at a 1:10 dilution. Each sample is now blue because it contains the loading dye.

Step 10 You will use a marker DNA sample that is commonly used in many laboratories as a size reference: lambda (λ) DNA that has been cleaved with the restriction endonuclease HindIII. Digesting lambda DNA with HindIII produces eight fragments of DNA, each with different

»»

LAB PERIOD I.1.4 CONT.

molecular weights that run at different rates in the agarose gel (the smallest fragments will migrate the fastest). Your instructor has already diluted the marker DNA and added 5x BJ so that your marker DNA is ready to load on your gel (the marker should be at a final concentration of 25 ng/µl). To prepare the marker DNA for electrophoresis, remove 6 µl (150 ng) of the marker DNA from the tube marked "lambda HindIII," and place it into a new tube labeled "lambda marker." The fragments produced by digesting lambda DNA with HindIII are presented in Appendix III.

Step 11   Heat *only* your "lambda marker" DNA tube at 65°C for 5 minutes. Spin briefly in your nanofuge to bring all of the liquid to the bottom of the tube. Place the tube on ice until you are ready to load the sample on your gel. You now have three samples to load: your undiluted "mouse DNA gel," your mouse DNA diluted 1:10 ("dil mouse gel"), and your "lambda marker" DNA.

PROCEDURAL NOTE

Heating the lambda markers is required for all of the bands to run according to their true molecular weight. This is because two of the lambda DNA fragments in each marker can hybridize together via their "*cos*" ends. These *cos* ends are 12 nucleotide, single-stranded overhangs that can hybridize together. This causes two of the fragments to join together (hybridize) and electrophorese as a single band unless they are denatured by the heating step.

Step 12   Load all of each sample into three wells in your gel underneath the buffer. This may take some practice; obtain some extra blue juice dye from your instructor. Practice loading samples of 5x BJ in extra wells on your gel before attempting to load your real samples! Record in your lab book which wells you loaded with which samples for future reference (see below).

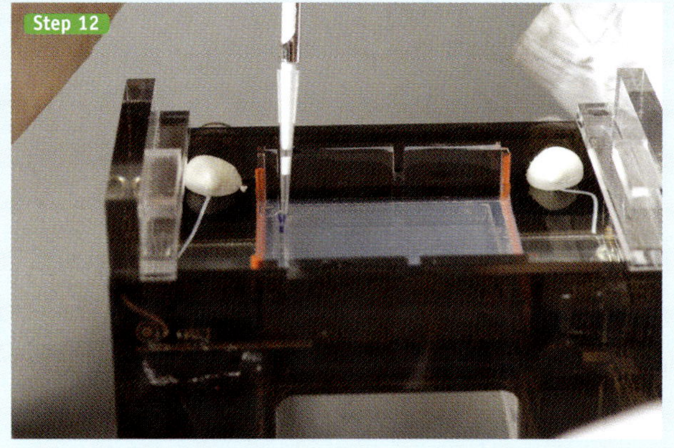

The recommended loading pattern is as follows:

Lane 1:   Practice

Lane 2:   Practice

Lane 3:   Practice

Lane 4:   "Lambda HindIII DNA" in BJ (150 ng)

Lane 5:   "Undiluted mouse DNA gel" in BJ

Lane 6:   "Diluted mouse DNA gel" in BJ

Lane 7:   Practice

Lane 8:   Practice

Lane 9:   Practice

Lane 10: Practice

LAB PERIOD I.1.4 CONT.

PROCEDURAL NOTE

Save your lambda HindIII marker and genomic mouse DNA tubes at 4°C. You will be using these again!

Wear gloves, lab coats, and safety glasses. Handle these gels with extreme care! They are slippery. Support the entire gel with the spatula to keep it from breaking.

Ethidium bromide is a known mutagen and carcinogen. Wear gloves, lab coats, and safety glasses! Handle gels carefully! Do not splash ethidium bromide! Always rinse the spatula in the water destain bath after it has been in contact with ethidium bromide! Always rinse the gel tray after sliding gels into the ethidium bromide bath. You may unknowingly contact the ethidium bromide! Be very careful not to drip the ethidium bromide anywhere!

Handle these gels with extreme care! Support the entire gel with a spatula to keep it from breaking.

Be careful not to cause unwanted contact with the ethidium bromide DNA gel stain by splashing or spreading this solution unnecessarily — always rinse the spatula and gel tray with water after each use.

Step 13 Connect the electrophoresis box to the power supply so that the DNA will migrate from the negative (–) anode to the positive (+) cathode. Otherwise, your DNA will migrate in the wrong direction (known affectionately as agarose gel retrophoresis)! Run the gel at 75 V for 2 hours.

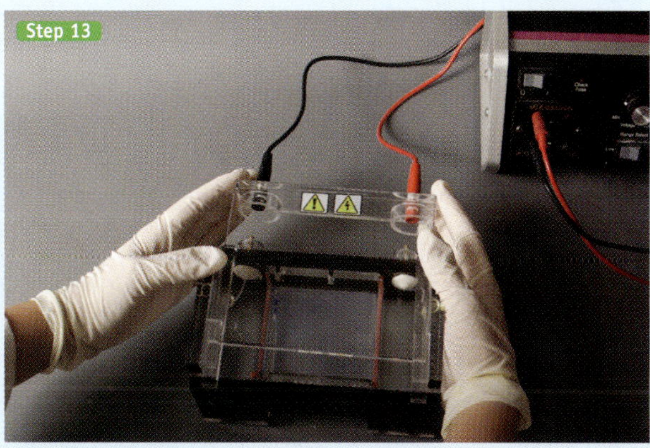

Step 14 After electrophoresis, turn off the power supply. Unplug the leads from the power supply, and carefully stain the gel for 8 to 10 minutes in an ethidium bromide bath using a spatula to transfer the gel from the electrophoresis box to the gel stain container. (The instructor will demonstrate this.) Using the spatula, transfer the gel to a water bath (to destain the gel) for a minimum of 15 minutes.

PROCEDURAL NOTE

Appendix IV describes an alternative DNA stain, SYBR Safe (Molecular Probes). SYBR Safe DNA gel stain has been shown to be much less toxic, mutagenic, and/or carcinogenic than the more commonly used ethidium bromide stain and therefore may be preferable for use in a laboratory course setting. Nevertheless, it is still advisable to use caution when using this reagent!!

Step 15 Place your gel on a UV transilluminator, and photograph the gel (or take a digital image with a gel documentation system). Your instructor will demonstrate this. After obtaining a high-quality photograph, discard the gel. An example of the expected pattern of the DNA fragments in this gel is shown in **Figure 1-3**.

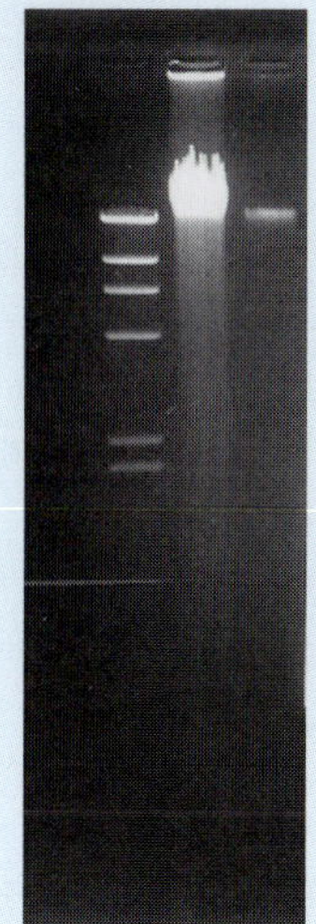

Fig 1-3 Expected pattern of DNA fragments in gel.

Step 16 You can estimate your mouse DNA concentration by comparing your mouse genomic DNA band (most likely from the diluted lane) to your marker DNA bands. Remember that the amount of fluorescence in each band is proportional to the amount of DNA in each band. Also remember that you can only get accurate quantitation from the gel only if you are comparing faint bands to faint bands.

»»»

LAB PERIOD I.1.4 CONT.

After the isolated genomic DNA has been (1) shown to be high molecular weight in the agarose gel, (2) shown to be free of contaminating RNA, protein, and other contaminants in the agarose gel and by spectrophotometry, and (3) quantified by spectrophotometry and electrophoresis, it is ready to be used for analysis as described in the following lab periods.

LAB PERIOD I.1.5.   RESTRICTION ENZYME DIGESTS OF GENOMIC DNA

PROCEDURAL NOTE

If you do not isolate your own mouse genomic DNA, your instructor will provide it in an appropriate concentration.

Step 1 Each group will set up five digests of their genomic mouse DNA with the restriction endonucleases AluI, EcoRI, SfiI, HpaII, and MspI. Each digest requires the addition of 1 µg of mouse genomic DNA. You prepared your own mouse genomic DNA and calculated the concentration as described previously. Calculate how many microliters of your mouse genomic DNA is necessary to give you 1 µg. In the digests below, this will be your volume "X." The amount of ddH$_2$O ("Y") that you add can then be calculated to bring the entire volume of the reaction to 50 µl. Each restriction enzyme requires addition of 10x buffer for optimal activity (see Appendix II).

AluI (AG'CT) Digest

X µl mouse DNA (1 µg)
5 µl 10x New England Biolabs (NEB) buffer 1
Y µl ddH$_2$O
3 µl AluI enzyme

———————
50 µl final volume
Incubate at 37°C, O/N

Uncut (Control)

X µl mouse DNA (1 µg)
5 µl 10x NEB buffer 2
Y µl ddH$_2$O
No enzyme

———————
50 µl final volume
Incubate at 37°C, O/N

EcoRI (G'AATTC) Digest

X µl mouse DNA (1 µg)
5 µl 10x EcoRI buffer
Y µl ddH$_2$O
3 µl EcoRI enzyme

———————
50 µl final volume
Incubate at 37°C, O/N

HpaII (C'CGG) Digest

X µl mouse DNA (1 µg)
5 µl 10x NEB buffer 1
Y µl ddH$_2$O
3 µl HpaII enzyme

———————
50 µl final volume
Incubate at 37°C, O/N

SfiI (GGCCNNNN'NGGCC) Digest

X µl mouse DNA (1 µg)
5 µl 10x NEB buffer 2
Y µl ddH$_2$O
3 µl SfiI enzyme

———————
50 µl final volume
Incubate at **50°C**, O/N

MspI (C'CGG) Digest

X µl mouse DNA (1 µg)
5 µl 10x NEB buffer 2
Y µl ddH$_2$O
3 µl MspI enzyme

———————
50 µl final volume
Incubate at 37°C, O/N

Step 2 Gently mix digests by stirring with a pipette tip. Store the restriction enzymes and buffers at −20°C.

»»»

LAB PERIOD I.1.5 CONT.

Step 3   After an overnight incubation, add 1 µl of additional enzyme to each of your five restriction digests to ensure complete digestion. After 4 to 8 hours of additional digestion, store the digests at −20°C for later preparation of the samples for running on an agarose gel. (If the agarose gel is to be run on the same day, simply store the tubes at 4°C).

LAB PERIOD I.1.6.   PREPARE RESTRICTION ENZYME DIGESTS FOR AGAROSE GEL ELECTROPHORESIS

Step 1   Retrieve your mouse genomic DNA restriction digest tubes (and the uncut DNA control tube) stored at −20°C or 4°C. If frozen, allow the reactions to defrost at room temperature.

Step 2   To ethanol precipitate the restriction digests, add 1/10th volume (5 µl) 3.0 M sodium acetate (pH 5.2) and 2 volumes (100 µl) of ice-cold 100% ethanol to each tube. Mix by inverting the tubes and place them on ice for 10 minutes.

Step 3   Remove your mouse digest tubes from the ice, and place them in the microfuge with the caps closed and the hinges pointing away from the center of the rotor, and centrifuge them for 10 minutes at 4°C at maximum speed (~16,000x *g*). This will bring the precipitated DNA to the bottom of the tube.

Step 4   Prepare a flame-drawn Pasteur pipette using the method shown. With gloved hands, hold each end of a long (9-inch) Pasteur pipette, and heat the upper portion of the thin section with the flame from a Bunsen burner until the glass begins to soften. Then remove the pipette from the flame, and quickly pull the pipette in both directions to extend the thin section of the pipette and reduce its diameter. At the same time, introduce a slight bend in the pipette. Carefully break off the excess portion of the thin section approximately 0.5 inches past the bend.

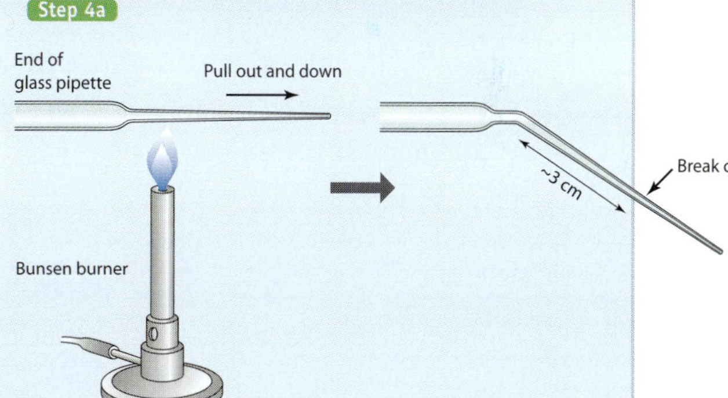

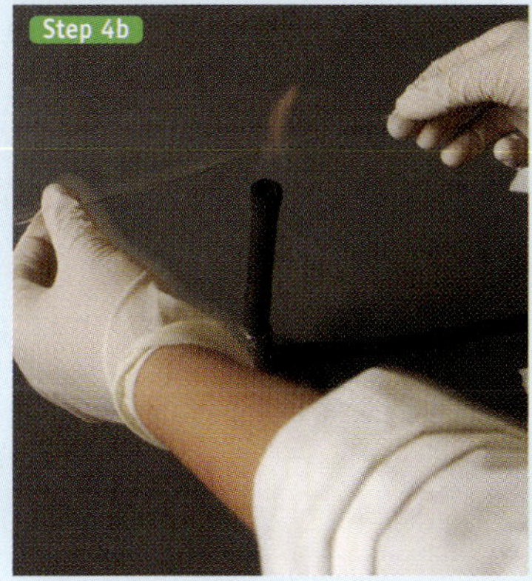

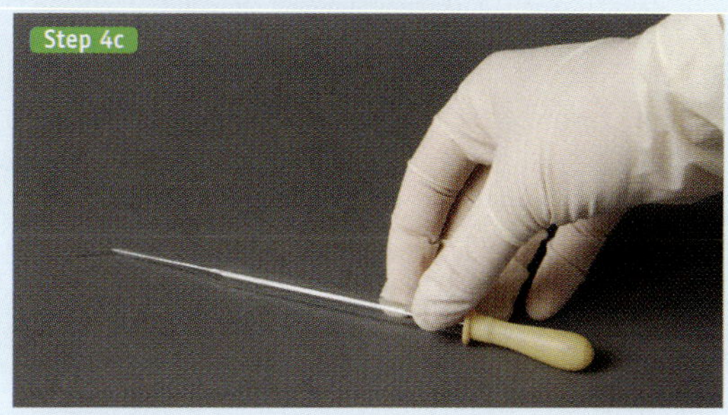

»»

LAB PERIOD I.1.6 CONT.

Step 5 Using only one flame-drawn Pasteur pipette for all of your supernatants, carefully remove the supernatant from each tube, and transfer each supernatant into a waste beaker. Be careful not to disturb the DNA pellets! The DNA pellet (which may not be visible) will be located directly under the hinge of the tube. Use the bent, flame-drawn pipette to remove supernatant from the bottom of the tube *across* from the side where the hinge is located. (Save the flame-drawn pipette.)

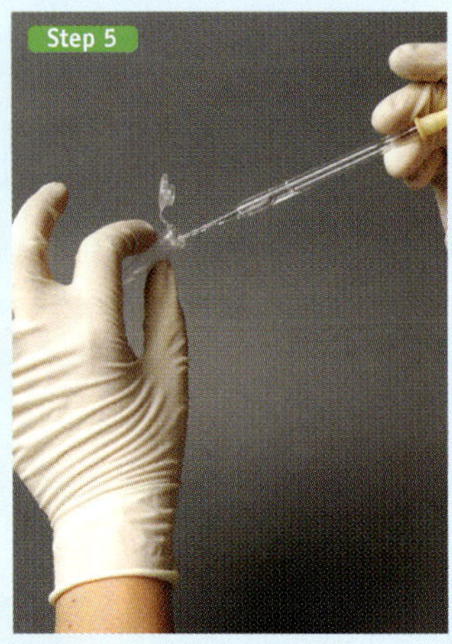

Step 6 Add 500 µl 70% ethanol to each tube. Invert the tubes several times to rinse the pellets.

Step 7 Centrifuge the tubes again for 5 minutes at 4°C in a microfuge at maximum speed (~16,000x *g*) to be sure that the DNA pellet is stuck securely to the bottom of the tube.

Step 8 Using your same flame-drawn Pasteur pipette, carefully transfer each supernatant into a waste beaker (be careful not to disturb the pellets!). Spin the tubes for 30 seconds in your nanofuge, and carefully remove residual ethanol with the same flame-drawn pipette.

Step 9 The instructor will collect the tubes and spin them briefly in a vacuum centrifuge ("speed-vac") to evaporate the ethanol. If there is no speed-vac available, the tubes can be placed in a temp block at 42°C for 10 to 15 minutes to evaporate the remaining ethanol.

Step 10 Add 12 µl of 1x TE buffer plus 3 µl of 5x BJ to each tube (*do not* mix by pipetting up and down). Place the tubes at 4°C and store overnight. During this time, the pellets will soften and make resuspension easier.

LAB PERIOD I.1.7.   PREPARE, LOAD, AND ELECTROPHORESE THE AGAROSE GEL

A VALUABLE analysis of your mouse genomic DNA digestions can be obtained by running them on an agarose gel: the size distribution will tell you whether you have succeeded in digesting your DNA with the restriction endonucleases. You will see dramatic differences in the cutting of your DNA with the four-cutter AluI and the six-cutter and eight-cutter enzymes. In addition, the digestion patterns provided by HpaII and MspI will provide information about the degree of methylation of the mouse genome.

Step 1 Each pair should pour a 1% agarose gel as described in Module I.1.4, Part B. Allow the gel to solidify for at least 20 minutes before adding the 1x TAE running buffer to the gel apparatus.

Step 2 Retrieve 5 µl (150 ng) of your λBstEII marker (in BJ) from 4°C storage, and pipette into a new microfuge tube. Heat the tube at 65°C for 5 minutes. Spin briefly in your nanofuge, and place on ice.

»»

PROCEDURAL NOTE

This molecular weight marker is made of lambda DNA that has been digested with the restriction endonuclease BstEII. The pattern produced by this digestion is presented in Appendix III.

Step 3 Retrieve your mouse restriction enzyme digests (stored at 4°C). Mix to complete the resuspension of the DNA pellets by pipetting the contents of each tube up and down. Load each of the restriction enzyme digests and your uncut control "digest" on the gel in the following order:

> Lane 2: 10 µl AluI-cut mouse DNA in 5x BJ
>
> Lane 3: 10 µl EcoRI-cut mouse DNA in 5x BJ
>
> Lane 4: 10 µl SfiI-cut mouse DNA in 5x BJ
>
> Lane 5: 10 µl HpaII-cut mouse DNA in 5x BJ
>
> Lane 6: 10 µl MspI-cut mouse DNA in 5x BJ
>
> Lane 7: 10 µl uncut control mouse DNA supplied in 5x BJ
>
> Lane 8: 5 µl λBstEII marker in 5x BJ (from Step 2 above)

Step 4 Electrophorese the gel at 50 V for 2 hours. Shut off the power supply, and disconnect the cables.

Step 5 Stain the gel with ethidium bromide gel stain for 10 minutes, followed by a 15-minute destain in a water bath (Lab Period I.1.4).

> **CAUTION!**
>
> Ethidium bromide is a known mutagen and carcinogen. Wear gloves, lab coats, and safety glasses! Handle gels carefully! Do not splash ethidium bromide! Always rinse the spatula in the water destain bath after it has been in contact with ethidium bromide! Always rinse the gel tray after sliding gels into the ethidium bromide bath. You may unknowingly contact the ethidium bromide! Be very careful not to drip the ethidium bromide anywhere!

PROCEDURAL NOTE

Appendix IV describes an alternative DNA stain, SYBR Safe™ (Molecular Probes).

Photograph the gel. An example of the expected pattern of the DNA fragments in this gel is shown in **Figure 1-4**. How does the photograph of your gel compare to this example?

Lanes 2 3 4 5 6 7 8

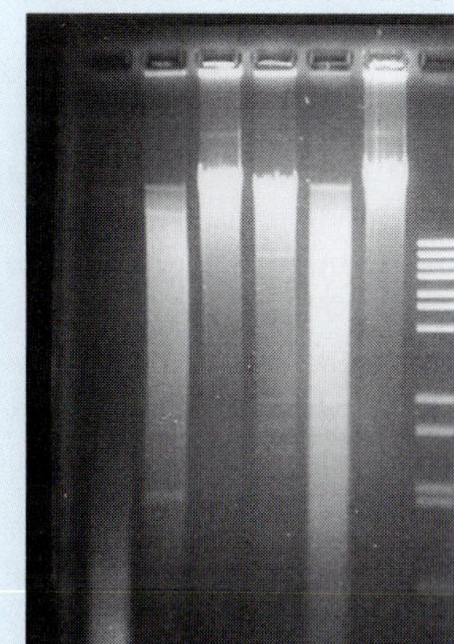

Fig 1-4 Expected pattern of DNA fragments in gel.

PROCEDURAL NOTES

Save this gel if proceeding to Module I.2 — you will need to use it for your Southern blot!

Lab Period I.2.1 is a direct extension of Lab Period I.1.7 and should be performed on the same day if the Southern blot is to be performed.

MODULE I.2

Southern Blot Experiment to Detect Gene-Specific Sequences in Genomic DNA

MODULE SUMMARY

Southern blot transfer of electrophoresed, digested genomic DNA. Nonradioactive PCR labeling of the *Rvt* gene for Southern blot hybridization to determine *Rvt* genomic organization. Southern hybridization and detection of the mouse *Rvt* gene. Analysis of the organization of the *Rvt* repeat sequences, and DNA methylation patterns seen in mouse genomic DNA.

MODULE BACKGROUND

The Southern blot method can be used to specifically detect an individual gene sequence within a sample of genomic DNA that contains all of the genes in the genome (at least 30,000 different genes in mammals!). To accomplish this, the DNA is digested with one or more restriction endonucleases, and the digestion products are separated by size by electrophoresis through an agarose gel (you have already done this as part of Module I.1).

The desired gene-specific fragment of DNA will have migrated to a specific position in the gel in response to electrophoresis depending on its size. This fragment can be detected by transferring the size-separated fragments of electrophoresed DNA to a hybridization membrane, hybridizing with a non-radioactively labeled, gene-specific probe, and then detecting the bound probe. This will allow you to confirm that the fragment is present in your sample of genomic DNA and to determine the exact size of the fragment.

For this project, you will use the gel from Module I.1 (Lab Period I.1.7) in which you have already electrophoresed several different genomic DNA samples digested with different restriction endonucleases. The "Southern transfer" of DNA from the gel to the membrane is accomplished by capillary action using a high-salt buffer. The membrane is placed directly on top of the gel, and the buffer is drawn through the gel and the membrane to a stack of dry filter papers and paper towels sitting on top of the membrane (see Figure 1-5b). The membrane is permeable to the high-salt buffer that passes directly through it. The membrane is not, however, permeable to the DNA. Thus, the DNA passes out of the gel and is deposited directly onto the membrane. During the transfer, the DNA stays in position according to its molecular weight. Thus, the membrane binds the DNA in the exact same position as it was in the gel. The membrane carrying the transferred DNA is called a "Southern blot."

To maximize the efficiency of the transfer process, the gel is first treated with mild acid (0.25 M HCl) that partially depurinates the DNA. Under these conditions, a purine base is removed from the DNA every few hundred nucleotides. This weakens the double helix so that when the DNA begins to transfer, it breaks into smaller pieces that transfer more easily out of the gel. After depurination, the gel is treated with a base (NaOH) to denature (separate) the double-stranded DNA into single-stranded DNA. This further aids transfer and prepares the DNA for subsequent hybridization with the labeled probe DNA. These treatments are followed by a neutralization step to return the gel to a neutral pH so that it will not degrade the blotting membrane during the transfer procedure. After transfer, the DNA is stably affixed to the blotting membrane by either baking the membrane or subjecting the membrane to UV irradiation.

You will be detecting the *Rvt* (LINE-1) repeated sequence in the mouse genomic DNA. You will use a cloned version of one of the *Rvt* repeats as a hybridization probe. You will use the polymerase chain reaction (PCR) to amplify and nonradioactively label the *Rvt*-specific probe fragment from a small amount of starting template DNA. PCR affords a simple and efficient method for labeling DNA. Refer to the introduction to Module I.3 for more information on PCR. In this reaction, one of the nucleotide triphosphates (dTTP) is partially replaced by fluorescein-labeled dUTP, which becomes incorporated into the amplified DNA as the new strands are synthesized. The dUTP is an unusual nucleotide for DNA but is incorporated during DNA synthesis just like dTTP and will hybridize with a complementary dATP just as dTTP will. Importantly, the specific sequence of the initial template DNA is faithfully replicated in each new strand of DNA produced by the PCR reaction. Thus, this method provides large quantities of a specific sequence of DNA that is labeled in a way that can be easily detected using an antifluorescein antibody and a chemiluminescent reaction after hybridization of this probe to the Southern blot membrane.

To detect the *Rvt* repeat sequence in your sample of genomic DNA, you will hybridize the labeled *Rvt* repeat gene probe to the genomic Southern blot. This will allow you to specifically detect the presence and organization of the *Rvt* repeat sequences in the mouse genome. For this hybridization experiment, you will use a labeled, full-length gene probe to detect the presence of the *Rvt* gene among all other sequences in the mouse genome. Thus, this is a very complex hybridization requiring high sensitivity to specifically detect the repeated 1.4-kb *Rvt* gene sequence among a total of approximately 3.1 billion bp of other sequence in the complete mouse genome!

LAB PERIOD I.2.1.   PROCESS THE AGAROSE GEL AND SET UP THE GENOMIC SOUTHERN BLOT TRANSFER

PROCEDURAL NOTE

Lab Period I.2.1 is an extension of Lab Period I.1.7 and should be performed on the *same* day.

Step 1 After photographing the agarose gel in Module I.1 (Lab Period I.1.7), place the agarose gel in a plastic container, and add enough Southern depurination solution to cover the gel. Depurinate the DNA for 15 minutes. Swirl the container occasionally.

Step 2 Pour off the Southern depurination solution and rinse the gel once in ddH$_2$0.

Step 3 Denature the DNA by incubating the gel in Southern denaturation solution for 20 minutes. Swirl the container occasionally.

Step 4 Pour off the Southern denaturation solution, and rinse the gel once in ddH$_2$0.

Step 5 Neutralize the gel by incubating the gel in Southern neutralization solution for 20 minutes. Swirl the container occasionally.

Step 6 While your gel is incubating in the various solutions, you can prepare your nylon membrane (Hybond™ N+ from GE Healthcare; a positively charged membrane). Be sure that you are wearing gloves whenever you touch nylon or nitrocellulose membranes because the oils on your hands will interfere with DNA transfer to the membrane. Cut one nylon membrane to the same size as your agarose gel (use your gel tray to get the proper dimensions). Write your group number or name and "DNA side" on the top of the membrane with a *pencil* (do not use ink; it may run when it gets wet). Using two pairs of forceps, bend the membrane in half without creasing it, and immerse the membrane at the bend into a tray of ddH$_2$0.

Step 7 As the membrane is immersed, allow it to slowly unfold into the ddH$_2$0. This allows the membrane to wet evenly and prevents air bubbles from being trapped beneath it.

Step 8 After 5 minutes in the water, transfer the membrane (using the same technique) to a tray containing 10x SSC for a minimum of 15 minutes.

»»

LAB PERIOD I.2.1 CONT.

Step 9 Components of the Southern transfer assembly and its arrangement are presented in **Figures 1-5a** and **1-5b**, respectively. To set up the Southern transfer, cut three sheets of Whatman 3MM paper to the exact dimensions of the gel (use your gel tray as a guide). Then cut two sheets of Whatman 3MM paper to the same width as your gel, but increase the length by 3 to 4 inches (these will serve as your wicks). Finally, cut paper towels to the same dimensions as your gel. You will need a stack at least 3 inches thick. The components to be assembled for the Southern transfer are shown in Figure 1-5a and 1-5b.

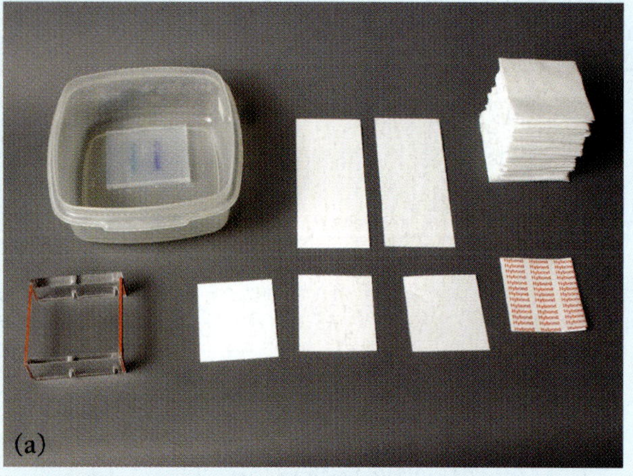

Fig 1-5 Southern transfer assembly. (a) Components. (b) Arrangement assembled.

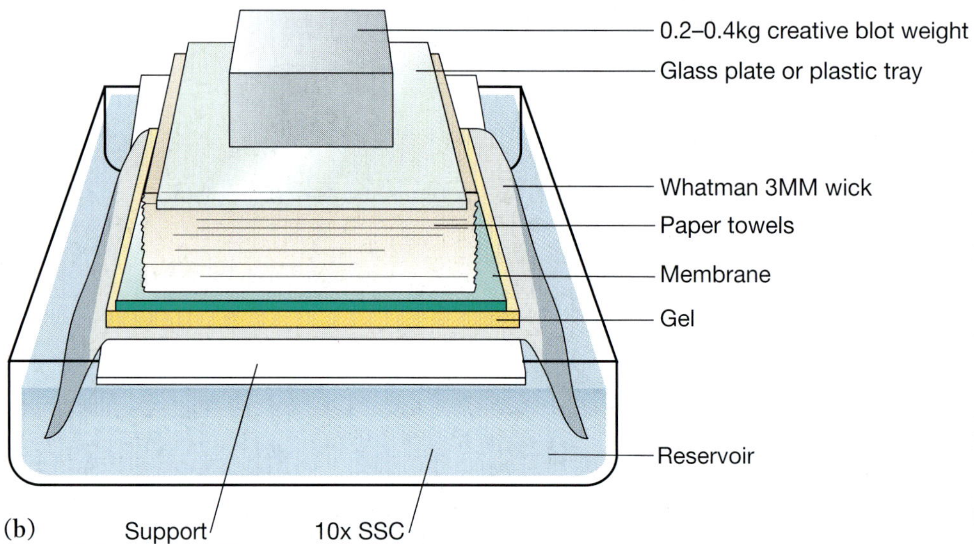

Step 10 Place your agarose gel tray upside down in a plastic container. Add enough 10x SSC so that the level is two thirds up the side of the gel tray.

Step 11 Saturate the filter paper wicks in 10x SSC, and align them on the back of the gel tray. Smooth out any air bubbles.

LAB PERIOD I.2.1 CONT.

Step 12 Use the spatula to carefully place the gel face down on the saturated filter papers, making sure no air bubbles are trapped between the gel and the wick.

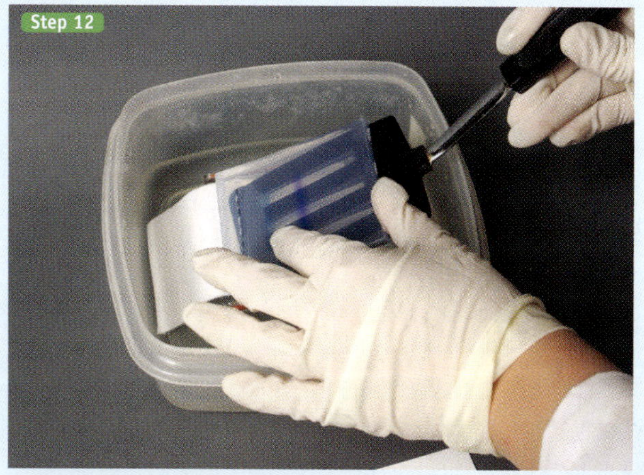

Step 13 Place the wetted nylon membrane (labeled "DNA side" down) on the gel. Do not let the membrane hang over the gel and come in contact with the wick. Again, be sure that no air bubbles are trapped between the gel and the membrane. You can remove trapped air bubbles by gently rolling a clean Pasteur pipette across the top of the membrane.

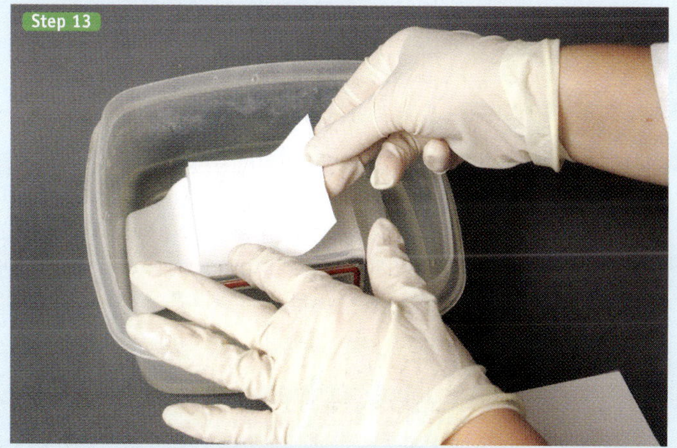

Step 14 Prewet one precut piece of filter paper in 10x SSC, and place it on top of the membrane. Add two dry pieces of filter paper to the stack. Be sure that these filter papers do not directly contact the gel or the wick! If they do, the transfer of buffer will go around the gel rather than through it and your DNA will not transfer.

Step 15 Cover the filter paper with the stack of precut paper towels at least 3 inches thick. Be sure that the paper towels do not directly contact the gel or the wick!

Step 16 Cover the paper towels with a rigid flat support (the hard plastic cover from your gel rig can be used for this).

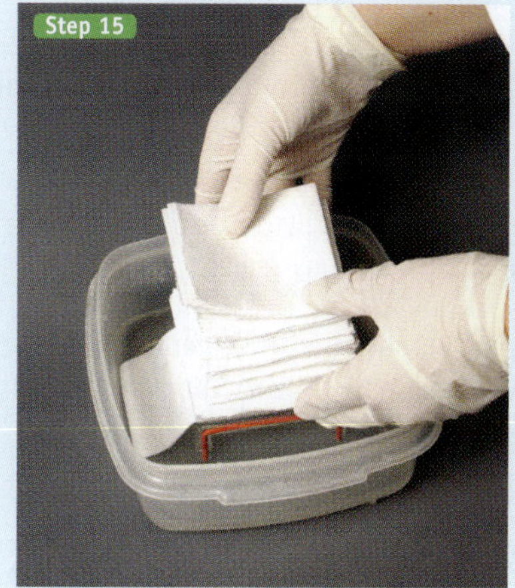

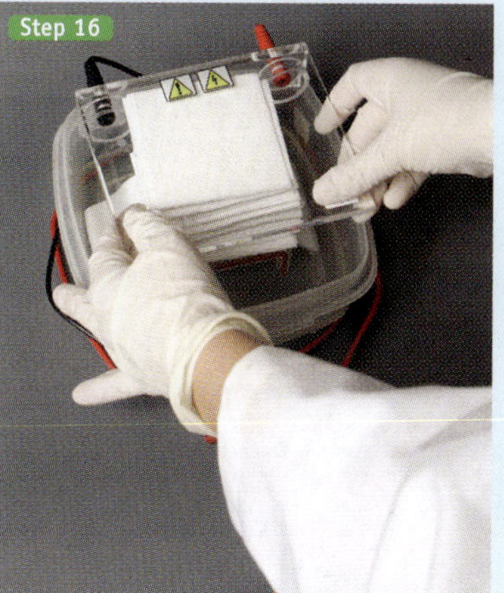

LAB PERIOD I.2.1 CONT.

Step 17 Cover the whole apparatus with plastic wrap to prevent evaporation, and place a "blot weight" of about 500 g (your gel box can be used for this) on the rigid support. The minimum transfer time is 6 hours, but overnight transfer is preferred.

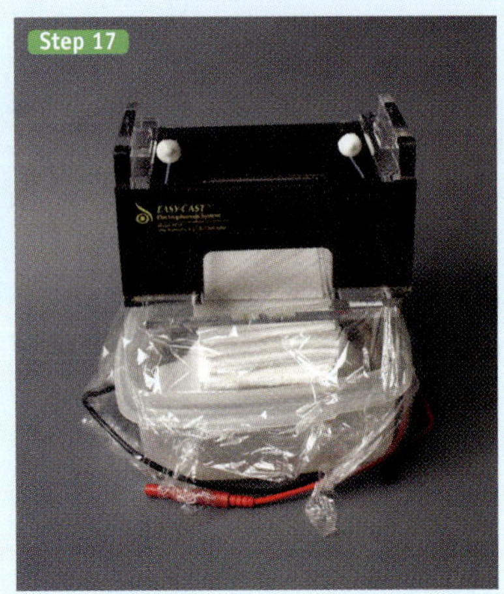

LAB PERIOD I.2.2.   DISASSEMBLE AND CROSS-LINK/BAKE THE GENOMIC SOUTHERN BLOT

Step 1 *Before* removing the membrane from the gel, cut a corner of the membrane to orient the blot (record which corner you cut). Next, mark the location of the wells on the back of the membrane with a pencil. Remove the membrane from the gel, and place it DNA side up on clean filter paper. Make sure that your group name/number are written on the DNA side of the membrane.

Step 2 Rinse the membrane for 2 to 5 minutes in 6x SSC in a group plastic container (one container for the entire lab is sufficient).

Step 3 Place your gel carefully on a UV light box. Lower the protective cover, and check to see that all of the DNA has been transferred out of the gel. If any fluorescent signal is still visible, contact your instructor. It is likely that some small amount of the highest molecular weight DNA may not have completely transferred.

Step 4 Lay the membrane "DNA side" up on a clean sheet of filter paper, and allow the membrane to air dry for at least 15 minutes.

Step 5 If a UV cross-linking instrument is available, UV cross-link, and then bake the filter at 80°C for 1 hour in an oven. If a cross-linker is not available, bake the filter at 80°C for 1 to 2 hours in an oven. The cross-linking and/or baking will fix the transferred DNA onto the membrane. Cross-linking times and settings will be determined by the instructor and will vary depending on the instrument that is being used.

Membranes can be stored at room temperature in sealed plastic bags for several days, or they can be stored in the refrigerator or freezer in sealed plastic bags for at least several weeks.

LAB PERIOD I.2.3.   PCR LABEL THE 1.4-KB *RVT* PROBE

EACH PAIR will do *one* fluorescein/PCR labeling reaction. You will use the PCR fluorescein labeling kit from Roche Diagnostics. The DNA that you will PCR amplify and label is one copy of the mouse 1.4-kb *Rvt* gene family that has been previously cloned into a plasmid. The labeled mouse *Rvt* DNA will be used as a hybridization probe to detect the highly repeated *Rvt* DNA sequence in the restriction digests of your mouse genomic DNA on your Southern blot.

Step 1 You will be provided with a 0.2-ml PCR tube containing a purified plasmid (pBluescript; Stratagene) with a cloned copy of the mouse 1.4-kb *Rvt* repeat DNA sequence (4 ng in 29.5 µl).

To this tube, add the following components:

5 µl 10x PCR Buffer

5 µl FL-dUTP PCR mix

8 µl 25 mM $MgCl_2$

1 µl *Rvt* repeat Forward Primer (10 pmol/µl)

1 µl *Rvt* repeat Reverse Primer (10 pmol/µl)

0.5 µl AmpliTaq™ Gold DNA Polymerase enzyme

50 µl total volume

> **CAUTION!**
>
> Because you are doing PCR, wear gloves while setting up this labeling reaction.
>
> The 0.2-ml tubes are very fragile — handle with care!

PROCEDURAL NOTE

The sequences of the *Rvt* primers are as follows:
Rvt repeat Forward Primer: 5' GAATTCTTTGTTCAGTTCTGAGCC 3'
Rvt repeat Reverse Primer: 5' CCCATATCTAAACATGATAAAAGC 3'

Step 2 Mix the components by stirring. Briefly spin the tube in a nanofuge (see Fig. A-10) to get all of the liquid to the bottom of the tube.

Step 3 Place the tube into the PCR instrument. The PCR program we will run for this labeling reaction is a 30-cycle program with the following parameters:

94°C for 12 minutes to denature the plasmid DNA and activate the AmpliTaq Gold DNA Polymerase and then 30 cycles of

a. 94°C for 60 seconds — Denaturation

b. 50°C for 60 seconds — Annealing

c. 72°C for 90 seconds — Extension

The reaction is completed with a final elongation step at 72°C for 7 minutes and then is held at 4°C.

Step 4 After the program is complete, remove your tube from the PCR instrument. Because the 0.2-ml tubes are so fragile, transfer the PCR product to a new 0.5-ml microfuge tube. Label this tube with your group numbers and "*Rvt* Label." Place the tube of labeled DNA on ice.

PROCEDURAL NOTE

The labeled probe can be stored at −20°C for at least 3 months.

LAB PERIOD I.2.4.   EVALUATE THE FLUORESCEIN-LABELED *RVT* PROBE ON AN AGAROSE GEL

PROCEDURAL NOTE

You will use a different DNA marker on this gel that is appropriate for comparison to low molecular weight PCR products. This marker, 100-bp ladder, includes a band at 100 bp, another at 200 bp, another at 300 bp, and so on (refer to Appendix III for a complete description). It will be provided to you with 5x BJ already added.

Step 1 Pour a 1% agarose minigel as described in Lab Period I.1.4, Part B.

CAUTION!

Ethidium bromide is a known mutagen and carcinogen. Wear gloves, lab coats, and safety glasses! Handle gels carefully! Do not splash ethidium bromide! Always rinse the spatula in the water destain bath after it has been in contact with ethidium bromide! Always rinse the gel tray after sliding gels into the ethidium bromide bath. You may unknowingly contact the ethidium bromide! Be very careful not to drip the ethidium bromide anywhere!

Step 2 Prepare a sample of your "*Rvt* Label" probe from Step 4 above (Lab Period I.2.3) to electrophorese on this gel by removing 8 µl of the probe to another microfuge tube and add 2 µl 5x BJ. Label this tube "*Rvt* Label Gel." Save the remainder of your "*Rvt* Label" probe for possible use in Module II.3.

Step 3 Load the agarose gel lanes with samples in the following order:

Lanes 1–4: Empty

Lane 5: 10 µl "Rvt Label" from Step 2 above

Lane 6: 10 µl 100-bp ladder in 5x BJ (50 ng/µl)

Lanes 7–10: Empty

Step 4 Run the agarose gel at 70 V for 1 to 2 hours.

Step 5 Look at the gel on the UV light box *before staining* with ethidium bromide or SYBR Safe stain. The lane that contains the *Rvt* probe you labeled with fluorescein (lane 6) should have a band visible at 1.4 kb. A visible band at this position indicates that the 1.4-kb *Rvt* repeat was successfully labeled! The large green blob below this band contains the unincorporated fluorescein-labeled dUTP. If there is no fluorescent band at a position equivalent to approximately 1.4 kb, then it means that the labeling reaction was not successful. In this case, the students should ask the instructor to provide some successfully labeled probe for the hybridization experiment described. An example of the expected result is presented in **Figure 1-6a**.

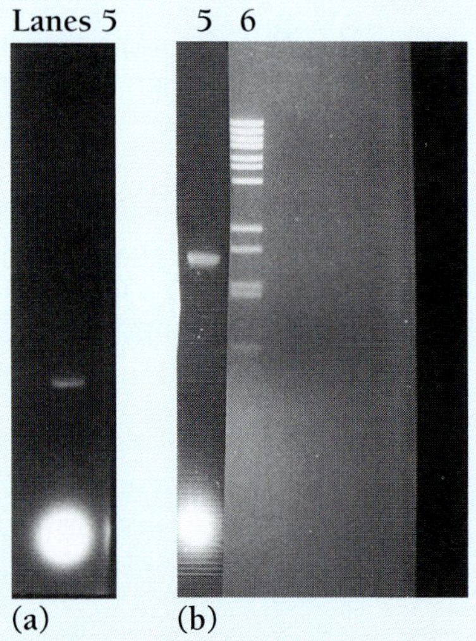

Fig 1-6 Expected pattern in gel.

Step 6 Stain the gel in ethidium bromide gel stain for 10 minutes, followed by 15 minutes destain in a water bath (Lab Period I.1.4), and then photograph the gel. Verify that the band you observed in Step 5 is indeed at 1.4 kb by comparing its position with those of the bands in the size standard lane (refer to the marker band positions presented in Appendix III). An example of the expected result is shown in **Figure 1-6b**.

»»»

LAB PERIOD I.2.4 CONT.

PROCEDURAL NOTE

Appendix IV describes an alternative DNA stain, SYBR Safe (Molecular Probes).

LAB PERIOD I.2.5.   **PREHYBRIDIZE (BLOCK) THE GENOMIC SOUTHERN BLOT**

EACH PAIR will prehybridize their Southern blots using a blocking solution containing nonfat dry milk and salmon sperm DNA. This step is referred to as the prehybridization step and is designed to "block" the filters before the addition of the DNA probe. The blocking agents bind to "sticky" spots on the membrane and prevent nonspecific binding of the DNA probe when it is added in the hybridization step. This results in a Southern blot that has a very low background signal.

Step 1 Transfer the genomic Southern blot membrane prepared in Lab Period I.2.2 into a sealable plastic bag. Add 4 ml of Probe-Amp hybridization solution to the bag. Carefully remove most of the air bubbles, and seal the bag as demonstrated by your instructor. (One method is shown here).

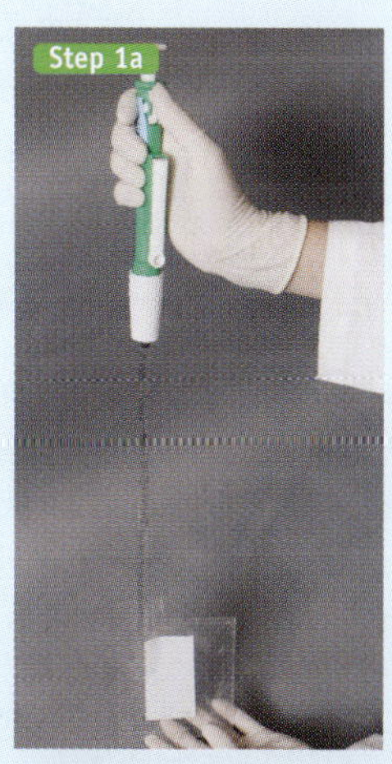

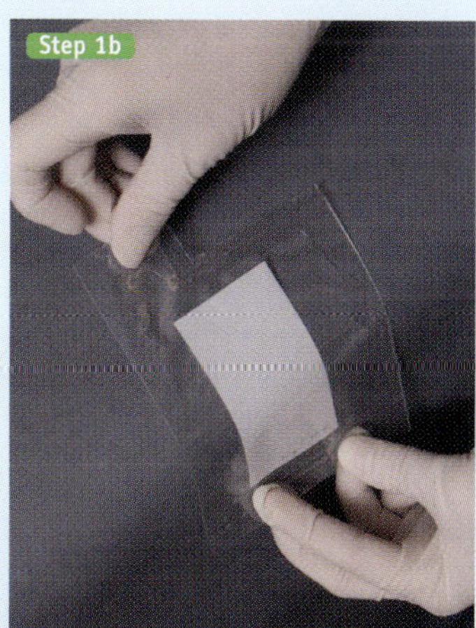

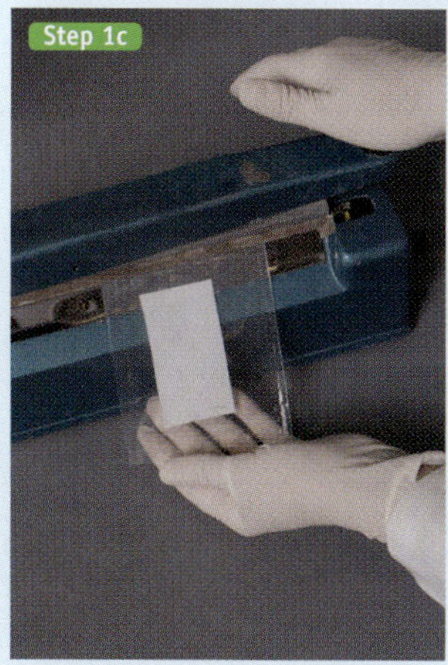

Step 2 Use a marker to write your group name/number and "*Rvt*/Genomic Hyb" on the bag.

Step 3 Incubate the prehybridizing filters in a shaking water bath or air incubator for 1 to 6 hours at 70°C.

PROCEDURAL NOTE

For scheduling convenience, this prehybridization can proceed overnight if necessary.

LAB PERIOD I.2.6.   HYBRIDIZE THE *RVT* REPEAT PROBE TO THE GENOMIC SOUTHERN BLOT

Step 1 Each group will transfer 25 µl of their "*Rvt* label" from Lab Period I.2.3 to a microfuge tube. Label this tube "*Rvt*/Genomic Probe."

Step 2 Place a "lid-lock" on the "*Rvt*/Genomic Probe" microcentrifuge tube to prevent the cap from opening during the subsequent heating step. Heat this tube of labeled probe at 95°C for 5 minutes to denature the DNA. Spin briefly in your nanofuge to collect all of the liquid at the bottom of the tube, and place immediately on ice.

Step 3 With a pair of scissors, snip off one corner of the bag containing your prehybridizing "*Rvt*/Genomic" Southern blot. *Do not* remove the hybridization solution in the bag. Pipette all 20 µl of your denatured fluorescein-labeled probe into the hybridization solution in the bag. Do not pipette the solution directly onto the membrane. Remove bubbles from the bag, and reseal carefully. Hybridize at 70°C overnight in a shaking incubator (if time is limited, an 8-hour hybridization is usually sufficient).

LAB PERIOD I.2.7.   WASH THE GENOMIC SOUTHERN BLOT AND DETECT THE LABELED PROBE

PROCEDURAL NOTE

For Steps 2 and 3 below, the entire class can use one plastic container for these washes.

Step 1 Carefully cut open the hybridization bag and remove the filter. Transfer the filter to a plastic container with Genomic Southern Wash solution I. Reseal the bags containing the hybridization solution, and discard in waste containers.

Step 2 Wash your Southern filter in a plastic container with Genomic Southern Wash I at 70°C for 15 minutes in a shaking incubator. Be sure to check the temperature of your wash solution with a thermometer.

Step 3 Pour off the first wash solution, and replace it with Genomic Southern Wash II. Wash your filter in this solution for 20 minutes at 70°C in plastic containers in a shaking incubator. Again, verify that your wash solution is at the proper temperature.

Step 4 Transfer your filter from the plastic container to a large plastic container. All of your remaining washes will be done in this container.

Step 5 Rinse your filter briefly in 50 ml Enhanced Chemiluminescence (ECL) buffer. Pour off the solution.

Step 6 Incubate your filter in 50 ml Antibody Blocking Solution for a minimum of 1 hour at room temperature with agitation on a rocking platform.

Step 7 Pour off the solution. Rinse filters briefly in 20 ml ECL buffer. Pour off the solution.

Step 8 Incubate your filter in 50 ml Antibody Solution for 1 hour at room temperature with agitation. Antibody solution is a 1:1000 dilution of the anti-fluorescein antibody in 0.5% BSA in ECL buffer.

Step 9 Pour off the Antibody Solution, and wash your filter in 100 ml 0.1% Tween in ECL buffer at room temperature twice for 10 minutes, followed by two additional washes for 5 minutes (all with rocking).

LAB PERIOD I.2.7 CONT.

PROCEDURAL NOTES

There are *four* total washes in this solution. Pour off the wash solution after each step, *except* the last one. Leave the filters in the final wash solution until Step 11.

The following steps will be done in the darkroom.

Step 10 Mix an equal volume of ECL detection solutions 1 and 2 (GE Healthcare) to give sufficient volume to cover your filter.

Step 11 Remove the filter from the last wash solution and drain off the excess.

Step 12 Place your filter into a clean plastic container, and pour the detection solution onto your filter. Incubate for 1 minute at room temperature.

Step 13 Drain off excess detection buffer, and place your filter into a clear acetate development folder.

PROCEDURAL NOTE

You can also use plastic wrap in place of a development folder.

Step 14 Tape your filter DNA side up in a development folder or, alternatively, between two pieces of plastic wrap. Place the covered filter in a film cassette, and place a sheet of X-ray film (Kodak BioMax) over the filter.

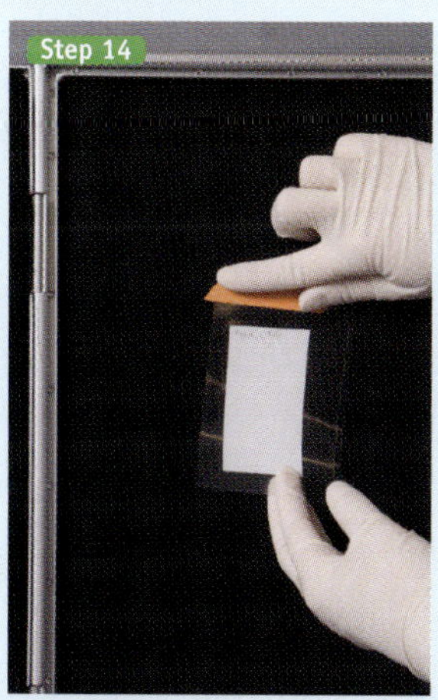

PROCEDURAL NOTE

After the x-ray film is developed and dried, it can be placed back over the blots so that you can trace the outline and position of each blot on the film.

Step 15 Close the cassette and expose the film for 30 minutes to 2 hours.

Step 16 Remove the X-ray film from the cassette in the darkroom with the red safelights on. Develop in Kodak GBX developer for 2 minutes. Rinse in tap water for 2 minutes. Fix in Kodak GBX fixer for 2 minutes, and finally rinse for 2 minutes in tap water.

Step 17 The first exposure should be done for a short period of time (3 to 5 minutes). When the first piece of film is removed from the cassette, immediately replace it with a second piece of film. If the first exposure is too faint, leave the second piece of film in cassette for 30 minutes to 2 hours.

»»

LAB PERIOD I.2.7 CONT.

Step 18 Read the lumigram and analyze the results. An example of the expected pattern of DNA fragments that hybridize to the *Rvt* probe on this blot is shown in **Figure 1-7**.

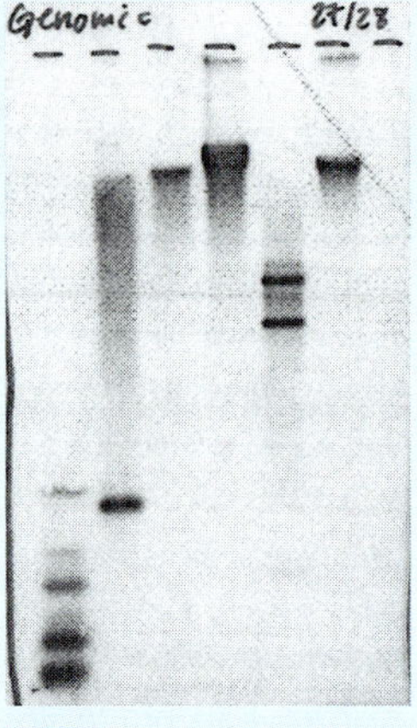

Fig 1-7 Expected pattern of DNA fragments on the lumigram.

PCR Amplify a Specific Gene from Mouse Genomic DNA and Verify Identity by Southern Blot Analysis

MODULE I.3

MODULE SUMMARY

PCR amplification of the *Ttr* gene from mouse genomic DNA and analysis by gel electrophoresis. Southern hybridization and detection of the oligonucleotide *Ttr* probe to verify the identity of the PCR amplification product.

MODULE BACKGROUND

Module I.3 describes the use of the polymerase chain reaction (PCR) to detect a specific DNA sequence/fragment within genomic DNA. PCR affords a much faster and much more sensitive method to detect a specific gene sequence than does Southern blotting or cloning. For this method, it is necessary to have initial DNA sequence information for two short (approximately 20 bp) regions that flank the gene-specific fragment to be amplified. These two short regions are then used to design DNA oligonucleotide primers that are complementary to each strand of the template DNA. When the double-stranded template DNA is denatured and the *Taq* polymerase and free nucleotides are added, two double-stranded copies can be made from each initial template molecule. This process is then repeated by cycling the DNA through a series of changes in temperature to facilitate sequential denaturation of the template DNA, annealing of the primers, and synthesis of the new DNA strands. The products of this reaction can then be visualized by electrophoresis in an agarose gel followed by staining with ethidium bromide. This allows you to visualize the new DNA that has been synthesized and to confirm the size of the amplification product.

In theory, 30 cycles of PCR will lead to the synthesis of 1 billion copies of each initial double-stranded molecule of DNA that contains the specific primer binding sites! In practice, however, the efficiency of this reaction is limited so that the amount of product produced generally plateaus after 25 to 35 cycles. Still, PCR can be used to convert a very small amount of starting material into a relatively large amount of product, and because amplification requires specific primer-binding sites, the product is gene specific. Therefore, this provides a convenient and sensitive method to detect and amplify a particular gene sequence starting with a very small amount of initial template DNA.

Because the production of a PCR product requires the presence of both primer binding sites on a contiguous piece of DNA and because the expected size of the PCR product is typically known in advance, a positive result is most often detected by simply visualizing the product on an agarose gel. Thus, after the PCR reaction, an aliquot of the product will be run on a gel to demonstrate the presence and size of the product. The chances that a product of the correct size will be produced by random annealing of the sequence-specific primers to sites flanking a fragment (of the correct size) other than the desired fragment are very low. Nevertheless, in some cases, it is desirable to confirm the identity of the PCR product further. A method for confirming the product by Southern blot hybridization is presented in this module. If the desired gene-specific fragment of DNA has been accurately amplified during PCR, it will contain the expected sequence of bases throughout its length. To confirm this, a single-stranded, *Ttr*-specific oligonucleotide probe, synthesized with a biotin-conjugated nucleotide, will be hybridized under stringent conditions to the membrane carrying the PCR products. You will then detect the hybridized probe using a chemiluminescent reaction. If this "internal oligonucleotide probe" hybridizes to the PCR product, then that product must contain the correct internal complementary sequence and, therefore, must be the desired gene-specific product. This result, along with the initial detection of an amplification product of the correct size on an agarose gel, should satisfy even the most skeptical reviewer (or laboratory instructor!) that the PCR product was indeed produced by amplification of the desired gene fragment.

LAB PERIOD I.3.1.   PCR AMPLIFY THE *Ttr* GENE FROM MOUSE GENOMIC DNA

IN THIS PCR experiment, you will amplify one 1100-bp sequence (part of the *Ttr* gene) from the entire 3 billion base pair genome of mouse. This will demonstrate the specificity of PCR and the power of PCR to amplify one region of genomic DNA from the entire genome. **Figure 1-8** shows the complete sequence of the *Ttr* cDNA and shows the organization of introns and exons in the genomic DNA. As you can see, the *Ttr* gene is divided into four exons and three introns. The *Ttr* PCR primers used in this PCR are "Forward 1 Primer" and "Reverse 1 Primer." When these primers are used to amplify genomic DNA, the size of the PCR product will be 1100 bp because it spans an intron. The amplification of a specific region of genomic DNA has very important applications in DNA diagnosis of genetic diseases, DNA forensics (DNA fingerprinting), population and evolutionary genetics, and many others.

Each pair will perform a PCR reaction on their mouse DNA using *Ttr* specific primers.

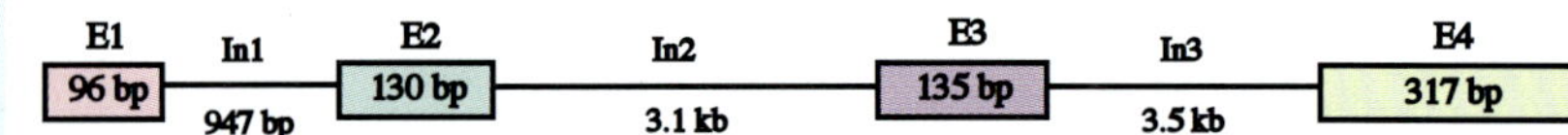

Ttr cDNA Sequence and Primer Design

Forward 1 and **Reverse 1:** Used to amplify the mouse transthyretin gene sequence. Yields a 150 bp product when using RNA or cDNA as template and a 1100 bp product when genomic DNA is used (due to amplification of intron 1)

Reverse 2: Used to make a 570 bp product (with **Forward 1**)

Ttr Oligo Probe: Designed to hybridize to exon 1 as the probe for the Southern Blot Experiment.

Fig 1-8 Mouse Transthyretin (*Ttr*) Gene Information

Step 1 In a 0.2-ml PCR tube, add the following:

30.5 µl ddH$_2$O

8 µl deoxynucleotide triphosphates (dNTP) mix

5 µl 10x PCR reaction buffer

3 µl MgCl$_2$ (25 mM)

1 µl *Ttr* Forward 1 Primer (10 pmol/µl)

1 µl *Ttr* Reverse 1 Primer (10 pmol/µl)

0.5 µl AmpliTaq Gold DNA polymerase

49 µl Final volume

Mix by stirring with your pipette tip.

»»»

PROCEDURAL NOTE

The *Ttr* Forward 1 Primer sequence is 5' ACACAGATCCACAAGCTCCTGACAG 3'. The *Ttr* Reverse 1 Primer sequence is 5' CGGACAGCATCCAGGACTTTGAC 3'.

Step 2 Add 1 µl of template DNA (your uncut mouse genomic DNA from Module I.1 or supplied by your instructor). Be sure to label this tube on the side with your group numbers and "*Ttr* DNA PCR" (do not label the cap of the tube because many thermal cycler block will remove any writing on the cap). One microliter of your mouse DNA will work in this PCR no matter what the concentration (as long as you recovered *some* mouse DNA!). Thus, there is no concern about adding a precise quantity of mouse genomic DNA into this reaction; however, if your gel in Module I.1 showed that you did not recover any mouse genomic DNA, borrow some DNA from another group (or from your instructor).

Step 3 Place your reaction tube in a chamber of the thermal cycler.

Step 4 The PCR program has been set as follows:

    a. An initial denaturing step at 94°C for 12 minutes is performed to insure that all of the DNA sample is single stranded and that the AmpliTaq Gold DNA polymerase is activated.

The following three steps (b–d) will be cycled 35 times:

    b. 94°C for 30 seconds to denature the double-stranded DNA.

    c. 58°C for 30 seconds to enable the primers to anneal to template DNA.

    d. 72°C for 1 minute to synthesize DNA from the primers (this is the temperature at which *Taq* polymerase is most active).

    e. After the 35 cycles, a final extension/annealing step at 72°C for 10 minutes is added to insure that all of the extensions are complete.

    f. The program will hold the tubes at 4°C until the instrument is turned off.

Later in the day, the tubes will be removed from the thermal cycler and will be stored at –20°C. To verify that the PCR reaction successfully amplified the target DNA, the samples will be run on an agarose gel alongside known molecular weight markers.

LAB PERIOD I.3.2.   ELECTROPHORESE THE *Ttr* PCR PRODUCT ON AN AGAROSE GEL. SOUTHERN BLOT TRANSFER TO CONFIRM THE IDENTITY OF THE PCR PRODUCT

IF THE desired gene-specific fragment of DNA has been amplified during PCR, the PCR product should be the length predicted by the distance between (and including) the two primers (1100 bp). If a PCR product of the correct length is identified by agarose gel electrophoresis, this is strong evidence that the correct sequence of DNA has been amplified. If further evidence of the identity of the PCR product is required, a Southern blot hybridization with a gene-specific oligonucleotide can be performed. In **Figure 1-9**, this oligonucleotide is indicated as "*Ttr* Oligo."

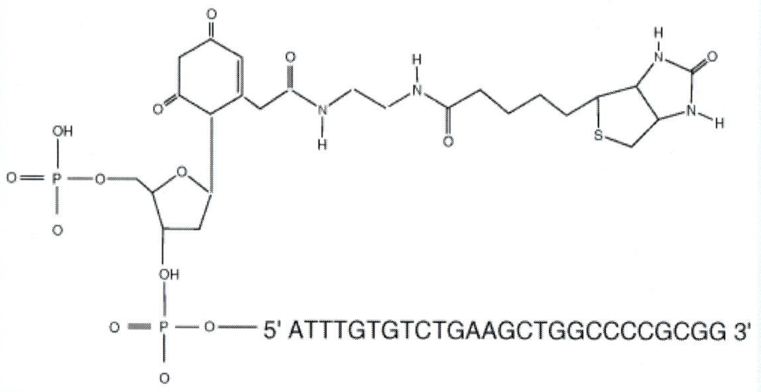

Fig 1-9 "*Ttr* oligo" oligonucleotide with a biotin molecule covalently linked to the 5' end.

5' ATTTGTGTCTGAAGCTGGCCCCGCGG 3'

Step 1 Prepare a 1% agarose gel as described in Lab Period I.1.4, Part B.

Step 2 Retrieve your genomic *Ttr* PCR reaction from −20°C (it is labeled "*Ttr* DNA PCR"). If this PCR reaction did not work for you, borrow what you need from another lab group! Carefully remove 5 µl of the PCR reaction product, and place into a new microfuge tube. Add 95 µl of 1x TE, and mix by vortexing. Spin briefly in your nanofuge to recover all of the liquid to the bottom. Label this as your "1:20 diluted *Ttr* PCR" product.

Step 3 Label two new microfuge tubes "PCR undil" and "PCR dil." Add 5 µl of the "*Ttr* DNA PCR" product to the "PCR undil" tube. Add 5 µl of the "1:20 diluted *Ttr* PCR" product to the "PCR dil" tube. Add 3 µl of 1x TE buffer and 2 µl of 5x BJ dye to each tube (total = 10 µl volume in each tube). Save these at your bench for loading on the agarose gel.

Step 4 Pipette 10 µl of 100 bp ladder directly from the marker tube to load directly on the gel (no heating required).

CAUTION!

Ethidium bromide is a known mutagen and carcinogen. Wear gloves, lab coats, and safety glasses! Handle gels carefully! Do not splash ethidium bromide! Always rinse the spatula in the water destain bath after it has been in contact with ethidium bromide! Always rinse the gel tray after sliding gels into the ethidium bromide bath. You may unknowingly contact the ethidium bromide! Be very careful not to drip the ethidium bromide anywhere!

Step 5 Load the gel in the following order:

Lanes 1–3: EMPTY

Lane 4: 10 µl 100-bp ladder in 5x BJ

Lane 5: "PCR dil" 10 µl from Step 3 above

Lane 6: "PCR undil" 10 µl from Step 3 above

Lanes 7–10: EMPTY

Step 6 Electrophorese the gel at 60 volts for 90 minutes.

»»

LAB PERIOD I.3.2 CONT.

Step 7 Stain 10 minutes in ethidium bromide DNA stain, followed by 15 minutes in a water bath to destain the gel (Lab Period I.1.4). Photograph the gel. An example of the expected pattern of the DNA fragments in this gel is shown in **Figure 1-10**.

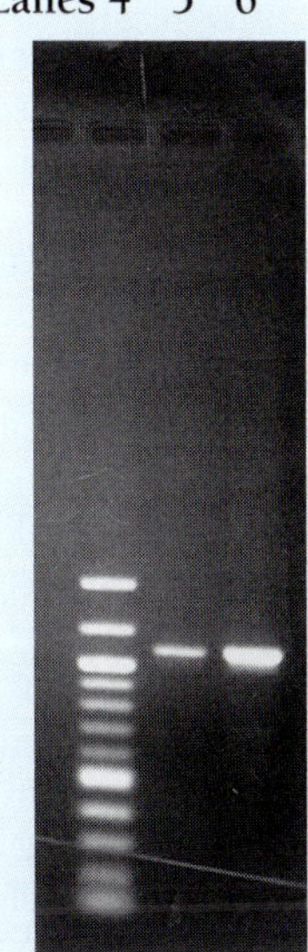

Fig 1-10 Expected pattern of DNA fragments in gel.

PROCEDURAL NOTES

Appendix IV describes an alternative DNA stain, SYBR Safe (Molecular Probes).

Do not discard this gel — you will need it for the Southern blot experiment that follows below!

Step 8 After photographing the agarose gel, place the agarose gel in a plastic container, and add enough Southern depurination solution to cover the gel. Depurinate for 15 minutes. Swirl occasionally.

Step 9 Pour off the Southern depurination solution, and rinse the gel once in ddH_2O. Pour off the ddH_2O.

Step 10 Denature by incubating the gel in Southern denaturation solution for 20 minutes. Occasionally, swirl the solution in the container.

Step 11 Pour off the Southern denaturation solution, and rinse the gel once in ddH_2O.

Step 12 Neutralize the gel by incubating it in Southern neutralization solution for 20 minutes. Swirl the solution occasionally.

Step 13 While your gel is incubating in the various solutions, you can prepare your nylon membrane (Hybond N+ from GE Healthcare; a positively charged membrane). Be sure you are wearing gloves whenever you touch nylon or nitrocellulose membranes because the oils on your hands will interfere with DNA transfer to the membrane. Cut one nylon membrane to the same size as your agarose gel (use your gel tray to get the proper dimensions). Write your group number or name and "DNA side" on the top of the membrane with a *pencil* (do not use ink; it may run when it gets wet). Using two pairs of forceps, bend the membrane in half without creasing it, and immerse the membrane at the bend into a tray of ddH_2O.

Step 14 As the membrane is immersed, allow it to slowly unfold into the ddH_2O. This allows the membrane to wet evenly and prevents air bubbles from being trapped beneath it.

Step 15 After 5 minutes in the water, transfer the membrane (using the same technique) to a tray containing 10x SSC for a minimum of 15 minutes.

»»

LAB PERIOD I.3.2 CONT.

Step 16 The arrangement of components of the Southern transfer assembly is schematically represented in Figure 1-5b. To set up the Southern transfer, cut three sheets of Whatman 3MM paper to the exact dimensions of the gel (use your gel tray as a guide). Then cut two sheets of Whatman 3MM paper to the same width as your gel but increase the length by 3 to 4 inches (these will serve as your wicks). Finally, cut paper towels to the same dimensions as your gel. You will need a stack at least 3 inches thick. The components to be assembled for the Southern transfer are shown in Figure 1-5a.

Step 17 Place your agarose gel tray upside down in a plastic container. Add enough 10x SSC so that the level is two thirds up the side of the gel tray.

Step 18 Saturate the filter paper wicks in 10x SSC, and align them on the back of the gel tray as done in Lab Period I.2.1, Step 11. Smooth out any air bubbles.

Step 19 Use the spatula to place the gel carefully face down on the saturated filter papers as done in Lab Period I.2.1, Step 12, making sure no air bubbles are trapped between the gel and the wick.

Step 20 Place the wetted nylon membrane (labeled "DNA side" down) on the gel as done in Lab Period I.2.1, Step 13. Do not let the membrane hang over the gel and come in contact with the wick. Again, be sure that no air bubbles are trapped between the gel and the membrane. You can remove trapped air bubbles by gently rolling a clean Pasteur pipette across the top of the membrane.

Step 21 Prewet one precut piece of filter paper in 10x SSC, and place it on top of the membrane. Add two dry pieces of filter paper to the stack. Be sure these filter papers do not directly contact the gel or the wick! If they do, the transfer of buffer will go around the gel rather than through it and your DNA will not transfer.

Step 22 Cover the filter paper with the stack of precut paper towels at least 3 inches thick, as done in Lab Period I.2.1, Step 15. Be sure that the paper towels do not directly contact the gel or the wick!

Step 23 Cover the paper towels with a rigid flat support (the hard plastic cover from your gel rig can be used for this).

Step 24 Cover the entire apparatus with plastic wrap to prevent evaporation, and place a "blot weight" of about 500 g (your gel box can be used for this) on the rigid support as done in Lab Period I.2.1, Steps 16 and 17. The minimum transfer time is 6 hours, but overnight transfer is preferred.

LAB PERIOD I.3.3.   DISASSEMBLE AND CROSS-LINK/BAKE THE PCR SOUTHERN BLOT

Step 1 Before removing the membrane from the gel, cut a corner of the membrane to orient the blot (record which corner that you cut). Next, mark the location of the wells on the back of the membrane with a pencil, as demonstrated by the instructor. Remove the membrane from the gel, and place it DNA side up on clean filter paper. Make sure that your group name/numbers are written on the "DNA side" of the membrane.

Step 2 Rinse the membrane for 2 to 5 minutes in 6x SSC in a group plastic container at the front bench.

Step 3 Place your gel carefully on a UV light box; lower the protective cover, and check to see that all of the DNA has been transferred. If any fluorescent signal is still visible, contact your instructor.

Step 4 Lay the membrane DNA side up on a clean sheet of filter paper, and allow the membrane to air dry for at least 15 minutes.

Step 5 If a UV cross-linking instrument is available, UV cross-link, and then bake the filter at 80°C for 1 hour in an oven. If a cross-linker is not available, bake the filter at 80°C for 1 to 2 hours in an oven. The cross-linking and/or baking will fix the transferred DNA onto the membrane. Cross-linking times and settings will be determined by the instructor and will vary depending on the instrument that is being used.

Membranes can be stored at room temperature in sealed plastic bags for several days, or they can be stored in the refrigerator or freezer in sealed plastic bags for at least several weeks.

LAB PERIOD I.3.4.   PREHYBRIDIZE THE PCR SOUTHERN BLOT

Step 1 Each pair should place their *Ttr*/PCR Southern into a plastic hybridization bag (sealable bag). Add 4 ml of *Ttr* Hyb Solution (see Lab Period I.2.5, Step 1).

PROCEDURAL NOTE

If desired, two groups can place their blots back to back in the same bag.

Step 2 Carefully remove most of the air bubbles, and seal the bag as demonstrated by the instructor (one method is presented in Lab Period I.2.5, Step 1). Use a marker to write your group numbers and "PCR Hyb" on the bag.

Step 3 Prehybridize these Southern blots at 55°C for at least 1 to 2 hours (overnight is fine).

LAB PERIOD I.3.5.   HYBRIDIZE THE *TTR* OLIGONUCLEOTIDE PROBE TO THE PCR SOUTHERN BLOT

PROCEDURAL NOTES

The hybridization probe for this experiment will be a 25-nt oligonucleotide that was synthesized by New England Biolabs with a single biotin molecule already attached to the 5′ end (see Figure 1-9). This will be provided to you for this hybridization experiment.

The oligonucleotide probe sequence is 5′ ATTTGTGTCTGAAGCTGGCCCCGCGGG 3′.

Step 1 Place a "lid lock" on the microcentrifuge tube containing 90 µl of *Ttr* Hyb Solution plus 10 µl of *Ttr* Oligonucleotide Probe (20 ng) to prevent the cap from opening during the subsequent heating step. Heat the tube at 95°C for 5 minutes (to ensure that the probe is in a single-stranded state (i.e., not hybridized to itself). Spin briefly in your nanofuge to collect all of the liquid at the bottom of the tube, and place the tube immediately on ice.

Step 2 With scissors, snip off one corner of the hybridization bag labeled "PCR Hyb," but do not remove the hybridization buffer. Add all 100 µl of the heated *Ttr* Oligonucleotide Probe (20 ng) solution from Step 1 above directly to the hybridization solution in the bag.

PROCEDURAL NOTE

Do not pipette the solution directly onto the filter.

Step 3 Remove air bubbles, and seal the bag carefully. Mix the probe with the *Ttr* Hyb solution in the bag and check the bag for leaks. Hybridize the *Ttr* Southern at 55°C for 2 to 17 hours in a shaking incubator (overnight is fine).

LAB PERIOD I.3.6.   WASH THE *TTR* PCR SOUTHERN BLOT, DEVELOP, AND ANALYZE THE RESULTS

PROCEDURAL NOTE

Steps 1 to 3 shown here can be performed for the entire class in one plastic container. After washes to remove excess probe and a "stringency wash" to ensure that only the correct (specific) hybridization will occur, a series of detection steps will be performed. These are summarized in **Figure 1-11**. These entail sequential additions of streptavidin (to bind to the biotinylated probe hybridized to the DNA on the membrane), biotinylated alkaline phosphatase to provide the enzymatic reaction to produce light (which is detected on X-ray film), and the detection reagent CDP-Star that is dephosphorylated to give off light. Throughout the protocol are a series of washes will be performed to remove excess reagents. This protocol is derived from the New England Biolabs Phototope Kit.

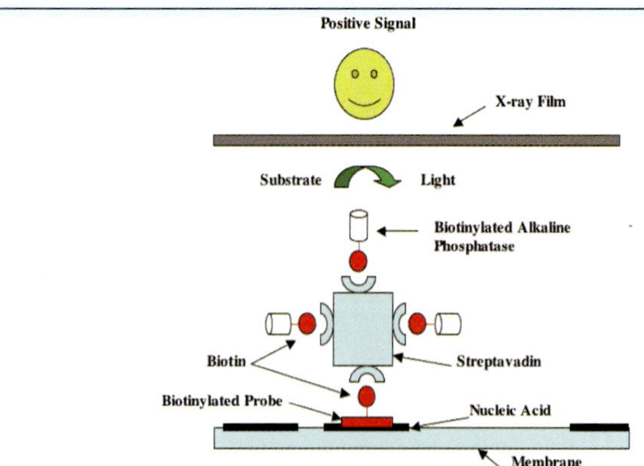

Fig 1-11 Sequence of detection steps.

LAB PERIOD I.3.6 CONT.

PROCEDURAL NOTE

Dump all waste from the washes in Steps 5 to 12 into a plastic container at your bench.

Step 1   Remove your filter from your hybridization bag, and transfer it to your plastic container. Reseal the bag containing the hybridization solution and discard.

Step 2   Wash the *Ttr* Southern blot filters in a plastic container with 100 ml Oligo Southern Wash I at room temperature for 5 minutes in a shaking water bath. Change the buffer, and then wash for another 5 minutes at room temperature.

Step 3   Wash the blot one time in 100 ml Oligo Southern Wash II, 15 minutes at 55°C in a plastic container in a shaking water bath. Change the buffer, and wash for another 15 minutes at 55°C.

Step 4   Transfer your Southern filter to your small plastic container and wash the filter for 15 minutes in 50 ml Phototope Blocking solution at room temperature by rocking the container.

Step 5   Discard the Phototope Blocking solution. Add 15 ml Streptavidin solution to your membrane in the container, and allow it to rock at room temperature for 5 minutes.

Step 6   Discard the Streptavidin solution, and add 25 ml Phototope Detection Wash solution I. Incubate, with rocking, for 5 minutes at room temperature.

Step 7   Discard the Phototype Detection Wash solution I, and repeat the 5-minute room temperature wash with 25 ml fresh Phototope Detection Wash solution I for three additional times (four washes total).

Step 8   Discard the solution, and add 15 ml Biotinylated Alkaline Phosphatase solution to your membrane. Incubate for 5 minutes at room temperature with rocking.

Step 9   Discard this solution, and add 50 ml Phototope Blocking solution to the membrane. Incubate for 5 minutes at room temperature with continuous rocking.

Step 10   Discard the Phototype Blocking solution, and add 25 ml Phototope Detection Wash solution II. Incubate for 5 minutes at room temperature with rocking.

Step 11   Discard this solution, and repeat the 5-minute wash with 25 ml fresh Phototope Detection Wash solution II for three additional times (four washes total). Do not discard the final wash until told to do so by the instructor.

Step 12   When instructed to do so, discard the last wash solution in your plastic container. Then add 10 ml Phototope Detection Mix to your membrane in the container. Incubate in Detection Mix for 5 minutes with rocking.

PROCEDURAL NOTE

The Phototope Detection Mix contains CDP* detection reagent in the correct buffer (see Appendix II).

Step 13   In the dark room with red safe lights on, place the blot with the DNA side up in a development folder or between two sheets of plastic wrap. Tape the blot inside an X-ray film cassette (see Lab Period I.2.7, Step 14). Place a sheet of autoradiography film (Kodak BioMax) over

»»

the blot. Place orientation marks on the film and the bag so that they can later be oriented properly. Close the cassette and expose the film for 1 to 5 minutes. It is recommended to first do a short exposure for 1 minute and then a longer exposure if necessary.

PROCEDURAL NOTE

Three blots can be placed side-by-side into a single X-ray cassette to conserve X-ray film.

Step 14 Remove the X-ray film from the cassette in the darkroom with the red safelights on. Develop in Kodak GBX developer for 2 minutes. Rinse in tap water for 2 minutes. Fix in Kodak GBX fixer for 2 minutes, and finally, rinse for 2 additional minutes in tap water. If you have an automatic X-ray film developer, you can use that instead.

Step 15 Dry the film (air drying works fine or you can use a hair dryer to speed up the process), and then replace the film into the X-ray cassette with the Southern membranes. Use the alignment marks to be sure that they are properly oriented. Now use a marking pen to outline the blots on the film. Trace the alignment lines from the filters onto the film, and write the group numbers next to each filter image on the film. Cut out the image of each filter from the film so that each student has a film (referred to as a lumigram!).

Step 16 Examine the lumigram, and analyze your results! The instructor will help you interpret your lumigram.

Step 17 Be sure to dump out and thoroughly rinse the plastic container at your bench that you used to discard all of your waste.

An example of the expected results is shown in **Figure 1-12**. How do your results compare with this example?

Lane 5   6

Fig 1-12 Expected pattern on lumigram.

MODULE MATERIALS AND REAGENTS (PER PAIR OF STUDENTS)

MODULE I.1

Frozen mouse liver (350 mg)
Liquid nitrogen (50 ml)
Qiagen buffer G2 (20 ml)
Qiagen genomic-tip DNA column
  (diethylaminoethyl [DEAE]) (1)
Qiagen buffer QBT (10 ml)
Qiagen buffer QC (30 ml)
Qiagen buffer QF (7 ml)
Isopropanol (5 ml)
70% Ethanol (5 ml)
100% Ethanol (600 µl)
TNE buffer (1 ml). See Appendix II.
0.1x TE buffer (pH 8.0) (600 ml). See
  Appendix II.
1x TE buffer (pH 8.0) (100 µl). See
  Appendix II.
1x TAE buffer (1000 ml). See Appendix II.
Agarose (2 g)
5x Blue Juice loading dye (BJ) (120 µl). See
  Appendix II.

Mouse genomic DNA (6 µg) (optional)
  (American Bioanalytical)
Lambda HindIII marker in 5x BJ (25 ng/µl)
  (6 µl)
Lambda BstEII marker in 5x BJ (25 ng/µl)
  (5 µl)
Ethidium bromide, DNA stain (200 ml)
AluI restriction endonuclease (3 µl) (New
  England Biolabs) [NEB]
EcoRI restriction endonuclease (3 µl) (NEB)
HpaII restriction endonuclease (3 µl) (NEB)
SfiI restriction endonuclease (3 µl) (NEB)
MspI restriction endonuclease (3 µl) (NEB)
10x NEB buffer 1 (10 µl)
10x NEB buffer 2 (15 µl)
10x NEB EcoRI buffer (5 µl)
3.0 M sodium acetate (pH 5.2) (30 µl). See
  Appendix II.
Mortar and pestle
Safety glasses (2)

45 mm Millipore Type VSWP, 0.025 mm (two
  dialysis membranes)
Spectrophotometer (UV-Vis), quartz cuvette
  Rocking platform or roller drum
Agarose gel electrophoresis apparatus, power
  supply
UV transilluminator
DNA gel photography unit
Lab coats (2)
Flame-drawn Pasteur pipette (1)
50-ml plastic tubes (polypropylene, conical) (3)
Plastic containers for staining and destaining
  gels (2)
15-ml glass Corex tube (1)
Disposable 50-ml syringe (1) and rubber
  stopper with hole (1)

MODULE I.2

Southern depurination solution (100 ml). See
  Appendix II.
Southern denaturation solution (100 ml). See
  Appendix II.
Southern neutralization solution (100 ml).
  See Appendix II.
ddH$_2$O (250 ml)
10x SSC (500 ml). See Appendix II.
6x SSC (200 ml). See Appendix II.
Ethidium bromide DNA stain (200 ml). See
  Appendix II.
Fl-PCR labeling kit (1 reaction worth) (Roche
  Diagnostics)
Rvt Repeat Forward Primer (10 pm/µl) (1 µl)
  (American Bioanalytical)
Rvt Repeat Reverse Primer (10 pm/µl) (1 µl)
  (American Bioanalytical)
Bluescript plasmid with 1.4 kb *Rvt* insert
  (4 ng in 29.5 µl) (American Bioanalytical)
Agarose (1g)

1x TAE buffer (500 ml). See Appendix II.
100-bp Ladder in 5x BJ (50 ng/µl) (10 µl).
  See Appendix II.
5x BJ (5 µl). See Appendix II.
Probe-Amp hybridization buffer (4 ml). See
  Appendix II.
Genomic Southern Wash solution I. See
  Appendix II.
Genomic Southern Wash solution II. See
  Appendix II.
Antibody blocking solution (50 ml). See
  Appendix II.
ECL Buffer (50 ml). See Appendix II.
0.1% Tween in ECL Buffer 1 (200 ml). See
  Appendix II.
Antifluorescein antibody/horseradish peroxi-
  dase conjugate (50 ml) (GE Healthcare)
ECL detection solution 1 (GE Healthcare)
ECL detection solution 2 (GE Healthcare)
GBX developer (Kodak)

GBX fixer (Kodak)
Spatula
UV cross-linker/80°C baking oven
DNA gel photography set up
500 g blot weight
UV transilluminator
Agarose electrophoresis apparatus and power
  supply
Plastic wrap
PCR thermocycler
X-ray cassette
X-ray film (Kodak BioMax™)
Darkroom access
Development folders or plastic wrap
Nylon transfer membrane (Hybond™ N+; GE
  Healthcare)
Whatman 3MM™ paper
Three inches of cut paper towels
Plastic containers (6)

MODULE I.3

PCR reagents (AmpliTaq Gold™ kit), Applied
  Biosystems (1 reaction)
Ttr Forward 1 Primer (10 pmol/µl), 1 µl
Ttr Reverse 1 Primer (10 pmol/µl), 1 µl
Prebiotinylated labeled *Ttr* Oligonucleotide
  Probe (20 ng)
Optional: 1 µl mouse genomic DNA
Agarose (1g)
1x TAE buffer (500 ml). See Appendix II.
Ethidium bromide gel stain (200 ml). See
  Appendix II.

Plastic staining and destaining trays.
1x TE (100 µl). See Appendix II.
5x Blue Juice Loading Dye (BJ) (5 µl). See
  Appendix II.
100-bp Ladder marker in 5x BJ (50 ng/µl)
  (10 µl) (American Bioanalytical). See
  Appendix II.
Southern depurination solution (100 ml). See
  Appendix II.
Southern denaturation solution (100 ml). See
  Appendix II.

Southern neutralization solution (100 ml).
  See Appendix II.
ddH2O (250 ml). See Appendix II.
10x SSC (500 ml). See Appendix II.
6x SSC (200 ml). See Appendix II.
Ttr Hyb solution (4 ml). See Appendix II.
Oligo Southern Wash I (200 ml), See
  Appendix II.
Oligo Southern Wash II (200 ml), See
  Appendix II.

MODULE MATERIALS AND REAGENTS (PER PAIR OF STUDENTS)

MODULE I.3 (CONT.)

Phototope Blocking Solution (100 ml), See Appendix II.

Streptavidin Solution (15 ml). See Appendix II.

Phototope Detection Wash Solution 1 (100 ml). See Appendix II.

Biotinylated alkaline phosphatase solution (15 ml). See Appendix II.

Phototope Detection Wash Solution II (100 ml). See Appendix II.

Phototope Detection Mix (10 ml). See Appendix II.

GBX developer (Kodak)

GBX fixer (Kodak)

PCR thermocycler

Agarose gel electrophoresis apparatus

DNA gel photography unit

Nylon transfer membrane (Hybond™ N+; GE Healthcare)

Whatman 3MM™ paper

Three inches of cut paper towels

Creative blot weight (500 g)

Plastic containers (6)

UV cross-linker or baking oven (80°C)

Development folder (or plastic wrap)

X-ray cassette

X-ray film (Kodak BioMax™)

Darkroom access

Pencil

Marking pen

Spatula

Plastic sealable hybridization bags

Lid locks

Heat block or water bath

Hybridization oven

Study Questions for Project I

Module I.1

1. In Lab Period I.1.1, Step 5, what is the purpose of the addition of the proteinase K, the detergents, and the RNase?

2. In Lab Period I.1.2, Step 8, did you see the mouse DNA precipitate out of solution? If so, describe what the precipitate looked like.

3. In Lab Period I.1.3, what is the purpose of dialyzing your mouse DNA sample against the 0.1x TE buffer?

4. What volume of liquid did you recover from the dialysis membranes in Lab Period I.1.4, Steps 1 and 2? How does this compare with the volume that was loaded onto the dialysis membranes? Did you recover more or less volume than you loaded onto the membranes? Whether your answer was more or less, try to explain why the volume recovered might be different than the volume that you loaded onto the membranes.

5. In Lab Period I.1.4, Part A, Steps 4 and 5, what were your absorbance readings for your mouse genomic DNA at 260 nm and at 280 nm? What is your 260/280 ratio? What is the concentration of your mouse genomic DNA? What is the total amount of mouse genomic DNA that you isolated? Does your DNA appear to be a fairly pure sample? If not, can you speculate on the possible nature of the contaminating material?

6. In Lab Period I.1.4, Part B, Step 17, what is the concentration and total amount of mouse genomic DNA that you estimated from the gel? How does this compare with the amounts you estimated by UV absorbance? Can you explain any difference that you see in the two measurements?

7. From the gel in question 6, was your genomic DNA of uniformly high molecular weight ($\geq$20 kb)?

8. From the gel in question 6, was your mouse genomic DNA free of contamination by RNA?

9. In Lab Period I.1.5, why is the SfiI digest done at 50°C instead of 37°C?

10. In Lab Period, I.1.6, what is the purpose of the ethanol wash done in Steps 6, 7, and 8?

11. In Lab Period I.1.7, what differences did you observe in the size patterns of the products of digestion of your genomic DNA with the different restriction endonucleases? How did these differences relate to the size of the recognition site of each endonuclease (four cutter, six cutter, or eight cutter)?

12. In Lab Period I.1.7, what differences did you observe in the size patterns of the products of digestion of your genomic DNA with the isoschizomers HpaII and MspI? What do these differences tell you about the status of DNA methylation of your mouse liver genomic DNA?

Module I.2

1. What is the purpose of the depurination step in Lab Period I.2.1, Step 1?

2. In Lab Period I.2.2, Step 3, did you successfully transfer most of your digested DNA samples from your gel to the blotting membrane? If not, what are some possible reasons for incomplete transfer?

3. In Lab Period I.2.3, Step 3, what is the purpose of the final elongation step at 72°C for 7 minutes?

4. In Lab Period I.2.4, Step 5, were you successful in labeling the *Rvt* gene probe nonradioactively? How do you know whether you were successful? If you were not successful, speculate on some possible reasons for the failure of the labeling procedure.

5. In Lab Period I.2.5, what is the purpose of the prehyrbridization step?

6. In Lab Period I.2.6, what is the purpose of hybridizing the Southern blot with the labeled *Rvt* probe?

7. In Lab Period I.2.7, what is the purpose of adding the antibody in Step 8?

8. In Lab Period I.2.7, Step 18, did you specifically detect the *Rvt* gene sequence in your DNA samples on your Southern blot? What is the molecular size (in bp or kb) of the *Rvt* gene sequence that you detected? What can you say about the status of DNA methylation of the *Rvt* gene sequence in mouse liver genomic DNA?

Module I.3

1. In Lab Period I.3.1, why are two different primers required for a successful PCR amplification?

2. In Lab Period I.3.1, why is it advantageous for the AmpliTaq Polymerase enzyme to be a thermostable polymerase?

3. In Lab Period I.3.2, does the result on your gel demonstrate that you successfully amplified the *Ttr* gene from total genomic DNA by PCR? What is your estimate of the size of this PCR product? Does this size make sense based on your knowledge of the gene?

4. In Lab Period I.3.3, what is the purpose of the UV cross-linking and baking of the membrane?

5. In Lab Period I.3.4, why do you think it is advantageous to keep the volume of the prehybridization/hybridization solution low (e.g., 4 ml instead of 40 ml)?

6. In Lab Period I.3.5, why is the hybridization of the *Ttr* probe to this membrane done at a much lower temperature (55°C) compared with the temperature of hybridization (70°C) of the *Rvt* probe to the genomic Southern blot membrane in Module I.2?

7. What is the difference in the way the *Ttr* oligonucleotide probe for the PCR blot and the *Rvt* gene probe for the genomic blot were labeled?

8. In Lab Period I.3.6, what is the purpose of the streptavidin addition in Step 5?

9. In Lab Period I.3.6, what is the purpose of the biotinylated alkaline phosphatase solution added in Step 8?

10. In Lab Period I.3.6, Step 16, did you specifically detect the *Ttr* PCR product on your Southern blot? How do you know that this hybridization signal is specific to the *Ttr* sequence?

11. What different lines of evidence do you now have to confirm the identity of the PCR product as representing the *Ttr* gene?

GENOMIC CLONING, DNA SEQUENCING, AND BIOINFORMATICS

PROJECT SUMMARY

Using the protocols described in this experiment, you will construct a mouse genomic DNA library in the bacteriophage cloning vector lambda ZAP Express (Stratagene). This lambda phage cloning vector is ideal for the construction of large genomic DNA libraries or of cDNA libraries (see Project IV). In addition to lambda ZAP Express, a variety of other lambda vectors from a variety of commercial suppliers are useful for making genomic and/or cDNA libraries.

The goal in this experiment will be to construct a mouse genomic DNA library from which you will isolate clones of a specific, highly repeated mouse DNA gene sequence. This highly repeated DNA sequence is 1.4 kilobases (kb) in length, is cleaved by the restriction endonuclease EcoRI (**Module II.1**), and includes a gene encoding reverse transcriptase (*Rvt*). The genomic library that you construct in lambda ZAP Express (Module II.1) will consist of millions of mouse DNA fragments, each ligated into an individual molecule of the lambda ZAP Express vector to form a "recombinant phage genome" referred to as a "clone." Each clone will be packaged into an individual infectious bacteriophage particle. These phage will be plated on a "lawn" of bacterial (*Escherichia coli*) cells on Petri plates where they will each form a "plaque." A plaque is a region where a single initial phage particle has propagated itself by infecting a single *E. coli* cell, replicated to form 100 to 200 progeny phage each carrying the same recombinant DNA molecule, and then lysed the cell so that each individual progeny phage particle can infect a neighboring bacterial cell. This process repeats and semiclear regions (plaques) appear where thousands of *E. coli* cells have been lysed, and millions of copies of phage carrying a unique clone have been produced. Some of these plaques will contain clones of the mouse *Rvt* sequence. To identify the *Rvt*-containing plaques, you will screen your mouse genomic library by hybridizing plaque lifts with a nonradioactively labeled DNA probe using a chemiluminescent detection system (**Module II.2**). The mouse *Rvt* gene sequence, originally cloned into lambda ZAP Express, will then be amplified from a positive clone using the polymerase chain reaction (PCR). The products of this PCR reaction will be subcloned into a plasmid vector (pDrive) (Qiagen) and propagated in bacterial cells (**Module II.3**). These plasmids will then be recovered from the bacterial cells and analyzed by gel electrophoresis. A *Rvt* gene PCR product derived from the plasmid DNA, produced in Module II.3, will then be subjected to dideoxy thermal cycle sequencing (**Module II.4**), and the DNA sequence data will be analyzed using bioinformatics methods (**Module II.5**).

PROJECT BACKGROUND

Isolation and characterization of genomic DNA were described in Project I. In Project II, you will use methods of recombinant DNA technology to "clone" a specific gene sequence. Recombinant DNA cloning involves the formation of new, heritable genetic material by insertion of foreign DNA into an appropriate "cloning vector." A cloning vector (e.g., plasmid DNA or bacteriophage DNA) facilitates the incorporation and propagation of genetic material in a convenient host organism (usually bacteria). This approach allows you to produce large quantities of a particular gene sequence and then to manipulate that sequence in a variety of ways that will allow you to study it much more extensively than is possible when it is present in its normal *in vivo* context. These analyses can include restriction endonuclease site mapping, DNA sequence analysis, *in vitro* mutagenesis, production of a specific hybridization probe, studies of gene structure, and studies of gene regulation and expression.

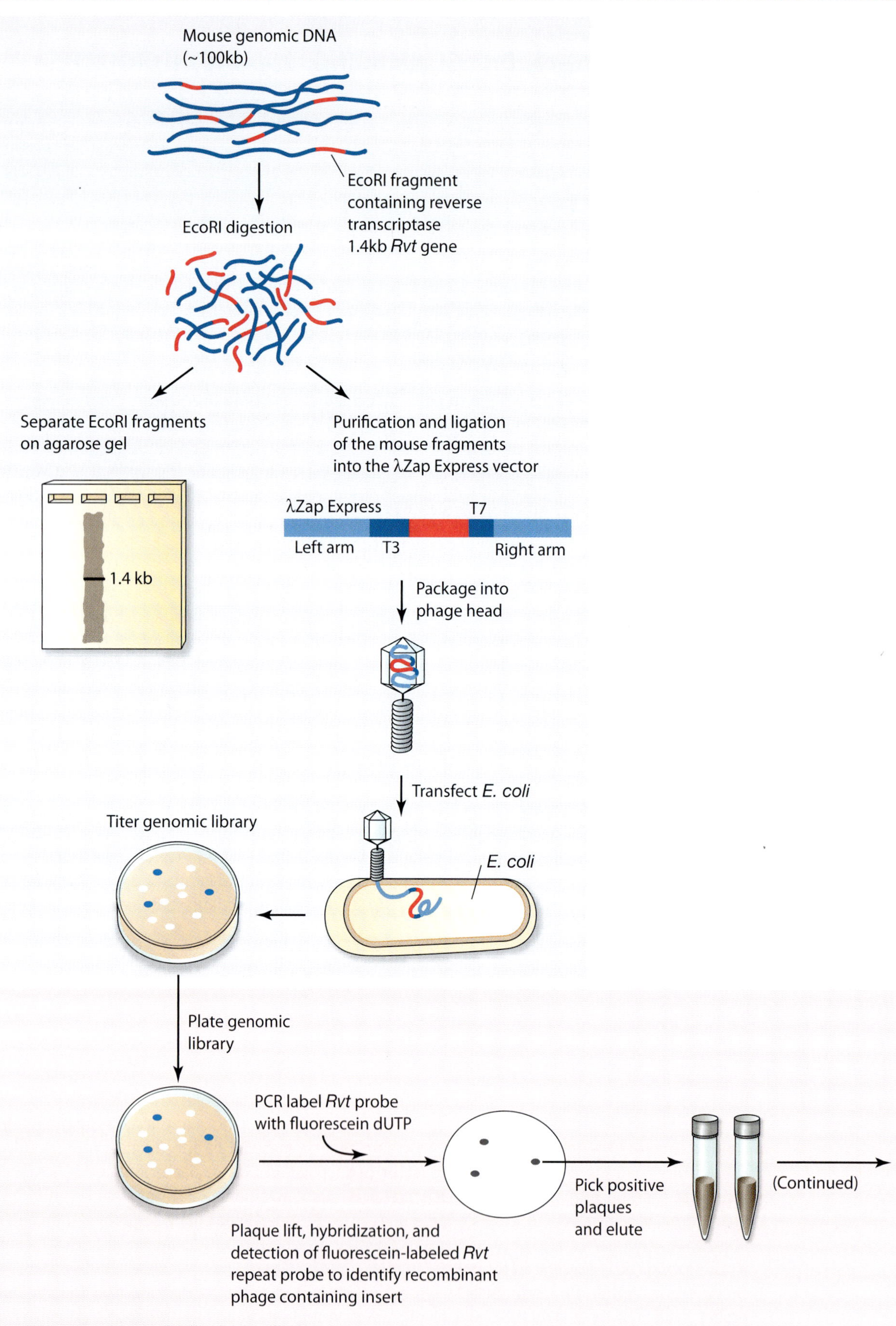

Project II flow diagram.

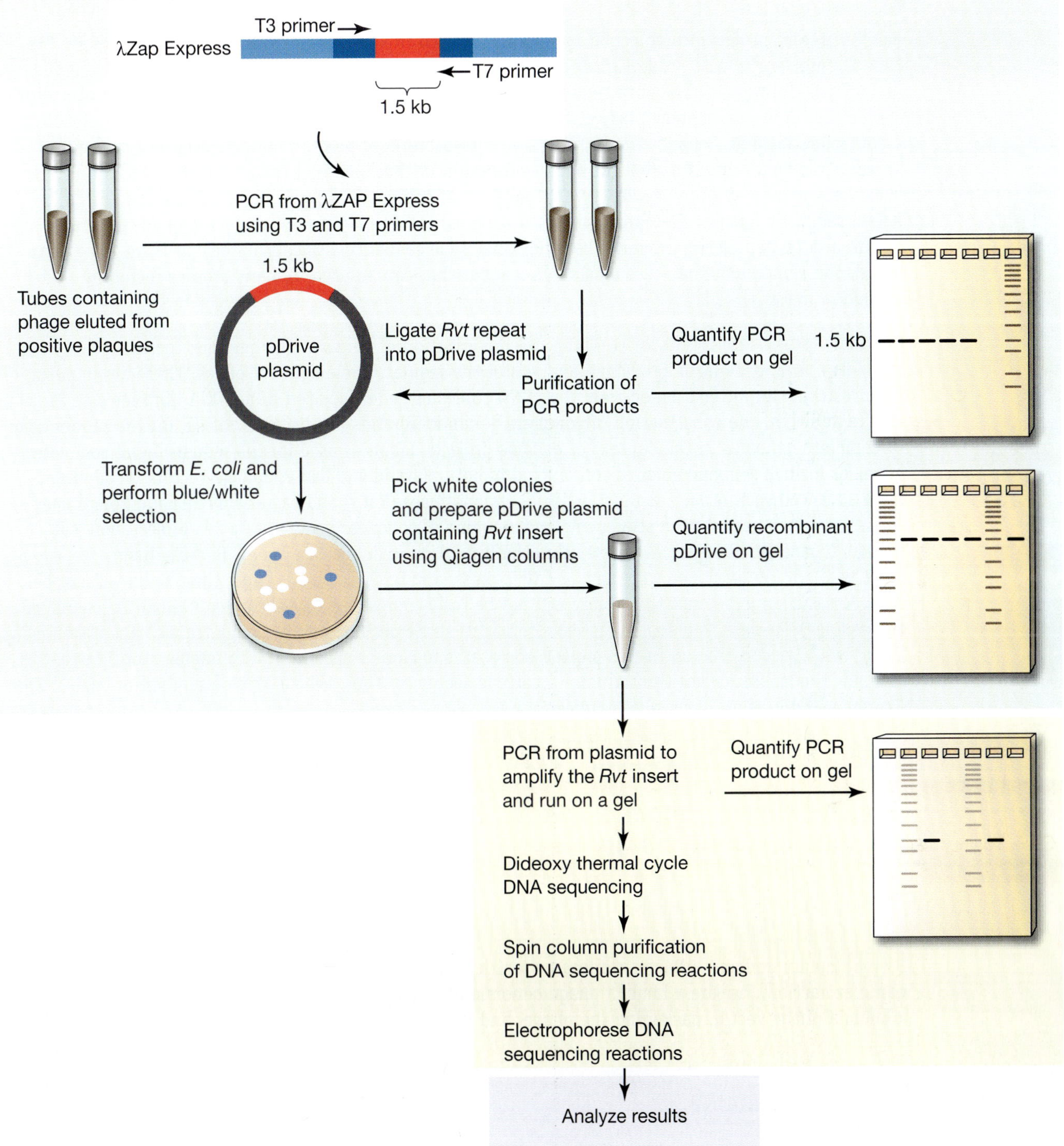

Project II flow diagram (continued).

BACKGROUND, continued

Many different projects in molecular biology begin with the molecular cloning of a specific gene. Three fundamental phenomena underlie our ability to use recombinant DNA technology to clone a specific gene. First, cleavage of DNA with sequence-specific restriction endonucleases and joining (ligation) of pieces of DNA with complementary ends enable the cutting and rejoining of different DNA molecules in a controlled fashion. Second, the introduction into bacteria of pieces of DNA ligated into a cloning vector carrying an origin of replication that function in that bacterial species allows us to propagate these recombinant DNA molecules to yield millions of copies of each. Third, hybridization with a labeled, sequence-specific DNA probe allows the identification of clones of a specific gene sequence within a large population of clones bearing other gene sequences. Once isolated, purified, and propagated, the cloned gene becomes a valuable reagent that can be used in many studies including DNA sequence analysis.

Genomic libraries are useful for studying the organization of genes in the genome, including the intron/exon arrangement of genes and analysis of the regulatory elements of genes. These are important features that cannot be studied from the cDNA copies of genes isolated from cDNA libraries (see Project IV). In addition, the construction of complete genomic libraries has become critical for the sequencing of complete genomes in the context of genome projects. Libraries used for the sequencing of complete genomes include genomic libraries constructed in both plasmid and bacteriophage vectors. In addition, more advanced vectors such as yeast artificial chromosomes (YACs) and bacterial artificial chromosomes (BACs) are often used for the sequencing and mapping of complete genomes. In this project, you will prepare a genomic library from mouse using the restriction endonuclease EcoRI. Because fragments as large as 12 kb can be cloned into lambda ZAP Express, any EcoRI fragment larger than 12 kb will not be cloned in this project. Thus, this library will not be truly complete (i.e., will not contain all fragments of the genome). In class, you should discuss alternative strategies for constructing a truly complete genomic library with your instructor. In this project, the goal is to clone a specific 1.4-kb fragment of DNA that contains the mouse reverse transcriptase repetitive sequence. Thus, the 12-kb cloning limit of lambda ZAP Express will not be an issue in this project.

MODULE II.1

Prepare Mouse Genomic DNA for Cloning; Production and Titering of a Lambda Genomic Library

MODULE SUMMARY

Prepare mouse genomic DNA for cloning by restriction enzyme digestion, gel electrophoresis, and organic extraction. Prepare a lambda phage genomic library by ligation of the prepared genomic DNA into a lambda phage vector, packaging the phage, and titering the library.

MODULE BACKGROUND

In order to produce a library of genomic DNA sequences, it is often useful to digest the genomic DNA with a sequence-specific restriction endonuclease in order (1) to yield DNA fragments with specific ends that can be efficiently ligated into a cloning vector bearing complementary ends and (2) to produce fragments of genomic DNA that are of an average size that, when incorporated into the cloning vector, can be productively propagated in bacterial cells. We will use the common restriction enzyme EcoRI, which digests DNA at the 6-bp sequence: 5'-GAATTC-3'. These sites occur essentially at random throughout the mouse genome such that digestion at each of these sites will yield fragments in a wide range of sizes, but the average size fragment produced will be approximately 4000 bp, which is an ideal size for cloning into a lambda bacteriophage cloning vector.

After the genomic DNA has been digested with EcoRI, it is ready to be ligated into the cloning vector. You will ligate the EcoRI-digested mouse genomic DNA into the lambda ZAP Express cloning vector from Stratagene. This vector has also been digested with EcoRI by the manufacturer; thus, it comes with ends that are directly compatible with the EcoRI ends on the digested mouse genomic DNA. However, unlike the ends on the insert genomic DNA, the vector ends have been dephosphorylated so that they will not religate to one another to form an empty vector or a double vector without any insert DNA. A 1:1 molar ratio of vector to insert DNA molecules is used in this ligation reaction. This ratio favors the formation of recombinant DNA molecules consisting of a single insert molecule ligated to both a left arm and right arm of the lambda ZAP Express vector.

The recombinant phage DNA (insert genomic DNA molecules ligated to cloning vector DNA molecules) can then be packaged into infectious phage particles by incubation with a commercially supplied "packaging extract." You will use Gigapack Gold packaging extract from Stratagene for this purpose. The packaging extract contains the proteins required to assemble phage heads and tails that can "package" the phage DNA and form a functional "phage particle." The formation of a functional phage particle requires the presence of a lambda genome or a recombinant lambda genome that becomes encapsulated within the head of the bacteriophage during the "packaging reaction." The packaging reaction takes place during a 2-hour period at room temperature. During this time, complete, infectious phage particles form spontaneously! As a result, you will produce your lambda genomic library (a collection of many phage particles, each containing different recombinant lambda genomes). After the packaging reaction, chloroform is added to stabilize the phage library. Your library can be stored at either 4°C or, after the addition of dimethyl sulfoxide (DMSO) to a final concentration of 7%, aliquoted and stored frozen at –70°C.

To assess the success of the recombinant DNA construction and packaging reaction, you will determine the number of bacteriophage in your lambda genomic library. This is known as the "titer" of the library. This procedure will provide you with two important pieces of information: (1) the number of infectious phage particles in your library (its "titer") and (2) the proportion of those phage particles that contain *recombinant* lambda genomes (i.e., the number that carry an insert molecule). To titer your library, you will combine an aliquot of the products of your packaging reaction with an aliquot of host bacterial cells. You will use XL1 Blue MRF′ *E. coli* bacterial cells for this purpose because they have been engineered by the manufacturer of the packaging extract to be an optimally efficient host for the phage formed during the packaging reaction. After plating and incubating at 37°C for about 8 hours to allow plaques to form, you can determine the titer of your entire library by simply counting the plaques formed from a small aliquot of your library and then calculating the number of plaque forming units (PFUs) in your total library. This can be done because each plaque is derived from a single initial phage (or PFU) that infects one bacterial cell. Thus, each phage results in one plaque.

In addition to determining the titer of your library, it is important to determine the percentage of recombinant plaques. Although the vector was dephosphorylated to avoid religation of vector ends to one another and hence avoid the formation of "empty" vectors during the ligation reaction, some vector molecules may not have been cut in the first place and thus may have been packaged into phage particles. These "vector-only" clones do not carry an insert and, therefore, form "nonrecombinant" plaques. Nonrecombinant plaques are distinguished from recombinant plaques on the basis of a colorimetric indicator, such that nonrecombinant plaques are blue and recombinant plaques are clear. The titer and the percentage of recombinant plaques are two important parameters that define the quality of any phage library.

LAB PERIOD II.1.1.   EcoRI RESTRICTION ENZYME DIGEST OF MOUSE GENOMIC DNA

Step 1   Each student will receive a tube in an ice bucket containing 40 µl of mouse genomic DNA (prepared as described in Project I and/or provided at a concentration of 6.25 ng/µl or 250 ng in 40 µl). Label this tube with your student number and "Mouse Eco" (this is important so that you can get your own tube back at the next lab period). Set up the following restriction endonuclease digest in a 1.5-ml microfuge tube containing the 40 µl (250 ng) of genomic mouse DNA. To this tube, add the following components in the order shown:

> 5 µl 10x EcoRI buffer
>
> 4 µl ddH$_2$O
>
> 1 µl EcoRI enzyme (add enzyme last)
>
> 50 µl final volume

PROCEDURAL NOTES

ddH$_2$O is sterilized double-distilled water.

Each restriction enzyme requires its own buffer for optimal activity.

Mix the reaction by pipetting up and down or by flicking the tube. Do not vortex! Spin briefly in your nanofuge to return all of the liquid to the bottom of the tube.

Step 2   Place the tube in an incubator at 37°C, and allow the digestion to proceed overnight.

PROCEDURAL NOTES

Do *not* remove enzymes from –20°C until you are ready to use them. Keep them on ice during use, and then return them to –20°C *immediately*. This is important because the enzymes will lose activity if they are allowed to warm.

Do not vortex enzymes or digests containing enzymes! Instead, always mix the contents by pipetting up and down or by flicking the tube several times.

Step 3   The following morning, add a second aliquot (1 µl) of EcoRI, and incubate at 37°C for 2 to 4 hours (longer is better). Mix the reaction by pipetting up and down or by flicking the tube. Spin briefly in your nanofuge to return all of the liquid to the bottom of the tube. If scheduling prohibits students from returning to the laboratory the day after the digestion is begun, the instructor may do Steps 3, 4, and 5.

PROCEDURAL NOTE

The reason for adding the second aliquot of the restriction enzyme is that the original enzyme will have lost some activity during the overnight digestion at 37°C. The additional EcoRI enzyme added in the morning will help to restore lost enzyme activity and will aid in digesting a greater proportion of the EcoRI sites in the mouse genomic DNA.

Step 4   Place your tube in a waterbath at 65°C for 20 minutes to heat-inactivate the EcoRI enzyme.

PROCEDURAL NOTE

Some restriction enzymes (such as EcoRI) are inactivated by heating to 65°C. Some enzymes require only 5 minutes for heat inactivation, whereas other enzymes cannot be inactivated at 65°C. Refer to New England Biolabs catalogue or webpages for information about each restriction endonuclease.

Step 5   Store the tube containing the digested DNA at 4°C until the next lab period.

LAB PERIOD II.1.2.   AGAROSE GEL ELECTROPHORESIS OF EcoRI-DIGESTED MOUSE GENOMIC DNA

Step 1 Prepare a 1.0% agarose gel as described in Lab Period I.1.4, Part B.

Step 2 Retrieve your EcoRI digested mouse DNA tube from storage at 4°C, and spin it briefly in the nanofuge to bring all of the liquid to the bottom.

Step 3 Pipette 12 µl from your "Mouse Eco" digest tube to a new microfuge tube. Add 3 µl of 5x Blue Juice loading dye (BJ) to this new tube, and label it "Mouse Gel." Stir with your pipette tip to mix. Store the remaining 38 µl of your "Mouse Eco" digest DNA at 4°C.

Step 4 You will load the samples into the wells underneath the buffer. If you have not done this before, this may take some practice. Practice loading 5- to 10-µl samples of 5x BJ in your extra wells (lanes 1, 2, 8, 9, and 10) before attempting to load your real samples!

Step 5 Each group will load the following samples:

a. Two control DNA size standards: lambda DNA previously cut with HindIII (lambda HindIII) and lambda DNA previously cut with BstEII (lambda BstEII). These are molecular weight markers premixed with 5x BJ at a final concentration of 25 ng/µl and provided to you by the instructor (refer to Appendix III for maps of these markers). Remove 6 µl of each marker, and transfer to two new microfuge tubes (label them lambda HindIII and lambda BstEII). Heat both tubes at 65°C for 5 minutes. Spin briefly in your nanofuge to bring all of the liquid to the bottom of the tube. Place the tubes on ice until you are ready to load these samples on your gel.

PROCEDURAL NOTE

Heating the lambda markers is required for all of the bands in the gel to electrophorese according to their true molecular weight. This is because two of the lambda DNA fragments in each marker can hybridize together via their "*cos* ends." These *cos* ends are 12 nucleotide, single-stranded overhangs that can hybridize together causing two of the fragments to join together and run as a single large band.

b. One uncut mouse DNA control (provided to you by the instructor at 10 ng/µl — premixed with 5x BJ, 15 µl total volume). *Do not* heat this marker!

c. Your 12 µl EcoRI digested mouse DNA to which you should have added 3 µl 5x BJ. Each student will now have 15 µl of EcoRI cut mouse DNA in BJ. Remember that each student has one 15-µl sample of digested mouse DNA (two samples total per group). Note the amount of each sample to load and the order in which they will be loaded.

Load the samples in lanes 3 to 7, as follows:

Lane 1: Empty

Lane 2: Empty

Lane 3: 6 µl (150 ng) lambda HindIII marker DNA in 5x BJ

Lane 4: 15 µl (150 ng) uncut mouse DNA in 5x BJ

Lane 5: 15 µl EcoRI digested mouse DNA in 5x BJ (approximately 50 ng) (partner 1)

Lane 6: 15 µl EcoRI digested mouse DNA in 5x BJ (approximately 50 ng) (partner 2)

Lane 7: 6 µl (150 ng) lambda BstEII marker DNA in 5x BJ

Lane 8: Empty

Lane 9: Empty

Lane 10: Empty

LAB PERIOD II.1.2 CONT.

PROCEDURAL NOTE

Save all of your DNA standards (markers) at 4°C. Also, save your remaining uncut mouse DNA at 4°C.

Step 6   Electrophorese the gel at 60 V for 1.5 to 2 hours.

Step 7   After electrophoresis, turn off the power supply. Unplug the leads from the power supply, and carefully stain the gels for 10 minutes in ethidium bromide DNA gel stain followed by 15 minutes in a destain water bath (refer to Lab Period I.1.4).

CAUTION!

Ethidium bromide DNA gel stain is toxic, mutagenic, and/or carcinogenic. Wear gloves, lab coats, and safety glasses when using this reagent!

PROCEDURAL NOTE

Appendix IV describes an alternative DNA stain, SYBR Safe (Molecular Probes).

Step 8   Place your gel on an ultraviolet (UV) transilluminator, and photograph the gel (or take a digital image). Does the EcoRI digest of your mouse DNA appear to have been successful? An example of the expected appearance of this gel after electrophoresis is shown in **Figure 2-1**. Compare the photograph of your gel with that in Figure 2-1.

CAUTION!

Always handle the gels carefully!

Step 9   Dispose of your gel in the appropriate waste containers for holding stained gels.

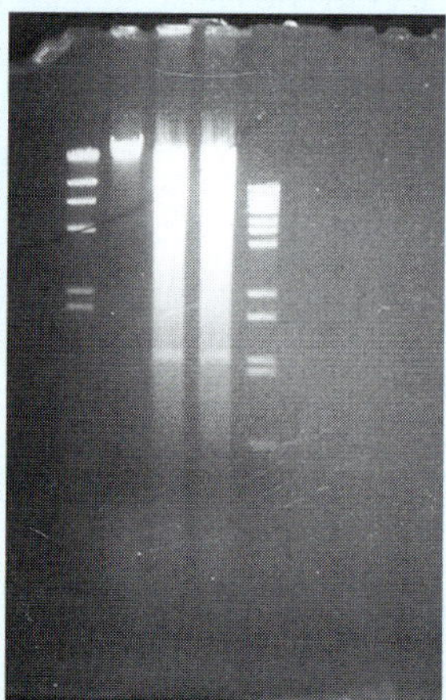

Fig 2-1 Expected appearance of EcoRI digest on mouse DNA gel.

LAB PERIOD II.1.3.   **PURIFY THE EcoRI-DIGESTED MOUSE GENOMIC DNA BY ORGANIC EXTRACTION AND DROP DIALYSIS**

PART A Organic Extraction

Organic extractions are used to remove proteins and some lipids from solutions containing a mixture of macromolecules, including nucleic acids. In the extraction, the proteins partition to the organic phase and the interface between the organic and aqueous phases. The nucleic acids partition to the aqueous phase. Phase-lock gel tubes used in this protocol are very useful because during centrifugation they form a partioning gel at the interface between the organic and aqueous phases. Thus, the nucleic acids become sequestered in the aqueous phase above the gel partition and are easily recovered by pipetting the aqueous phase containing the nucleic acids, leaving the undesired proteins sequestered in the organic phase below the gel partition.

»»»

LAB PERIOD II.1.3 CONT.

Step 1 Transfer both partners' DNA samples (38 µl each; 76 µl total) into *one* prespun green Phase Lock Light tube (Eppendorf) provided by the instructor. These tubes contain a gel material that enables the physical separation of organic and aqueous solutions, after centrifugation, due to differences in their densities. These tubes are "prespun" by microcentrifugation at 16,000x *g* for 1 minute).

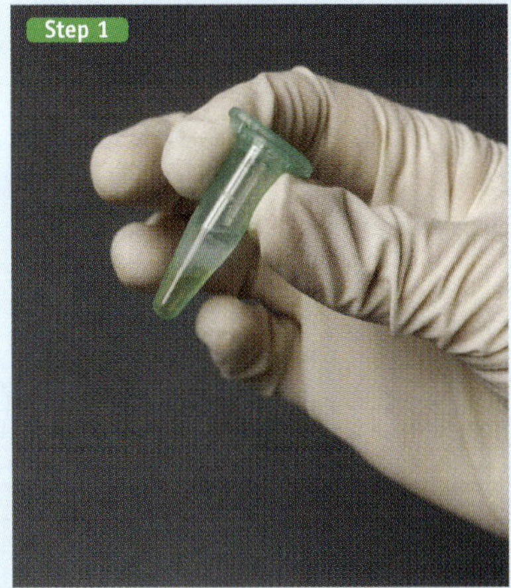

> **CAUTION!**
>
> These steps require the use of phenol and "chisam" (24:1 chloroform/isoamyl alcohol vol/vol). Please note the following precautions.
>
> ---
>
> Phenol and chloroform (in chisam) are both toxic and produce noxious fumes. Because phenol and chisam are solvents that are caustic and will burn skin and clothes, be sure to always wear a lab coat and gloves. Keep tubes closed at all times when not in use. Work in a fume hood and open the tubes containing phenol and/or chisam for brief periods only.
>
> ---
>
> Handle the tubes containing phenol and chisam carefully!

Step 2 Add 24 µl of ddH$_2$O to the tube to bring the total volume to 100 µl.

Step 3 Add 50 µl buffered phenol to your green phase lock tube. Close the cap. Hold the cap securely, and thoroughly mix by inverting the tube several times to form a transiently homogeneous suspension. Do not vortex.

Step 4 Add 50 µl chisam to the green phase lock tube. Close the cap. Hold the cap securely, and thoroughly mix by inverting the tube several times. Do not vortex.

Step 5 Spin this tube and a balance tube for 5 minutes in a microfuge at 16,000x *g*. This will separate the phases such that the organic phase will be at the bottom of the tube. The phase lock gel partition will form a middle layer, and the aqueous phase containing the DNA will be on top, as shown in Step 1.

Step 6 Using a P100 or P200 pipette, transfer the upper aqueous phase to a new microfuge tube. Label this tube with your group number and "Ext DNA." You will drop dialyze the sample on a membrane as described below.

PART B Drop Dialysis

Each pair will need one Petri dish, one 250-ml beaker, and one filter membrane to drop dialyze their EcoRI-digested mouse genomic DNA. Be sure to set up the dialysis in a protected area of your bench! This dialysis will purify your mouse genomic DNA by removing trace amounts of phenol, chisam, and other low molecular weight contaminants. The pore size of the membrane used in this protocol allows low molecular weight contaminants to pass through the membrane but does not allow the DNA to move through the membrane.

Step 1 Put 200 ml sterile ddH$_2$O into the 250-ml beaker.

Step 2 Float a 45-mm diameter Millipore dialysis membrane filter *shiny side up* on the buffer surface. Allow the filter to prewet for about 5 minutes (*do not submerge*). This is shown in Project I, Lab Period I.1.3, Step 2.

Step 3 *Carefully* pipette all of the "Ext DNA" from Step 6 (between 50 and 100 µl) onto the center of the dialysis filter. This is shown in Project I, Lab Period I.1.3, Step 6.

>>>

LAB PERIOD II.1.3 CONT.

Step 4   Cover the beaker with the Petri dish, and *do not disturb* for 3 to 8 hours or more (longer is better!). If it is convenient for your schedule, overnight dialysis is also excellent.

PROCEDURAL NOTE

The following two steps need to be performed on the day after Lab Period II.1.3. They can either be performed by the instructor or by the students if they can return on the day after Lab Period II.1.3.

Step 5   After 8 hours or more of dialysis, *carefully* retrieve as much of the DNA sample as possible with a P200 pipette, and transfer it into a new microfuge tube (label it with your group number and "pure mouse Eco"). It is important that you keep the pipette tip at a 90° angle to the membrane. This will prevent you from submerging the membrane and losing your sample.

Step 6   It will be necessary for you to reduce the volume of your DNA before ligation to the lambda vector. Centrifuge your dialyzed sample in a vacuum centrifuge ("speed-vac concentrator") to 6 µl or less (if less, adjust the volume of your sample to 6 µl by adding ddH$_2$O). Store the sample at 4°C until the next lab period.

LAB PERIOD II.1.4.   **LIGATE THE DIGESTED GENOMIC DNA TO A LAMBDA (λ) BACTERIOPHAGE VECTOR**

BEGINNING WITH this lab period, you will use the mouse genomic DNA that you prepared in the previous lab periods to construct a genomic library in a lambda bacteriophage vector (lambda ZAP Express; Stratagene). **Figure 2-2** shows a diagram of the pBK-CMV plasmid located within the lambda ZAP Express vector and the sequence of the vector around the cloning site. We will use this diagram when we assess our cloning and sequencing results.

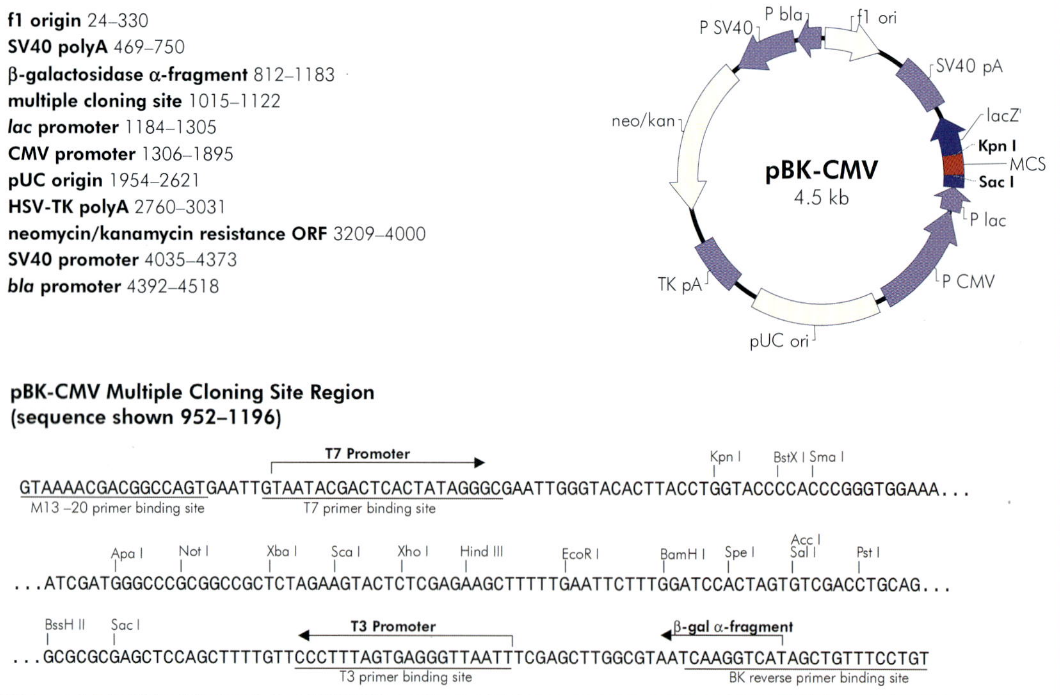

f1 origin 24–330
SV40 polyA 469–750
β-galactosidase α-fragment 812–1183
multiple cloning site 1015–1122
lac **promoter** 1184–1305
CMV promoter 1306–1895
pUC origin 1954–2621
HSV-TK polyA 2760–3031
neomycin/kanamycin resistance ORF 3209–4000
SV40 promoter 4035–4373
bla **promoter** 4392–4518

pBK-CMV Multiple Cloning Site Region
(sequence shown 952–1196)

Fig 2-2   Diagram of the lambda ZAP Express vector and its sequence around the cloning site. (Courtesy of Stratagene; reprinted with permission.)

»»

LAB PERIOD II.1.4 CONT.

Each pair will set up *one* ligation reaction using their purified mouse genomic DNA. At this point, each pair has *one* tube containing 6 µl EcoRI digested mouse DNA.

Step 1 Retrieve your tube containing 6 µl EcoRI digested mouse DNA (the microfuge tube labeled "pure mouse Eco").

Step 2 Add all 6 µl of the EcoRI cut mouse DNA (about 200 ng) to a different microfuge tube containing 1 µl of EcoRI-cut, dephosphorylated lambda ZAP Express vector DNA (1.0 µg) (this tube will be provided by the instructor). Label this tube with your group number and "Lig."

Step 3 Place each of the following reagents into the microfuge tube containing your insert and vector DNA:

> 1 µl 10 mM ATP (pH 7.5) (This is NOT <u>d</u>ATP)
>
> 1 µl 10x ligase buffer
>
> 1 µl T4 DNA ligase (add the enzyme last) (4 units/µl)
> ____
>
> 10µl final volume

Mix by stirring with your pipette tip or by tapping the tube gently. Spin briefly in your nanofuge. *Do not vortex!*

Step 4 Incubate the ligation reaction overnight at 14°C.

Step 5 The next day, remove the reaction tube from the 14°C incubator, and store it at 4°C until Lab Period II.1.5 (this can be done by the instructor if necessary).

LAB PERIOD II.1.5.   **PACKAGE THE LAMBDA PHAGE GENOMIC LIBRARY**

TO PACKAGE the mouse lambda ZAP Express ligation into lambda phage particles, *each pair* will do *one* packaging reaction. Retrieve your ligation reaction from the 14°C incubator or from storage at 4°C if it has been more than one day since the ligation reaction was performed.

Before use, the packaging extract (Gigapack Gold; Stratagene) is stored at –70°C. These tubes *must be kept on dry ice until immediately before use. Do not thaw the extract tube until you are ready to use it!*

Step 1 Quickly warm a packaging extract tube between your fingers until it *just* begins to thaw (this takes just a few seconds).

Step 2 *Immediately* add 5 µl of your mouse lambda ZAP Express ligation reaction to the extract tube.

Step 3 Stir gently with your pipette tip. Do not pipette up and down. Do not introduce any air bubbles.

PROCEDURAL NOTES

It is difficult not to introduce a few air bubbles when you do the stirring. It has been shown that production of bubbles in the reaction causes the packaging to be less efficient. Do not create a "foam" as you stir.

The packaging extract contains all of the lambda proteins necessary for the spontaneous assembly of lambda phage particles when the ligated DNA is added.

»»

LAB PERIOD II.1.5 CONT.

Step 4 Spin the extract tube 5 seconds in your nanofuge, and incubate the reaction at room temperature for *2 hours*. Do not exceed 2 hours! Label this pale blue tube with your group number and "Genomic Library."

Step 5 After the 2-hour incubation of the ligated DNA with the packaging extract, add 500 µl of SM buffer (phage storage buffer) and 20 µl of *chloroform (not chisam!)* to the "Genomic Library" tube. Mix gently by inverting several times.

CAUTION!

Wear gloves while using chloroform and work in a fume hood!

Step 6 Spin the "Genomic Library" tube briefly in your nanofuge to sediment the debris. Save this pale blue tube on ice at your bench, or if storage needs to be longer than for a few hours, store the tube at 4°C.

CAUTION!

Save this tube! Do not discard! Do not put the library tube at –20°C or you will inactivate your phages!

PROCEDURAL NOTE

After the packaging reaction is complete, this tube now contains your mouse genomic library! At this point, you are ready to titer your genomic library. You could, however, also choose to store this library for short-term periods of up to 1 month at 4°C or for longer periods at –70°C after addition of dimethyl sulfoxide (DMSO) to a final concentration of 7%. If you do store your library for any significant amount of time, you should always "retiter" it before beginning any other procedure because the titer can change (decrease) during long periods of storage.

LAB PERIOD II.1.6.   **TITER THE LAMBDA PHAGE GENOMIC LIBRARY**

AFTER THE recombinant phage genomes are packaged into infectious lambda particles, they are ready to be used to transfect *E. coli* cells to determine the "titer" of the library (the number of PFUs). In order to have cells ready for a transfection reaction, it is necessary to initiate a culture of these cells the night *before* this reaction will be performed. This is done as follows:

a. Inoculate 50 ml of LBMM medium with an appropriate strain of host *E. coli* designed to foster growth of the recombinant lambda bacteriophage optimally. For lambda ZAP Express, the optimized host bacterial strain is XL-1 Blue MRF' (Stratagene). These cells are grown overnight at 30°C with active rotation in a bacterial shaker/incubator; 50 ml will typically provide sufficient cells for titering at least 40 lambda libraries.

PROCEDURAL NOTE

These cells come from Stratagene as a glycerol stock. The cells are best revived by scraping crystals with a sterile loop onto a LB + tetracycline plate, streaking them out (see Appendix IV), and then growing overnight at 37° C. A single colony is then picked to start the LBMM culture in this step.

»»

LAB PERIOD II.1.6 CONT.

b. The next morning, pellet the bacteria by centrifuging them at 2500 rpm for 10 minutes in a tabletop centrifuge.

Resuspend the pelleted bacteria in freshly prepared 10 mM $MgSO_4$ to an OD_{600} of 0.5 (based on spectrophotometric measurements using a 1.0-cm path length cuvette). The cells can then be kept on ice for up to 12 hours before use.

Step 1 Dilute an aliquot of your genomic library (from Step 6 above) 1:10 in SM buffer by removing 1 µl from the library tube to a new tube and adding 9 µl of SM buffer. Label this tube "Dil Genomic Library."

Step 2 You will receive two tubes containing 200 µl of XL-1 Blue MRF' *E. coli* cells prepared as described above. Because the cells settle over time, flick these tubes to resuspend the *E. coli* cells shortly before you use them.

Step 3 To one of two tubes containing 200 µl of XL-1 Blue MRF' *E. coli* cells, add 2 µl of your "Dil Genomic Library." Label this tube "Dil." Add 2 µl of the undiluted library ("Genomic Library") to the other tube of XL-1 Blue MRF' cells (label this tube "Undil").

Step 4 Store the original "Genomic Library" tube at 4°C. Discard the "Dil Genomic Library" tube.

Step 5 Mix and incubate the bacteria and phage for 15 minutes at 37°C to allow the phage to adsorb (attach) to the host bacterial cells. After this incubation, keep the tube at room temperature (up to 30 minutes is fine) until you are ready to plate out the contents of your tubes.

> **CAUTION!**
>
> Store the pale blue "genomic library" tube at 4°C. Save this tube! Do not discard! Do not put the library tube at −20°C or you will inactivate your phage! Do not add chisam (chloroform/isoamyl alcohol) to your library tube or you will inactivate your phage!

PROCEDURAL NOTES

For Steps 6 through 8, prepare one plate with your *undiluted* XL-1 Blue MRF' mixture and one plate with your 1:10 *diluted* XL-1 Blue MRF' mixture. Label your plates along the side.

Nonrecombinant plaques are distinguished from recombinant plaques on the basis of a coloimetric indicator (using X-gal and IPTG) such that the nonrecombinant plaques are blue and recombinant plaques with inserts are clear. The blue color will indicate cells containing vectors with no insert in the ß-galactosidase gene. The intact gene encodes a protein (ß-galactosidase) that can cleave the compound X-gal, resulting in a "blue" plaque. Clear plaques are those that have inserts.

Step 6 Add 15 µl 0.5-M IPTG to a disposable 15-ml tube containing 3 ml molten NZY top aga**rose** in a water bath at 55°C.

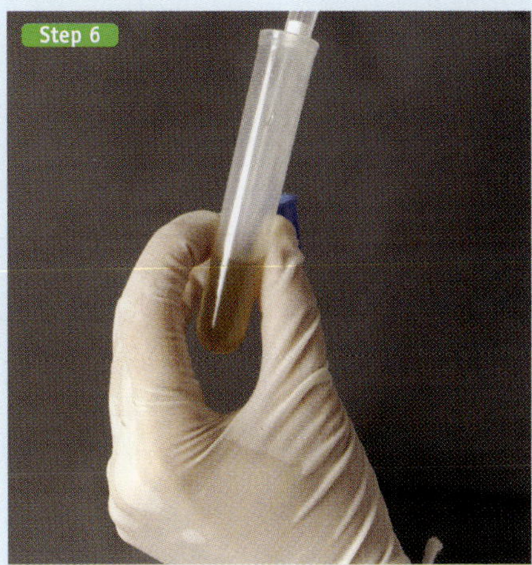

Step 7 To one tube, add 50 µl 250-mg/ml X-gal (made up in dimethyl formamide); vortex briefly, and place the tube back into the water bath. Next, flick the tube of cells/phage to mix, and add the entire "undil" mixture (about 200 µl) to the tube of top agarose.

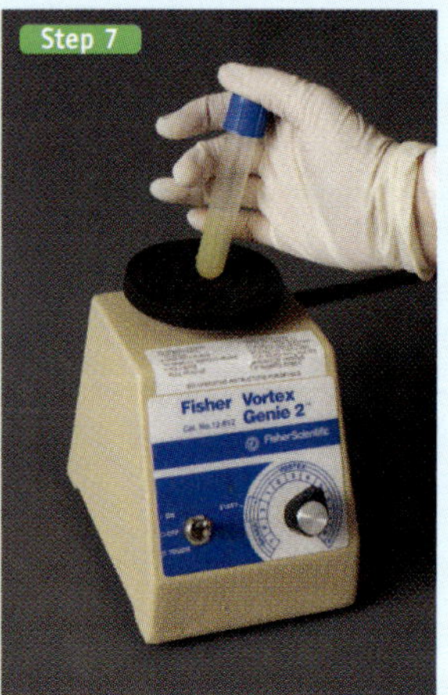

Step 8 Vortex this mixture, and pour *immediately* onto your warm NZY agar plate labeled "undil." Swirl the plate to evenly distribute the liquid agarose on the top of the agar on the plate.

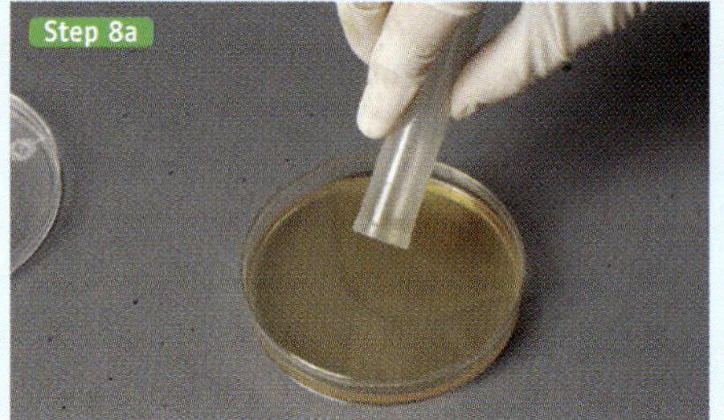

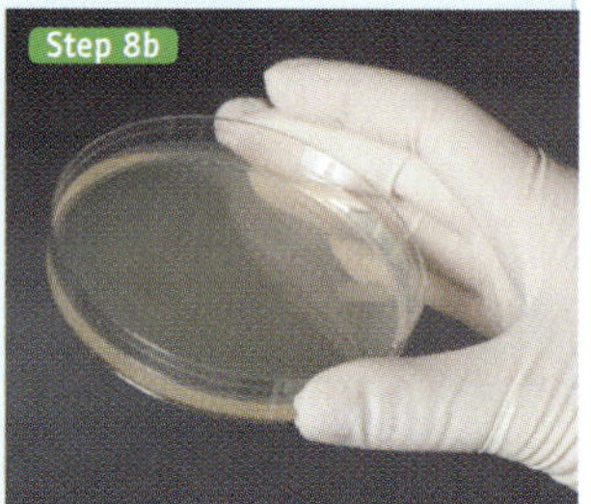

Step 9 Repeat Steps 6 through 8 for your tube containing the 1:10 diluted cells ("dil" tube).

PROCEDURAL NOTE

The top agarose is 0.6% aga**rose** in NZY medium. It is autoclaved and placed in a 55°C water bath. This agarose will quickly solidify upon removal from the water bath. To avoid problems with solidification, take your plates to the water bath, and plate your libraries next to the water bath. In each case, *immediately* after adding the cell/phage mixture to the top agarose and vortexing, pour the contents of this tube onto the agar plate, and gently swirl to distribute the top agarose evenly over the bottom agar. Be sure to make up the top agarose using the same agarose you used to make up gels. Do not use agar!

Step 10 Let the plates sit for at least 5 to 10 minutes to solidify.

Step 11 Incubate the plates *inverted* at 37°C for 8 to 24 hours.

PROCEDURAL NOTE

After the incubation period, the plates should be removed from the incubator and stored at 4°C until the next lab period.

LAB PERIOD II.1.7.   CALCULATE THE TITER OF THE LAMBDA PHAGE GENOMIC LIBRARY

AFTER AN incubation period of 8 hours or more, phage plaques should be easily visible, as shown in **Figure 2-3**. Each student will determine the titer of his or her lambda phage genomic library (i.e., the total number of plaques that could be obtained if all of the phage were used to infect *E. coli* and plated out).

Step 1 To determine the titer of your library, choose a plate with at least 20 plaques, but not so many that you cannot distinguish or count individual plaques (i.e., 1000 plaques on the undiluted plate would be difficult and time consuming to count, whereas 100 plaques on the diluted plate would be much easier to count and would still provide an accurate estimate of the titer).

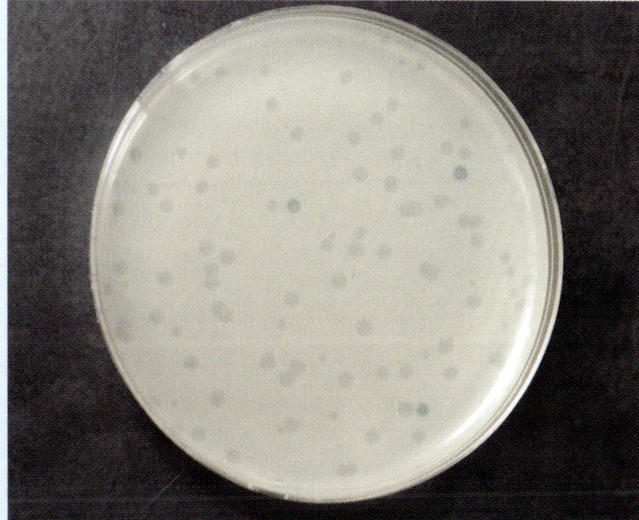

Fig 2-3 Phage plaques after 8 hours incubation.

Step 2 Count all of the plaques on the most appropriate plate (see above). The titer is the total number of plaques divided by two (because you used 2 µl of library); times 10 (*if* you counted the "dil" plate); times the total volume of the library (530 µl). (If you counted the "undil" plate, the titer is the total number of plaques on the plate times 530). This calculation will give you the total number of plaque forming units or phage in your library. If you do not multiply by 530 µl, you will have the number of plaque forming units per microliter of your library, which is also a useful measurement.

Step 3 Determine the percentage of recombinant plaques. To determine the percentage of recombinant plaques on your titer plate, divide the number of clear plaques by the number of total plaques, and multiply by 100. This number is the percentage of recombinant plaques in your library. Examples of clear and blue plaques are shown in Figure 2-3.

Step 4 Compare your data to that from the rest of the class. Record the range of titers and percentage of recombinant plaques for the entire class.

PROCEDURAL NOTE

After you have determined the titer and percentage of recombinant plaques in your library, it is ready to be used in any of several types of procedures (e.g., hybridization screening as described in Module II.2). Alternatively, your library can be stored refrigerated at 4°C for up to a month or more or stored frozen at –70°C following addition of DMSO to a final concentration of 7%. It is often useful to prepare aliquots of your library and store one at 4°C for immediate use and store the rest at –70°C for future use. Aliquots of libraries that have been stored at –70°C must be diluted 10-fold with SM buffer *immediately* after thawing to reduce the concentration of DMSO. As noted above, it is advisable to retiter a library after long-term storage at –70°C before using it for other procedures. Libraries stored properly at –70°C in DMSO can be a useful resource for many years.

MODULE II.2

Screening a Lambda Genomic Library to Detect Clones of a Specific Gene

MODULE SUMMARY

Plate the lambda phage library and perform plaque lifts. Hybridize with a nonradioactively labeled DNA probe, followed by chemiluminescent detection of the plaques containing the specific gene of interest (*Rvt*). Isolate phage clones containing the specific gene.

MODULE BACKGROUND

A genomic library of the kind you produced in Module II.1 contains clones of fragments from throughout the mouse genome. To identify the clone(s) that carry a genomic fragment containing the *Rvt* gene, you will screen the library with a hybridization probe that will detect the specific sequence of the *Rvt* gene. This is accomplished by plating a sufficient number of recombinant phage to ensure a greater than 99% likelihood of representation of the gene of interest. Thus, the number of plaques that must be screened to detect a clone of a specific gene depends on the representation of that gene in the genome. For unique genes (one copy per genome), it is often necessary to screen up to 1,000,000 or more plaques to find a positive clone. This requires multiple plating steps to initially detect and subsequently purify a clone of the desired gene. Because this is not practical in a short laboratory course, we will screen the library to detect a highly repeated gene sequence — the *Rvt* gene. This gene is repeated hundreds of thousands of times in the mouse genome and is part of the much longer LINE-1 repeat that is found dispersed throughout the mouse genome. The LINE-1 repeat is about 7 kb in length and contains two EcoRI sites that flank the region of the repeat that contains the *Rvt* gene. These two EcoRI sites are 1.4 kb apart, and cleavage of the mouse genome with EcoRI results in a 1.4 kb band that is visible in agarose gels because of the high copy number of the repeat. Although most copies of the *Rvt* gene in the mouse genome are nonfunctional, it is believed that some copies are functional and produce the reverse transcriptase enzyme that is critical for the ability of the LINE-1 repeat to move in the genome as a retrotransposon (a movable genetic element that moves via an RNA intermediate).

Because of the high copy number of the *Rvt* gene, you can plate a relatively small number of plaques (as few as 50) and still have a very high likelihood that one or more plaques will contain the *Rvt* gene. Detection of specific plaques carrying the *Rvt* gene will be accomplished by hybridizing a probe of the *Rvt* gene sequence to "plaque lifts" on nylon membranes that carry phage from the original plates. This will allow you to identify one or more *Rvt*-positive plaques from which you can then recover phage carrying a clone of the *Rvt* gene.

LAB PERIOD II.2.1.   PLATE THE LAMBDA PHAGE GENOMIC LIBRARY

EACH STUDENT will plate approximately 50 plaques of their lambda ZAP Express-mouse library onto one NZY plate in preparation for doing plaque lifts on these plates. You do not want so few plaques that you may not find any positive clones on the plate; however, you also do not want the plaques so crowded together that you cannot pick a positive plaque without having contamination with phage from an adjacent negative plaque (the phage will diffuse short distances through the top agarose). Typically, a library is screened at a much higher plaque density than this (thousands of plaques per plate). For this exercise, however, you screen the plates at low density so that it is easier to identify positive plaques among the negative plaques and to be sure that the phage picked from a positive plaque are all of one type (i.e., not contaminated with phage from a nearby plaque).

Step 1 You will use an aliquot of the lambda phage from your "Genomic Library" tube to transfect XL-1 Blue MRF' *E. coli* cells for the screening process. In order to have these cells ready for transfection, it is necessary to begin cultures the night before the transfection step. This is done as described in detail in Lab Period II.1.6 and briefly below (refer to Figures 2-3 through 2-7).

a. Grow XL-1 Blue MRF' bacterial cells overnight at 30°C in LBMM medium.

PROCEDURAL NOTE

These cells come from Stratagene as a glycerol stock. The cells are best revived by scraping crystals with a sterile loop onto an LB+ tetracycline plate, streaking them out (see Appendix IV), and then growing them overnight at 37°C. A single colony is then picked to start the LBMM culture in this step.

b. Pellet the bacteria by centrifuging them at 2500 rpm for 10 minutes in a tabletop centrifuge.

c. Resuspend the bacteria in freshly prepared 10 mM $MgSO_4$ to an OD_{600} of 0.5 (1.0 cm path length). The cells can be kept on ice for up to 12 hours before use.

Step 2 Based on the titer of your mouse genomic library, calculate the volume of the library necessary to give 50 plaques per plate. Dilute some of your packaged genomic library with SM buffer if necessary.

Step 3 Flick a tube containing 200 µl XL-1 Blue MRF' cells to disperse them and then add an appropriate aliquot of your library containing 50 plaque forming units (PFUs) to the cells.

Step 4 Flick the tube again, and incubate for 15 minutes at 37°C (without agitation) to allow the phage to adsorb (attach) to the bacteria. Place the tube at room temperature until it is your turn to plate (no longer than 30 minutes).

Step 5 Add 15 µl 0.5 M IPTG to each tube containing 3 ml of molten NZY top agarose in a water bath at 55°C. Mix by vortexing and return the tube to the 55°C water bath.

Step 6 To a tube of the top agarose, add 50 µl 250 mg/ml X-gal, and vortex briefly. Place the tube back into the water bath. Next, flick the tube containing the bacteria and phage mixture, and then add the entire mixture (about 200 µl) to the tube of top agarose. Quickly mix by vortexing and plate the contents of the tube on a NZY agar plate.

PROCEDURAL NOTE

The top agarose will solidify soon after it is removed from the water bath; thus, plate your mixture *quickly*.

»»»

LAB PERIOD II.2.1 CONT.

Step 7 Allow the plates to sit undisturbed for 10 to 15 minutes and then place them inverted in a 37°C incubator overnight.

PROCEDURAL NOTES

Return your "Genomic Library" tube to 4°C.

After an 8- to 24-hour incubation period, the plates can be removed from the incubator and stored at 4°C until the next lab period.

LAB PERIOD II.2.2.   PREPARE PLAQUE LIFTS FROM THE LAMBDA PHAGE GENOMIC LIBRARY

IT IS possible to transfer phage from plaques in top agarose to nylon filters by a procedure called a "plaque lift." The transferred phage can subsequently be lysed to release their DNA. After immobilizing the DNA onto the nylon filters, the DNA can be used as a substrate for hybridization with a DNA probe.

Each student will do *one* plaque lift from *one* of the lambda ZAP Express-mouse library plates. The plaque lift procedure will transfer some of the phage from each plaque on the Petri plate to the nylon filter (also referred to as a nylon membrane or a "plaque lift").

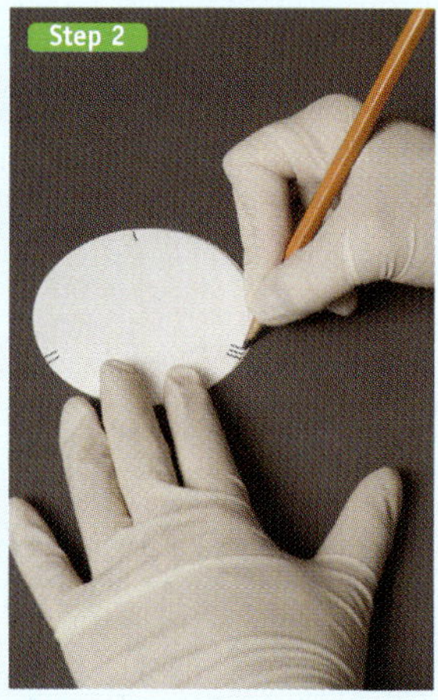

Step 1 Refrigerate the plates for *at least* 1 hour before lifting. This will help prevent the agarose overlay from separating from the agar in the Petri dish during the plaque lift procedure.

Step 2 Wearing gloves and using a pencil, label two nylon filters "A" and "B" (the filters are Schleicher and Schuell Nytran +). Be sure to write your group name/numbers on both filters. Write "plaque side" on the side of the filter that will contact the plaques. Also, draw three sets of lines on the filter asymmetrically as shown here. These asymmetric marks will enable you to align your Petri plates with the plaques to your results on the X-ray film later on.

PROCEDURAL NOTE

Use only #2 pencils to label the filters! Do not use ink! Most types of ink will not stay on the membranes during the wash and transfer procedures and will smear or disappear entirely.

Step 3 Retrieve your plates from the refrigerator. Label one plate "A" and one plate "B."

Step 4 Wearing gloves, carefully place the "A" nylon filter on the "A" plate with the writing-side face down (the side that says "plaque side"). Take care to place the membrane so that it is centered on the plate (if you have not done this before, the instructor will demonstrate this technique). After you have placed the

»»

membrane on the plate, *do not* move or attempt to adjust the membrane in any way. Also try to avoid getting air bubbles under the filter. Now do the same for the "B" filter on the "B" plate.

Step 5 Mark the bottom of the Petri plate asymmetrically with lines corresponding to those on the filter (use a marker pen). After marking the Petri plates, place them in the refrigerator for 5 minutes.

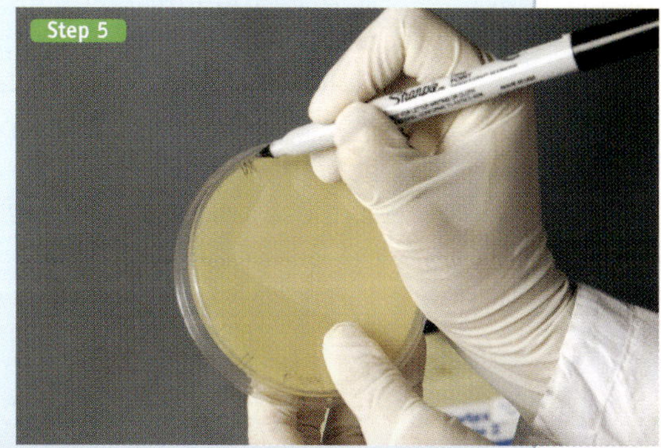

Step 6 Retrieve the plates from the refrigerator, and carefully lift the nylon filters off the plates with a pair of forceps to avoid disturbing the agarose overlay and place the filters *plaque side up* to dry on absorbant paper for 5 minutes (the side with the writing should be the plaque side).

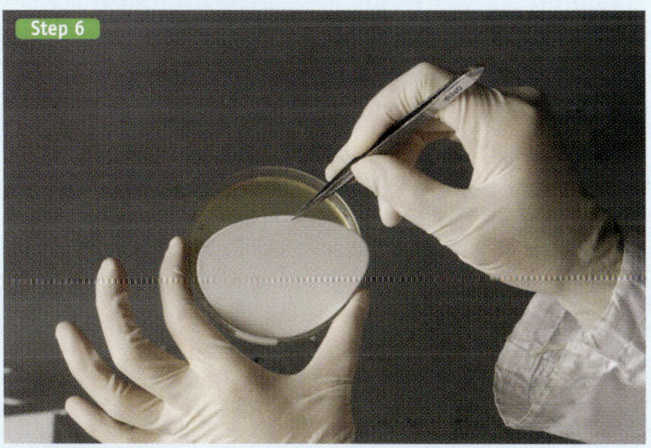

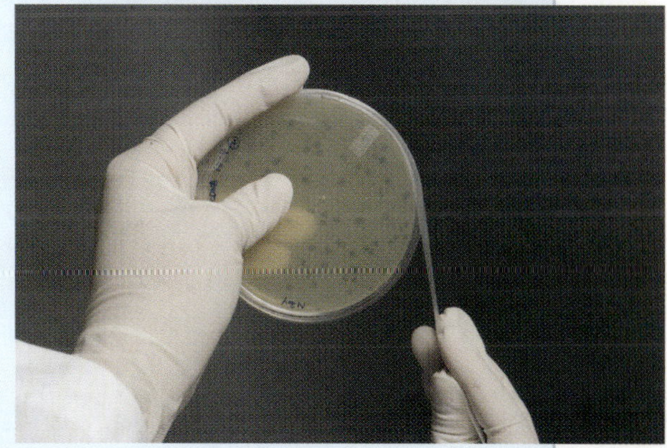

PROCEDURAL NOTE

After removing the plaque lift filter from your plate, seal the plate with parafilm as demonstrated by the instructor and shown here, and return it to the refrigerator to be stored at 4°C. You will use this plate again in Lab Period II.2.6.

Step 7 Stack two pieces of Whatman 3MM filter paper on a sheet of plastic wrap. Each piece of filter paper should be large enough to hold a nylon filter. Repeat so that there are three stacks of filter papers on plastic wrap. Saturate the first stack of filter papers with plaque lift denaturation solution. Saturate the second stack with plaque lift neutralization solution. Saturate the third stack with 2x SSC. The filter papers should be thoroughly saturated but should not have pools of buffer on the surface.

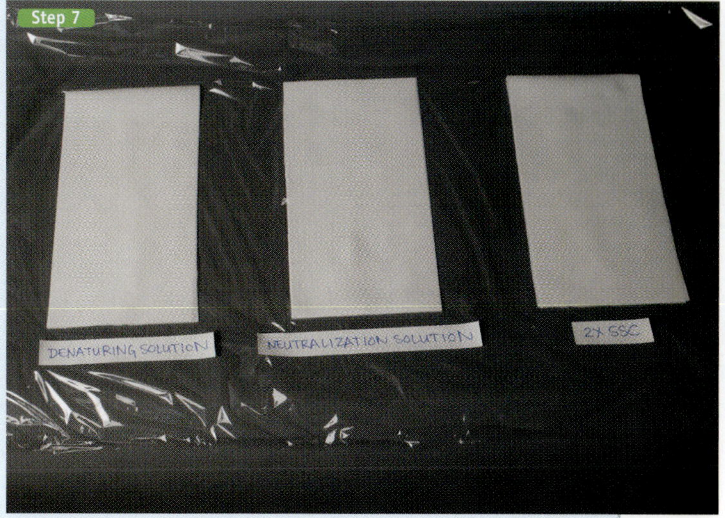

LAB PERIOD II.2.2 CONT.

Step 8 Denature the filters by placing them on the first stack of Whatman 3MM paper saturated with the plaque lift denaturation solution for 2 minutes with the plaque side-up. This step breaks open the phage on the filters and releases the phage DNA. The NaOH also causes the double-stranded phage DNA to be denatured to single-stranded DNA.

Step 9 Neutralize the filters in the same manner by placing filters plaque side up on the stack of Whatman 3MM paper saturated with plaque lift neutralization solution for 5 minutes. This step is necessary to reduce the pH of the membranes. If this is not done, the membranes will disintegrate (not a desirable result!).

Step 10 Place the filters plaque side up on the stack of Whatman 3MM paper saturated with 2x SSC for 2 minutes.

Step 11 Air dry the filters on an additional sheet of absorbant Whatman 3MM paper for about 15 minutes (plaque side up).

Step 12 UV cross-link the filters in a UV cross-linking instrument. Then bake the filters at 80°C for 1 hour in a vacuum oven.

PROCEDURAL NOTES

A vacuum oven is not necessary if nylon filters are used; any oven will be fine. If no UV cross-linker is available, then bake the filters for 2 hours instead. If you use a UV cross-linker, the instructor will guide you as to the settings to use.

Filters can be used immediately or stored at room temperature in plastic bags for later hybridization (if the filters are completely dry and stored in sealed plastic bags, they can be used even years later).

LAB PERIOD II.2.3.   PCR REACTION TO NONRADIOACTIVELY LABEL THE MOUSE *RVT* GENE DNA PROBE

PROCEDURAL NOTE

If you have already performed the *Rvt* PCR-labeling procedure described in Module I.2 (Lab Period I.2.3), then an aliquot of that labeled probe can be used for the hybridization procedure in Lab Period II.2.4, and it is not necessary to carry out a separate labeling procedure for this project. In that case, you may proceed directly to Lab Period II.2.4. However, in the event you have not performed the labeling procedure described in Lab Period I.2.3, that procedure is described again in detail below.

YOU WILL do one fluorescein PCR labeling reaction to prepare a probe for hybridizing to the mouse genomic library plaque lifts. The DNA you will PCR amplify and label is one copy of the mouse 1.4-kb EcoRI fragment that contains the *Rvt* repeated gene sequence cloned into a plasmid (this plasmid will be provided by the instructor). This labeled copy of the mouse *Rvt* gene will be used as a hybridization probe to detect other copies of this gene family in the plaque lift hybridization. The fluorescein labeling method is described in Module I.2 and the general process of PCR is described in Module I.3.

PROCEDURAL NOTE

Because you are performing PCR, wear gloves while setting up this labeling reaction!

Step 1   You will be provided with a 0.2-ml PCR tube containing 4 ng (29.5 µl) of a purified plasmid (pBluescript) with one copy of the *Rvt* gene cloned into the EcoRI site. The 0.2-ml tubes are very fragile — handle with care! They can be cracked by squeezing them too hard or by poking them with a pipette tip. To this tube, add the following components:

   5 µl 10x PCR buffer

   5 µl FL-dUTP PCR mix

   8 µl 25-mM $MgCl_2$

   1 µl *Rvt* Repeat Forward Primer (10 pmol/µl)

   1 µl *Rvt* Repeat Reverse Primer (10 pmol/µl)

   0.5 µl AmpliTaq Gold DNA Polymerase enzyme

   50 µl Total volume

PROCEDURAL NOTE

The *Rvt* forward primer is 5' GAATTCTTTGTTCAGTTCTGAGCC 3'. The *Rvt* Reverse Primer is 5' CCCATATCTAAACATGATAAAAGC 3'.

Step 2   Mix the components by stirring with your pipette tip. Place the tube into the PCR instrument.

Step 3   The PCR program we will run for this labeling reaction is a 30-cycle program with the following parameters:

94°C for 12 minutes to denature the plasmid DNA and activate the AmpliTaq Gold DNA Polymerase. This is followed by 30 cycles of

   a. 94°C for 60 seconds — Denaturation

   b. 50°C for 60 seconds — Annealing

   c. 72°C for 90 seconds — Extension

»»

LAB PERIOD II.2.3 CONT.

The reaction is completed with a final elongation step at 72°C for 7 minutes and then is held at 4°C.

Step 4 When the PCR cycling is complete, transfer the PCR product to a new 0.5-ml microfuge tube. Label this tube with your group numbers and "*Rvt* Label," and store this tube at 4°C until the next lab period.

PROCEDURAL NOTE

This step can be done after the end of the lab period by the instructor if necessary.

Your labeled *Rvt* DNA probe can be stored at –20°C for at least 3 months. This is the same labeled probe as that used in Module I.2 (Lab Period I.2.3) for hybridization to the genomic Southern blot. This single PCR labeling reaction will provide more than enough probe for both of these experiments.

LAB PERIOD II.2.4.   **VISUALIZE THE *Rvt* GENE PROBE FOLLOWED BY PREHYBRIDIZATION AND HYBRIDIZATION OF THE PLAQUE LIFT FILTERS**

PART A Visualization of the *Rvt* Gene Probe on an Agarose Gel

PROCEDURAL NOTE

If you previously labeled the *Rvt* probe in Project I (Lab Period I.2.3) and checked it on a gel (Lab Period I.2.4), then you do not need to run this gel!

Step 1 Pour a 1.2% agarose minigel (refer to Lab Period I.1.4, Part B). Allow it to solidify for at least 20 minutes before adding 1x TAE buffer into the electrophoresis chamber.

Step 2 Prepare a sample of your "*Rvt* Label" probe from Step 4 (Lab Period II.2.3) by removing 8 µl of the probe to another microfuge tube and adding 2 µl 5x BJ. Label this tube "*Rvt* Label Gel." Save the remainder of your "*Rvt* Label" probe at 4°C for use in the hybridization to the plaque lifts later in this experiment.

Step 3 Load the agarose gel with samples in the following order:

> Lanes 1–4: Empty
>
> Lane 5:   10 µl 100-bp ladder in 5x BJ (50 ng/µl)
>
> Lane 6:   10 µl "*Rvt* Label" from Step 2 above
>
> Lanes 7–10: Empty

PROCEDURAL NOTE

The 100-bp ladder marker is made up of DNA fragments that run at 100 bp, 200 bp, 300 bp, etc. (see Appendix III for the sizes of all the bands in this marker).

»»

> **CAUTION!**
>
> Ethidium bromide is a known mutagen and carcinogen. Wear gloves, lab coats, and safety glasses! Handle gels carefully! Do not splash ethidium bromide! Always rinse the spatula in the water destain bath after it has been in contact with ethidium bromide! Always rinse the gel tray after sliding gels into the ethidium bromide bath. You may unknowingly contact the ethidium bromide! Be very careful not to drip the ethidium bromide anywhere!
>
> Always handle the gels carefully!

Step 4 Run the agarose gel at 80 V for 1 hour.

Step 5 Look at the gel on the UV light box *before staining* with ethidium bromide. The lane that contains the *Rvt* probe you labeled with fluorescein (lane 6) should have a band visible at 1.4 kb. This should be the only band visible. A band at this position indicates that the 1.4 kb *Rvt* gene was successfully labeled! The large green blob below this band contains the unincorporated fluorescein-labeled dUTP. Figure 1-6a (Lab Period I.2.4) presents the expected result.

Step 6 Stain the DNA in ethidium bromide DNA stain for 10 minutes. Destain the gel in a water bath for 15 minutes, and then photograph the gel (see Figure 1-6b in Lab Period I.2.4).

PROCEDURAL NOTE

Appendix IV describes an alternative DNA stain, SYBR Safe (Molecular Probes).

Step 7 Verify that the band you saw above in step 5 is indeed at 1.4 kb (use the marker in lane 5 for reference). Figure 1-6b in Lab Period I.2.4 presents the expected result.

PART B Prehybridization of the Plaque Lift Filter

This step is referred to as the "prehybridization" step. It is designed to "block" the filters before the addition of the DNA probe. The blocking agents (nonfat dry milk and the salmon sperm DNA) will bind to sticky spots on the membrane and prevent the probe DNA from sticking to these random spots on the membrane. This blocking step dramatically reduces the background signal on the X-ray film that is obtained following detection of the probe.

PROCEDURAL NOTES

This prehybridization can be done while the gel from Part A is running or can be done immediately without doing Part A if the gel results are already known.

Two plaque lifts filters can be prehybridized together (back-to-back with the plaque sides facing out) in one hybridization bag.

Step 1 Place all of the plaque lift filters from the class into a large container of Plaque Wash buffer. The exact volume of the buffer is not critical, but approximately 50 ml buffer per filter is sufficient. Incubate the plaque lift filters for 2 hours to overnight at 42°C in a shaking incubator (2 to 6 hours is fine, but if it works better for your schedule, overnight is also good).

»»

LAB PERIOD II.2.4 CONT.

Step 2 Transfer the plaque lift filters back-to-back (DNA side facing outward) into a "seal-a-meal bag."

Use a 10-ml pipette to add 6 ml Probe-Amp hybridization solution.

Carefully remove most of the air bubbles as shown here, and seal the bag using a heat sealer (this will be demonstrated by the instructor). Write your number on the outside of the bag.

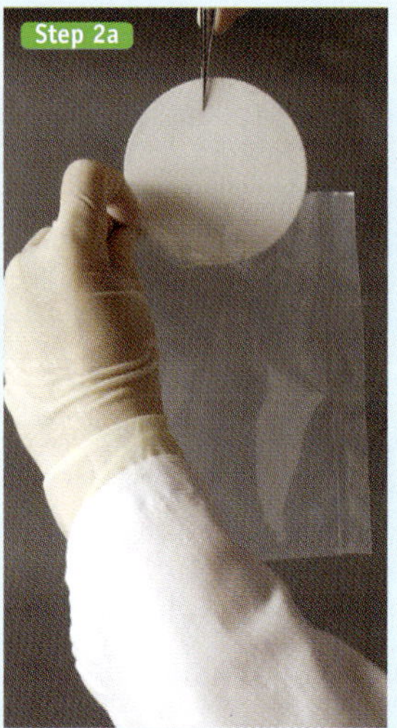

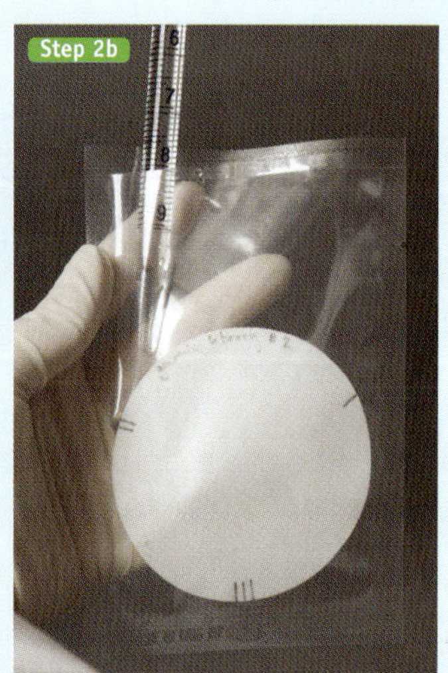

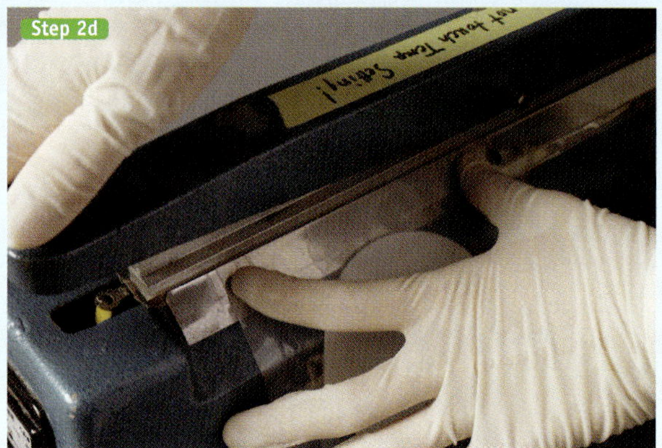

Step 3 Incubate the filters in a shaking water bath or rocking air incubator at 65°C for at least 1 hour (up to overnight is fine if that works best for the laboratory schedule).

PART C Hybridization of the Plaque Lift Membrane with the *Rvt* DNA Probe

Step 1 Remove 10 µl of PCR-labeled *Rvt* probe from your tube labeled "*Rvt* Label" (produced either in Project I or Project II). Add the 10 µl to a new microfuge tube containing 40 µl of Probe-Amp hybridization buffer (label this tube "*Rvt* probe"). Store the remainder of your *Rvt* DNA probe at −20°C. The probe can be stored for at least 3 months for use in subsequent experiments.

Step 2 Place a "lid lock" on the microcentrifuge tube that contains the *Rvt* probe to prevent the cap from opening during the subsequent heating step. Heat this *Rvt* probe tube at 95°C for 5 minutes to denature the DNA. Spin the tube briefly in your nanofuge to collect all of the liquid at the bottom of the tube, and place the tube immediately on ice.

»»

LAB PERIOD II.2.4 CONT.

Step 3 With scissors, snip off one corner of the bag containing your prehybridizing plaque lift filters. *Do not* remove the hybridization solution in the bag. Pipette all 50 µl of your denatured fluorescein-labeled *Rvt* probe (from Step 1) into the hybridization. Take care not to pipette the probe directly against the membrane. Remove bubbles from the bag and reseal carefully.

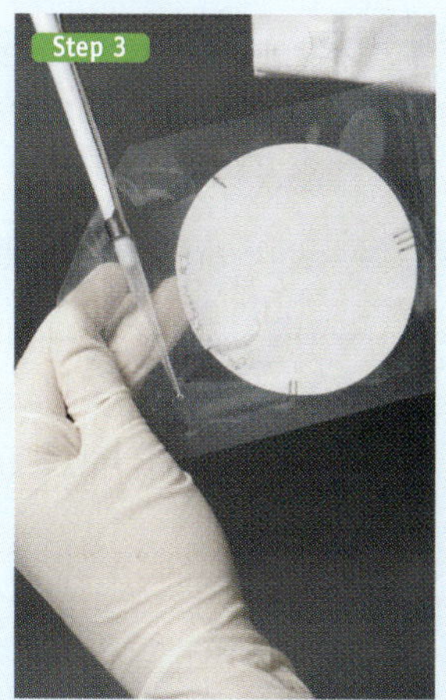

Step 4 Hybridize the filters at 65°C overnight in a shaking water bath or in a rocking air incubator.

PROCEDURAL NOTE

If possible, it is best to proceed directly from the hybridization step to the washing and detection steps as described for Lab Period II.2.5. If necessary, however, the hybridization can be continued at 65°C for 2 days maximum.

LAB PERIOD II.2.5. ■ WASH THE PLAQUE LIFT FILTERS AND CHEMILUMINESCENT DETECTION OF THE HYBRIDIZED PROBE

PART A Wash the Plaque Lift Filters

Step 1 Cut open the hybridization bag and remove the filters. Transfer the filters to a plastic container with 1x SSC, 0.1% SDS. Reseal the bag containing the hybridization solution, and discard in a waste container. The filters from an entire class can be washed in a single container for 5 minutes at room temperature.

Step 2 Transfer the plaque lift filters to a second plastic container with 1x SSC, 0.1% SDS at 65°C for 15 minutes in a rocking incubator. Be sure that the wash solution is actually at the proper temperature by checking with a thermometer.

Step 3 Transfer your filters from the 1x SSC, 0.1% SDS wash solution to a new container with 0.5x SSC, 0.1% SDS, and wash for 20 minutes at 65°C in a rocking incubator.

PROCEDURAL NOTE

This is the highest stringency wash, and thus, the temperature of this wash is critical. Be sure that the wash solution is actually at the proper temperature by checking with a thermometer.

PART B Detection of Hybridized Probe on the Plaque Lift Membrane

Step 1 Rinse your filters briefly in ECL buffer in a plastic container (the filters from the entire class can be washed together in several hundred ml of the buffer).

Step 2 Transfer your filters to the lid of a P1000 pipette tip box (called a "large tip top") or another plastic container. If you are washing two filters together, be sure that your filters are oriented back to back so that they wash efficiently. All of the remaining washes will be done in these containers.

》》

LAB PERIOD II.2.5 CONT.

Step 3 Incubate your filters back to back in 50 ml Antibody Blocking Solution for 1 hour or longer at room temperature with agitation on a rocking platform. Note: The blocking solution is 0.5% nonfat dry milk in ECL buffer.

Step 4 Pour off the blocking solution, and rinse the filters briefly in 50 ml ECL buffer. Pour off the solution.

Step 5 Incubate filters in 50 ml Antibody Solution for 1 hour at room temperature with agitation on a rocking platform.

PROCEDURAL NOTE

Antibody solution is a 1:1000 dilution of the anti-fluorescein antibody in a solution of 0.5% BSA in ECL buffer. The antibody is conjugated to horseradish peroxidase. Detection will be accomplished using the horseradish peroxidase enzyme in a chemiluminescent reaction (GE Healthcare).

Step 6 Pour off the Antibody Solution, and wash the filters in 50 ml ECL buffer/0.1% Tween at room temperature twice for 10 minutes followed by two additional washes for 5 minutes (all with agitation and 50 ml buffer). Leave the filters in the final wash solution until Step 8.

PROCEDURAL NOTE

There are *four* total washes in the ECL buffer/0.1% Tween solution. Pour off the wash solution after each step *except* the last one.

Step 7 Mix an equal volume of ECL Detection Solutions 1 and 2 to give sufficient volume to cover the filters (1 ml of each solution is usually sufficient for each filter). These solutions contain the molecule luminol, which when modified by horseradish peroxidase, becomes unstable and emits light as it decays.

Step 8 Remove the filters from the last wash solution and drain off the excess liquid.

Step 9 Place the filters on clean plastic wrap or in a plastic container, and pour the detection solution onto the filters. Incubate for 1 minute at room temperature.

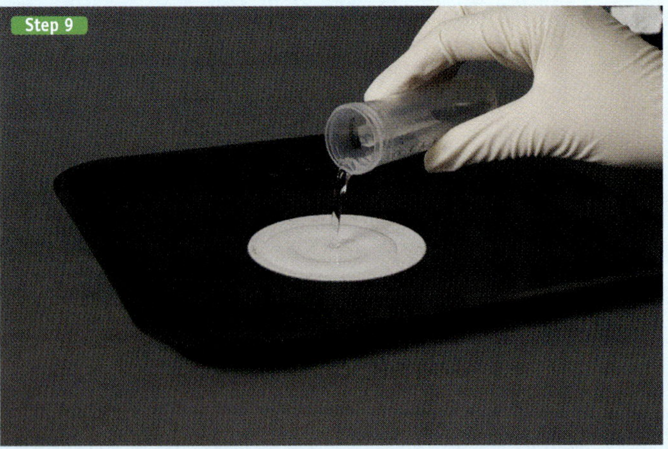

Step 10 Drain off excess detection buffer and place your filters into a clear acetate development folder as shown here.

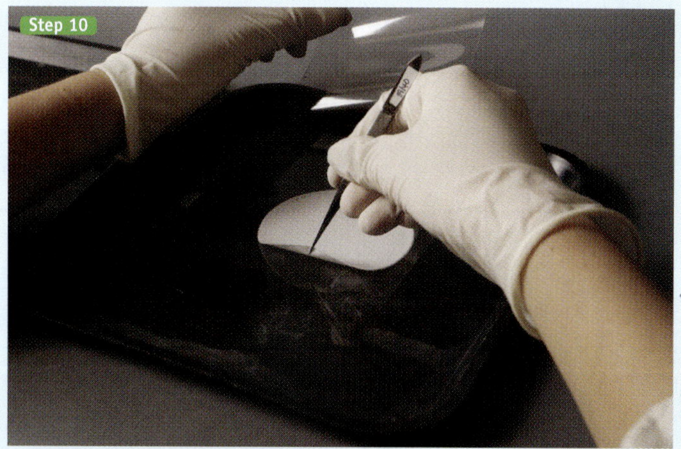

PROCEDURAL NOTE

Any clear plastic folder will do; you can also use plastic wrap in place of a development folder.

Step 11 In a darkroom with red safe lights, tape the filters plaque side up in a film cassette, and place a sheet of X-ray film (Kodak BioMax) onto the blots. Be sure to mark the film and the nylon filters so that the film can later be oriented properly to the plaque lift filters. This can be done using a marking pen or by pushing pin holes through the film and the filters.

Step 12 Close the cassette, and expose the film for 3 minutes to 60 minutes. We recommend first doing a short exposure for 3 to 5 minutes and then a longer exposure if necessary. You should put a new piece of film in the cassette immediately after removal of the first film from the cassette. This is important because the reaction runs efficiently for less than 2 hours, and if you require a longer exposure, you may need to maximize the exposure time for the second film.

Step 13 Remove the X-ray film from the cassette in the darkroom with the red safelights on. Develop in Kodak GBX developer for 2 minutes. Rinse in tap water for 2 minutes. Fix in Kodak GBX fixer for 2 minutes, and finally, rinse for 2 additional minutes in tap water. If you have access to an automatic X-ray film developer, you can use that instead.

»»

LAB PERIOD II.2.5 CONT.

Step 14 Dry the film (air drying works fine or you can use a hair dryer to speed up the process), and then place the dry, developed film back into the X-ray cassette holding the development folder with the plaque lift filters. Use the pin holes or the orientation marks to align the film properly with the plaque lift filters. Now, use a marking pen to trace an outline of the circular filters onto the film, and trace the alignment lines from the filters onto the film (these alignment lines were the ones that you marked on the original Petri plate and on the filter). Also, write your identifying numbers next to each filter image on the film. Cut out the image of each filter from the film (with the identifying number) so that each student has a piece of X-ray film with the image from his or her filter(s).

Step 15 Examine the film (called a "lumigram"), and look for black spots that indicate positive plaques! An example of the appearance of positive black spots on a lumigram is shown in **Figure 2-4**.

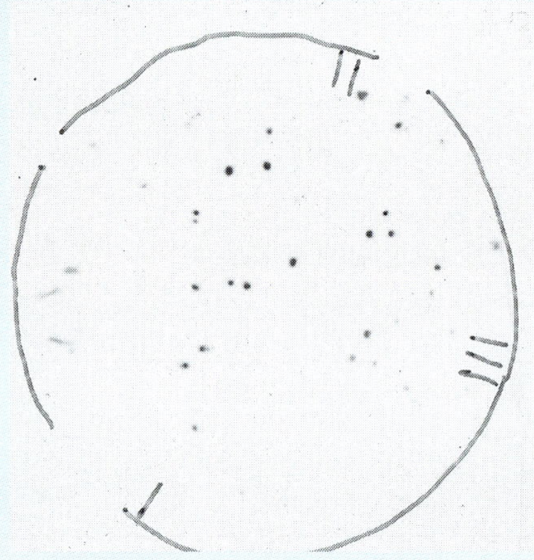

Fig 2-4 Lumigram showing positive black spots.

PROCEDURAL NOTE

Your spots will, of course, be in a different pattern, and you may have a significantly different number of spots.

LAB PERIOD II.2.6.   **ALIGN LUMIGRAMS TO PLATES AND PICK POSITIVE LAMBDA PHAGE PLAQUES**

IN THIS lab period, students will align their own lumigrams to the individual lambda ZAP Express plate from which their plaque lift was made, and select *two* positive plaques *each* for PCR amplification. To pick lambda ZAP Express library plaques:

Step 1 Retrieve the plates from which you made your plaque lifts.

PROCEDURAL NOTE

The plates have been stored in the refrigerator at 4°C since Lab Period II.2.2.

»»

LAB PERIOD II.2.6 CONT.

Step 2 Align the lumigram of the plaque lifts with the lambda ZAP Express plate using the asymmetric alignment marks; simply place the film on the back of the Petri plate, and hold it up to the light as shown here.

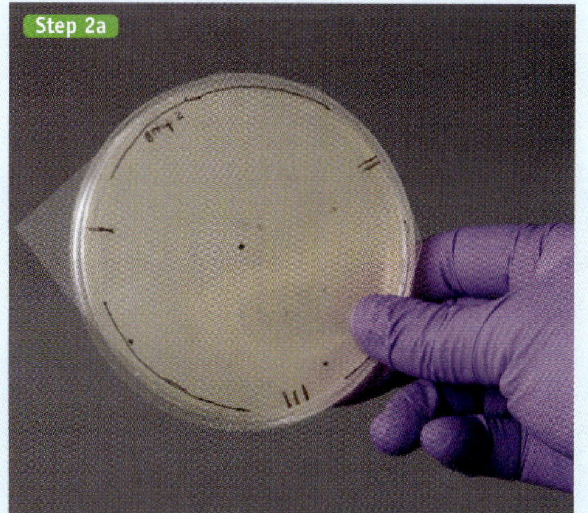

Rotate the film until the alignment marks match up. Based on the signals on the lumigram, identify a single, well-isolated plaque giving a strong positive signal. Mark this plaque "A" with a marking pen on the back of your Petri dish so that the "circles" can be seen from the top (agar side) of the plate, as can be seen here. Select a second positive plaque, and mark it "B" in the same way. Each student should pick two plaques. If you are unable to pick two from your plate, "borrow" plaques from a classmate.

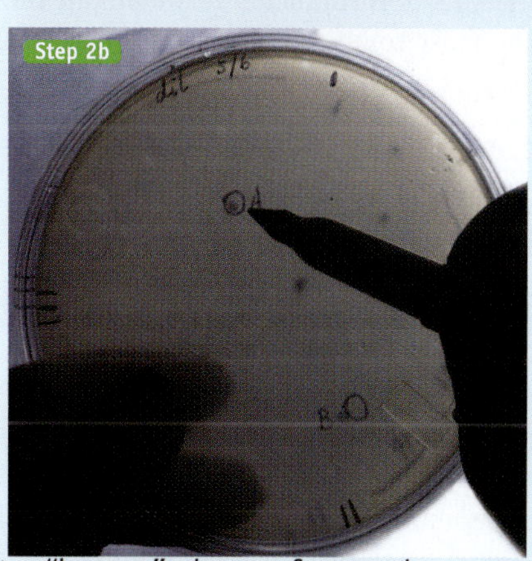

Step 3 Pipette 50 µl SM buffer into each of four 1.5 ml microfuge tubes.

Step 4 Pick each positive plaque by transferring the plaque into a P1000 tip and expelling it into 50 µl of SM buffer in a blue microfuge tube as shown here. It may be helpful to attach the tip to a 1000-µl pipette to expel the agar plug containing the plaque. Vortex the tube to release the phage particles into the SM buffer. Label these tubes with your number and "*Rvt* Phage Stock-A" and "*Rvt* Phage Stock-B." Place the *four* phage stock tubes at 4°C overnight to further elute phage from each agar plug. Note that you and your partner have two tubes each.

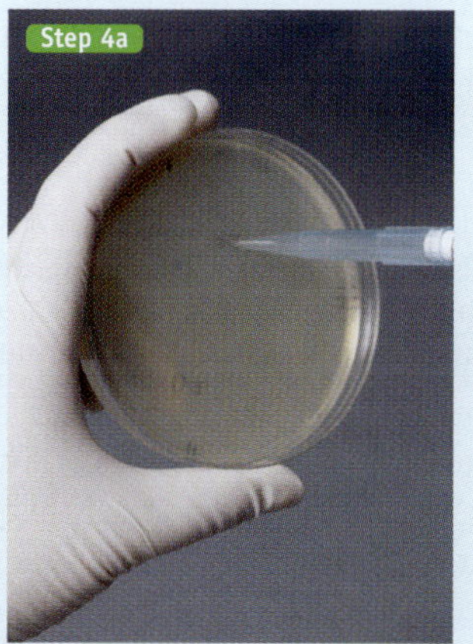

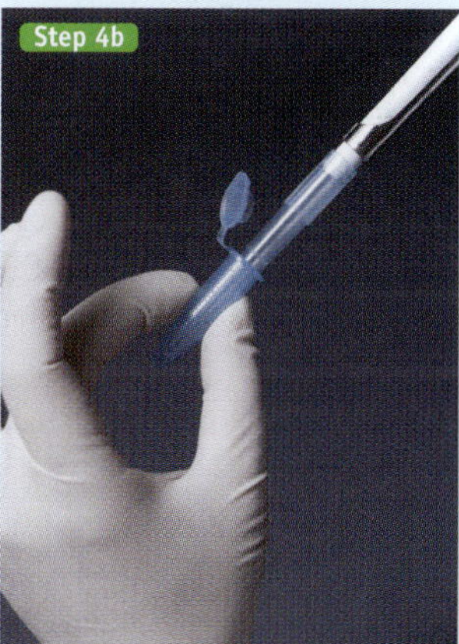

Congratulations! You have now isolated a clone of the mouse Rvt *gene!* You may now tell your friends that you have cloned a gene! It is typically desirable, however, to manipulate and characterize your cloned gene further, as described in Modules II.3 and II.4.

MODULE II.3

PCR Subclone into a Plasmid Vector, Propagate in Bacterial Cells, and Prepare Purified Recombinant Plasmid DNA

MODULE SUMMARY

PCR amplify, purify, and characterize the positive lambda phage insert. Ligate the *Rvt* PCR product into the pDrive plasmid vector, and transform the plasmids into *E. coli* cells. Pick, prepare, and quantify the pDrive/*Rvt* recombinant plasmid.

MODULE BACKGROUND

Gene libraries are typically constructed in lambda vectors because bacteriophage vectors provide the most efficient means of introducing recombinant DNA molecules into bacteria for subsequent propagation to yield the most complete library possible. Plasmids, however, provide a more stable vector in which to maintain a particular clone and are easier to manipulate and propagate in bacterial cells. Thus, it is often preferable to generate an initial library in a lambda vector and to screen that library to identify an individual clone of interest and then to transfer that clone into a plasmid vector for subsequent propagation and manipulation. Transferring a specific cloned recombinant DNA molecule from one vector to another is known as "subcloning." Subcloning involves the liberation and recovery of an insert DNA fragment from one vector and ligation of that fragment into a different vector to facilitate further propagation of that particular clone. In this experiment, you will subclone the insert from your positive *Rvt* lambda clone (selected on the basis of screening as performed in Module II.2) into a plasmid vector.

Plasmids are small circular molecules of DNA that are maintained and replicated independently of the "host" chromosome of a bacterial cell. Depending on the particular plasmid vector used, up to 200 or 300 copies can be maintained and propagated in a single bacterial host cell. In addition, there are many different commercially available plasmid vectors into which a particular cloned fragment can be inserted. The different plasmid vectors offer different unique advantages that allow you to manipulate your clone of interest in different ways. One common use of plasmid vectors involves the propagation of an individual clone in *E. coli* to yield large amounts of plasmid carrying a particular insert. Other manipulations using plasmids involve expression of the cloned DNA fragment in *E. coli* to yield either encoded RNA or protein. Different plasmid vectors are available to facilitate each of these goals.

Subcloning involves isolating a cloned insert from one vector and ligating it into a different vector. One quick method for doing this is to use the polymerase chain reaction (PCR) to amplify the insert using primers complementary to vector sequences that flank the insert on each side in the original vector. The PCR product is then to ligated into the new vector. To do this, you will take advantage of a unique quirk of the PCR process — when amplification is performed using *Taq* DNA polymerase (or any other nonproofreading polymerase), an adenine (A) is added onto the 3' end of each strand of the amplicon (an amplicon is the double-stranded amplification product of PCR). This produces a single nucleotide "A" 3'-overhang on each end of each amplicon. To take advantage of this phenomenon, commercial suppliers now provide dephosphorylated, linearized plasmid vectors with single nucleotide "T" or "U" 3'-overhangs. These 3'-overhangs are complementary to the 3'-overhangs on the PCR amplicons and, therefore, enhance the efficiency of ligation of the linear PCR product to the linear plasmid to form a circular recombinant plasmid. These recombinant plasmids can then be transformed into bacterial cells and propagated, enabling the recovery of large quantities of pure, recombinant plasmid for use in many different types of experiments.

Theoretically, because the vector has 3' "T" or "U" overhangs, vector molecules should not be able to ligate to themselves or to other vector molecules. This should avoid one problem often encountered in ligations: the formation of empty vectors without inserts (nonrecombinant clones). When the companies prepare these vector molecules, however, some are not cut in the initial digest, whereas others may not have the "T" or "U" overhangs successfully added. Thus, nonrecombinant "vector-only" clones that do not carry an insert are still recovered in these subcloning experiments. Nonrecombinant colonies are

BACKGROUND, continued

distinguished from recombinant colonies on the basis of a colorimetric indicator (using X-gal and IPTG; see Appendix II), such that nonrecombinant colonies are blue and recombinant colonies with inserts are white. In addition, because the PCR products (insert molecules) have the 3' "A" overhangs, insert molecules should not be able to ligate to themselves or to other inserts. In practice, however, addition of the 3' "A" nucleotide is not 100% efficient, and therefore, some ligation of inserts is seen; however, vector/vector and insert/insert ligations are much less frequent with this PCR cloning system compared with more typical cloning experiments with compatible sticky ends from restriction digests.

LAB PERIOD II.3.1.   PCR AMPLIFY THE POSITIVE LAMBDA PHAGE INSERT

YOU WILL perform *two* PCR reactions (one on each of your "positive" lambda phage in SM buffer isolated in Lab Period II.2.6). If you did not perform Module II.2, then your instructor will provide you with phage templates for the following lab period. This PCR reaction will amplify the *Rvt* insert DNA in the positive lambda ZAP Express plaques that you identified in Module II.3 using hybridization screening. In this module, you will transfer ("subclone") the PCR products into the "pDrive" plasmid vector (Qiagen).

Step 1 Retrieve your "*Rvt* Phage Stock-A" and "*Rvt* Phage Stock-B" tubes from 4°C (or they will be given to you by your instructor). Vortex each tube for 30 seconds, and then centrifuge each tube for 2 minutes in your nanofuge (this disperses the phage and sediments unwanted agarose particles).

Step 2 Label two 0.2-ml PCR tubes on the side of the tubes with your number and "A" or "B." Do not write on the top of the tubes, as the heat block will cause your writing to disappear! You will next prepare the following "PCR Master Mix" in a 1.5-ml microfuge tube by mixing sufficient reagents for two PCR reactions:

> 63 µl DdH$_2$O
>
> 16 µl dNTPs mix
>
> 10 µl 10x PCR reaction buffer
>
> 6.0 µl MgCl$_2$ (25 mM)
>
> 2.0 µl T3 Primer (10 pmol/µl)
>
> 2.0 µl T7 Primer (10 pmol/µl)
>
> 1.0 µl AmpliTaq Gold DNA polymerase
>
> 100 µl Total volume

PROCEDURAL NOTES

If you are doing this with a lab partner, you can double the previously mentioned volumes to make sufficient "PCR Master Mix" for four reactions.

T3 primer: 5' AATTAACCCTCACTAAAGGG 3'. T7 primer: 5' GTAATCAGACTCACTATAGGGC 3'.

Step 3 Mix your "PCR Master Mix" by setting your P200 to 100 µl and pipetting up and down several times.

LAB PERIOD II.3.1 CONT.

Step 4 Aliquot 46 µl of the "PCR Master Mix" to each of the two 0.2-ml PCR tubes that you have already labeled.

Step 5 Add 4 µl of template DNA ("*Rvt* Phage Stock-A" or "*Rvt* Phage Stock-B") to the appropriate PCR tube (A or B), and mix again by stirring with your pipette tip.

Step 6 Place your reaction tubes into the heat block of the PCR instrument. The PCR program has been set as follows:

a. An initial denaturing step at 94°C for 12 minutes to insure that the lambda phage are lysed, that all of the double-stranded DNA has been converted to single-stranded, and that the *Taq* Gold is activated.

b. 55°C for 5 minutes to anneal primers to the template.

The following three steps (c–e) will be cycled 35 times:

c. 72°C for 90 seconds for DNA synthesis to extend from the primers (this is the temperature at which *Taq* polymerase is most active).

d. 94°C for 45 seconds to denature the double-stranded DNA.

e. 55°C for 45 seconds to allow primers to anneal to the template DNA.

After the 35 cycles, the instrument will perform the following steps:

f. A final extension step at 72°C for 10 minutes is added to insure that the extensions are complete.

g. The program will then maintain 4°C until the instrument is turned off.

The PCR will take about 4 hours for this program. The tubes can then be removed from the PCR instrument and stored at 4°C until the next lab period.

LAB PERIOD II.3.2.   **PURIFY AND CHARACTERIZE THE *RVT* PCR PRODUCT**

PART A Purify the PCR Product

You will purify your PCR amplified *Rvt* PCR products to remove unincorporated dNTPs, primers, and salts. You will use the QIAquick PCR Purification Kit from Qiagen (a silica-based system). In high salt and at a pH less than or equal to 7.5, the double-stranded DNA PCR product will bind to the silica in the column, whereas the short, single-stranded oligonucleotide primers and unincorporated dNTPs will not.

Step 1 Transfer each of your PCR products from the fragile 0.2-ml PCR tubes to new 1.5-ml microfuge tubes. Label these tubes with your number and "*Rvt* PCR insert-A" or "*Rvt* PCR insert-B."

Step 2 To each tube containing your 50-µl *Rvt* PCR products add 250 µl (5 volumes) Buffer PB, and mix by pipetting up and down.

CAUTION!

Wear gloves! The PB buffer is toxic!

»»

Step 3 Apply each 300-µl sample to a separate QIAquick spin column supported in a 2-ml collection tube. In this step, the PCR amplified DNA will bind to the spin column, whereas primers and dNTPs will pass through the column during the subsequent centrifugation.

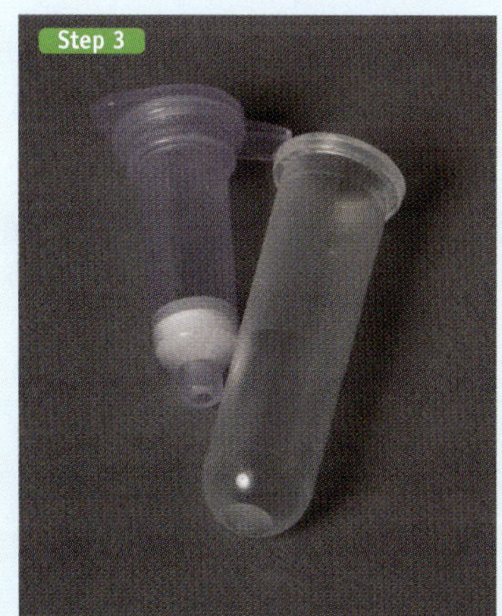

Step 4 Centrifuge the spin columns for 1 minute at 10,000x *g* in a microfuge.

Step 5 Discard the flow-through from the 2-ml collection tubes, and place the spin columns back into the *same* 2-ml collection tubes.

Step 6 Add 750 µl Buffer PE to each column. This buffer contains mostly ethanol and is used to wash the columns to remove any residual dNTPs, primers, and salts. The PCR-amplified DNA will remain bound to the column.

Step 7 Centrifuge the spin columns again for 1 minute at 10,000x *g* in a microfuge.

Step 8 Discard the flow-through from the 2-ml collection tubes, and place the spin columns back into the *same* 2-ml collection tubes.

Step 9 Centrifuge the spin columns again for 1 minute at 10,000x *g* in a microfuge to remove any residual Buffer PE.

Step 10 Place the spin columns into *new* 1.5-ml microfuge tubes. To elute the DNA from the columns, add 30 µl 0.1x TE, pH 8.0, to each column. The PCR products that were previously bound to the silica in the column under high-salt/low-pH conditions will now be eluted in low salt at pH 8.

PROCEDURAL NOTE

The pH of the 0.1x TE is critical! Using 0.1x TE with a more acidic pH (7 to 7.5) prevents elution of the DNA.

Step 11 Let the columns stand for 1 minute at room temperature then centrifuge for 1 minute at 10,000x *g* in the microfuge.

Step 12 Discard the columns, and close the lids of the 1.5-ml blue microfuge tubes. Label these tubes with your number and "Pure *Rvt* PCR-A" and "Pure *Rvt* PCR-B."

PART B Characterize the PCR Product by Agarose Gel Electrophoresis

Step 1 Pour a 1.0% agarose gel (as described in detail in Lab Period I.1.4, Part B) to be used to check the size and concentration of their PCR products. Allow the gel to solidify for 20 minutes before adding 1x TAE buffer to the electrophoresis chamber.

»»

LAB PERIOD II.3.2 CONT.

Step 2 Remove 5 µl of each "Pure *Rvt* PCR" product, and transfer to new microfuge tubes. To each of these tubes, add 2 µl 5x BJ and 3 µl ddH$_2$O. Label these tubes with your number and "undiluted PCR-A" or "undiluted PCR-B."

Step 3 Remove another 5 µl of each of the PCR products, and transfer to another set of new microfuge tubes. To these tubes, add 95 µl ddH$_2$O. Mix by vortexing.

Step 4 Remove 5 µl of each of these diluted PCR products to separate new microfuge tubes, and add 2 µl 5x BJ and 3 µl ddH$_2$O to each. Label these tubes with your number and "diluted PCR-A" or "diluted PCR-B."

PROCEDURAL NOTE

Store the remainder of your PCR products in your 4°C box until you use one of them for the ligation procedure in Lab Period II.4.

Step 5 Remove 6 µl of lambda BstEII marker DNA from your stock kept at 4°C into a separate microfuge tube, and heat at 65°C for 5 minutes (refer to Appendix III for a description of this molecular weight marker). Keep the sample on ice before loading the gel.

Step 6 Load the following samples on your agarose gel in this order:

    Lane 1:   10 µl "undiluted PCR-A" from Step 2 (partner 1).

    Lane 2:   10 µl "undiluted PCR-B" from Step 2 (partner 1).

    Lane 3:   10 µl "undiluted PCR-A" from Step 2 (partner 2).

    Lane 4:   10 µl "undiluted PCR-B" from Step 2 (partner 2).

    Lane 5:   10 µl "diluted PCR-A" from Step 4 (partner 1).

    Lane 6:   10 µl "diluted PCR-B" from Step 4 (partner 1).

    Lane 7:   6 µl λBstEII DNA in BJ (150 ng) from Step 5

    Lane 8:   10 µl "diluted PCR-A" from Step 4 (partner 2).

    Lane 9:   10 µl "diluted PCR-B" from Step 4 (partner 2).

    Lane 10: Empty

Step 7 Electrophorese the gel at 50 to 60 V for 1 to 2 hours. After electrophoresis, turn off the power supply; unplug the leads from the power supply.

Step 8 Carefully stain the gel for 10 minutes in ethidium bromide DNA gel stain, followed by a 15-minute gel destain in a water bath.

PROCEDURAL NOTE

Appendix IV describes an alternative DNA stain, SYBR Safe (Molecular Probes).

》》

LAB PERIOD II.3.2 CONT.

> **CAUTION!**
>
> Ethidium bromide is a known mutagen and carcinogen. Wear gloves, lab coats, and safety glasses! Handle gels carefully! Do not splash ethidium bromide! Always rinse the spatula in the water destain bath after it has been in contact with ethidium bromide! Always rinse the gel tray after sliding gels into the ethidium bromide bath. You may unknowingly contact the ethidium bromide! Be very careful not to drip the ethidium bromide anywhere!
>
> Always handle the gels carefully!

Step 9 Place your gel on a UV transilluminator, and photograph the gel (or take a digital image). An example of the expected appearance of this gel following electrophoresis is shown in **Figure 2-5**. Compare the photograph of your gel with that in Figure 2-5. Select a PCR product to be ligated into the pDrive plasmid vector during the next lab period. Use the lambda BstEII marker band sizes, as shown in Appendix III, to determine the approximate sizes of the PCR bands. If possible, pick a PCR product that is approximately 1.5 kb and shows only a single band on your gel. Remember that other size clones are possible, but if one of your PCR products is 1.5 kb, that is the best one to use for subcloning. How can you explain products that are sizes *other* than 1.5 kb? There are many short copies of the *Rvt* repeat with deletions and some longer copies with insertions; thus, many sizes are possible. Also, it is possible to have two EcoRI fragments ligated together to give a large dimer insert (2.9 kb).

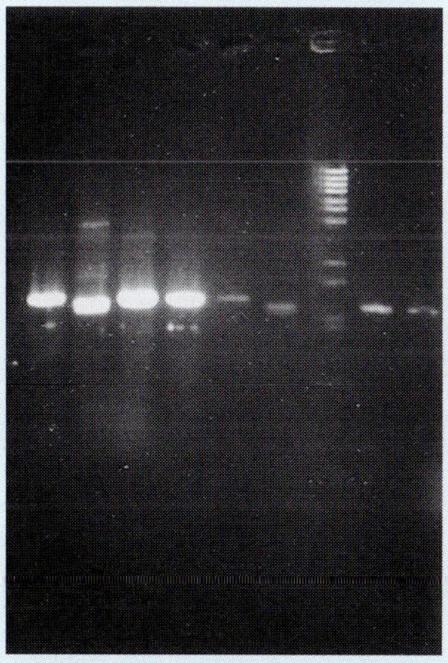

Lane 1 2 3 4 5 6 7 8 9

Fig 2-5 Expected appearance of gel.

PROCEDURAL NOTE

The PCR product has an additional 100 bp because the PCR primers hybridize to the lambda vector and amplify about 100 bp of lambda vector along with the insert. Thus, your PCR products actually contain some lambda DNA at each end and will be 1.5 kb. It is also possible that you may have cloned two or more inserts in the lambda vector. For example, two 1.4-kb inserts would yield a 2.9-kb fragment with the addition of the 100 bp of lambda vector sequence. Other sizes are possible because of the additions and deletions often found in these repeat sequences.

Step 10 Estimate the concentration of your "Pure *Rvt* PCR" products (A and B) by comparing your samples to that of known amounts of DNA in the marker lanes (refer to the chart detailing amounts of DNA in each fragment of lambda BstEII DNA marker in Appendix III). Calculate the total amount of each of your *Rvt* PCR DNA fragments. Calculate how much volume of your PCR product contains 50 to 100 ng DNA. Your instructor will provide instructions on how to do this and will help you with your estimates and concentrations (also see page 17). Write the concentration of each of your PCR products on the tubes.

Step 11 Dispose of your gel in the appropriate waste container.

LAB PERIOD II.3.3.   LIGATE THE *RVT* PCR PRODUCT INTO THE pDRIVE PLASMID VECTOR

EACH PAIR of students will prepare a ligation reaction using *one* of your "Pure Rvt PCR" products and an aliquot of the pDrive plasmid vector. A schematic of the linear pDrive plasmid with its 3' "U" overhangs is shown in **Figure 2-6**.

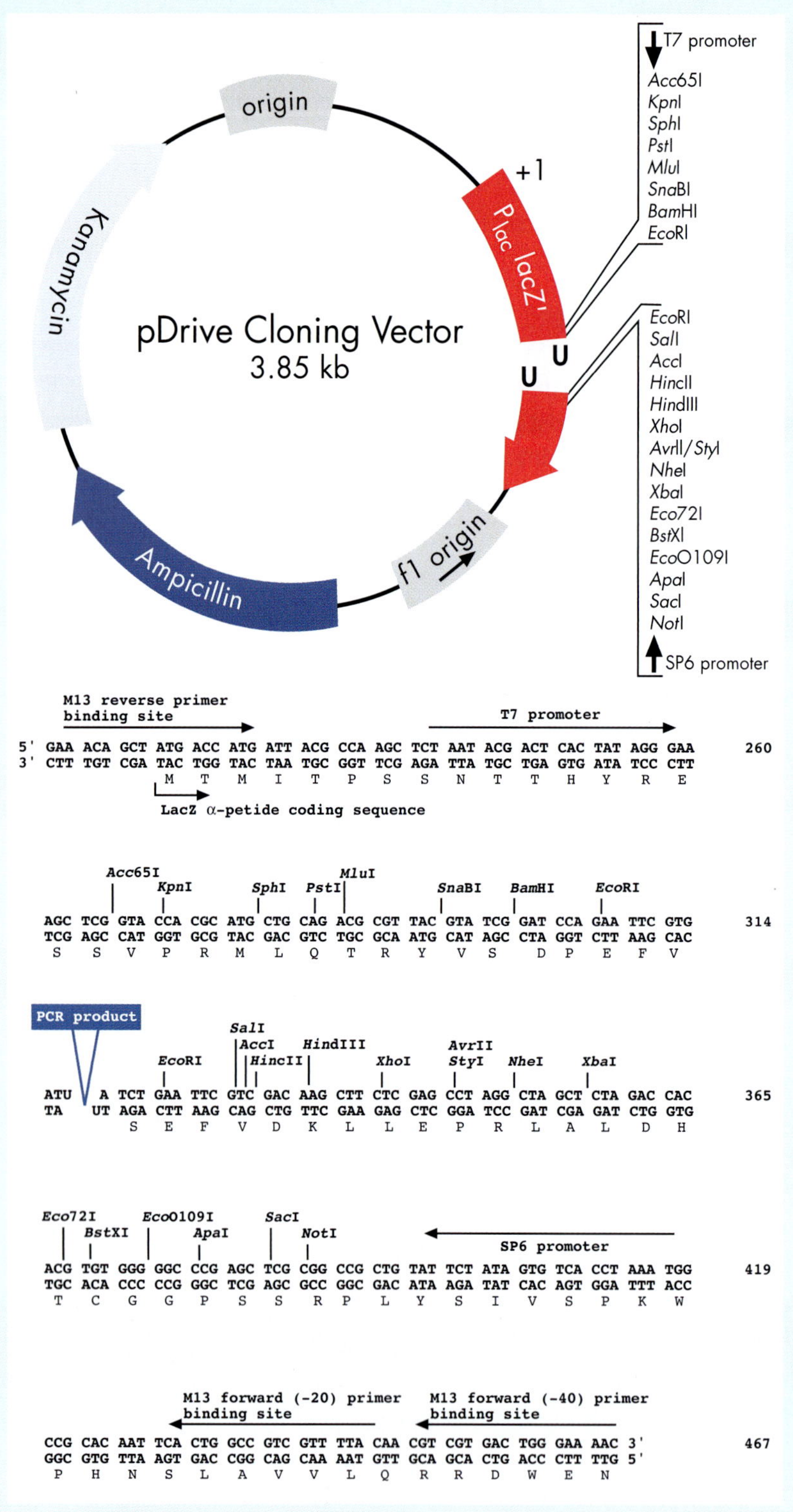

Fig 2-6 Schematic of pDrive plasmid with its 3' "U" overhangs. (Courtesy of and reprinted with permission of Qiagen, Inc.)

LAB PERIOD II.3.3 CONT.

Step 1 You will be provided with a premeasured aliquot (25 ng) of cut vector DNA (pDrive) for the ligation. Label the tube with your number and pDrive.

Step 2 Transfer a volume containing 50 to 100 ng of the selected "Pure *Rvt* PCR" in a volume of less than 4 µl to a new 1.5-ml microfuge tube.

PROCEDURAL NOTES

Remember that you selected one "Pure *Rvt* PCR" product based on the gel analysis you performed in Lab Period II.3.2 to use for this subcloning experiment, and you also determined the concentration of that PCR product from this gel as well. Add ddH_2O to adjust the volume to 4 µl.

Store all of the tubes containing your remaining "Pure *Rvt* PCR" products at 4°C. You may sequence these purified PCR products later; thus, label them carefully, and store them where you can find them easily at 4°C.

Step 3 Add all 4 µl of "Pure *Rvt* PCR" DNA to the microfuge tube containing 1 µl (25 ng) of linear pDrive vector (Qiagen) provided by the instructor. Then add the following:

> 5 µl 2x Ligation Master Mix (contains ligase buffer, ATP, and ligase) (Qiagen)
>
> ─────────────────────
>
> Final volume = 10 µl

Mix by stirring the solution with your pipette tip.

Step 4 Incubate your ligation 30 minutes to 2 hours in the 12°C incubator. Two hours will give more recombinant molecules than will 30 minutes. After the ligation period, your tubes should be stored at –20°C.

PROCEDURAL NOTE

The ratios of insert to vector DNA are *not* arbitrary. In fact, the molar ratios of the two types of DNA and their concentration in solution are critical parameters that must be considered in order to achieve successful ligations. A typical ligation is done with an insert to vector ratio of 1:1. The insert to vector ratio in this ligation is between 2:1 and 10:1. Why is this higher ratio acceptable in this type of ligation? Hint: Think about the ends of the PCR products that are being ligated to the vector.

LAB PERIOD II.3.4.   TRANSFORM THE pDRIVE PLASMID INTO *E. COLI* CELLS

A COMMON method used to introduce foreign DNA into bacterial cells is by the method of transformation. This is a method by which naked DNA is introduced into bacterial cells that have been prepared to make them "competent" for transformation. Bacterial cells to be transformed are rendered competent by their growth and preparation in selected media usually containing Mg^{2+} and/or Ca^{2+} ions. This method requires that the naked DNA be mixed with the competent *E. coli* cells and then given a "heat pulse" that is required to get the DNA into the cells. The method is not very efficient at getting DNA into *E. coli* but is perfectly acceptable when high efficiency is not needed (such as in this experiment). Competent cells are very fragile because of the treatment in the Mg^{2+} and Ca^{2+} solutions.

Each pair of students will transform competent *E. coli* cells with an aliquot of their pDrive/*Rvt* ligation reaction from the previous lab period (Lab Period II.3.3). The competent cells are provided in the pDrive Cloning Kit (Qiagen EZ Competent Cells). A protocol for making your own competent cells can be found in Appendix IV.

»»

LAB PERIOD II.3.4 CONT.

Step 1 You will be given a microfuge tube containing 50 µl of competent cells that have been thawed on ice. To this tube, add 2 µl of pDrive/*Rvt* ligated DNA. Again, mix *gently* by tapping the tubes. (Store the remainder of your ligation product at 4°C.) *Do not* centrifuge the competent cells in your nanofuge—they are fragile because of their treatment and you will kill them!

Step 2 Leave this tube on ice for 5 minutes or longer.

Step 3 Heat shock the cells by placing the samples in a 42°C water bath for *exactly* 30 seconds.

Step 4 Place immediately back on ice for 2 minutes.

Step 5 Add 250 µl of room temperature SOC medium to the tube.

Step 6 Label two LBA (LB agar + ampicillin) plates containing X-gal and IPTG with your number and "20 µl" or "100 µl." (Ampicillin will allow selection of cells containing the pDrive plasmid because the plasmid contains the gene for ampicillin resistance. The IPTG and X-gal facilitates the differentiation of recombinant colonies [white] from nonrecombinant colonies [blue].)

Step 7 Flick the tube to mix the cells. Onto the "20" plate, pipette 20 µl of the SOC culture. Then use one plastic cell spreader (per student) to spread the cells on the plate. Repeat the process with 100 µl on the other ("100") plate. This technique is shown in the two photos here and will be demonstrated by your instructors.

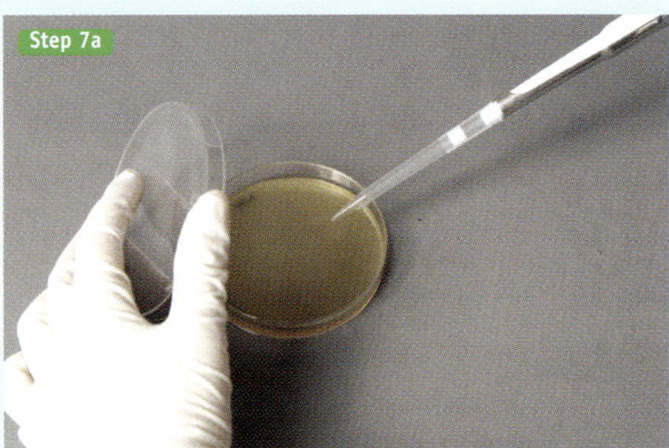

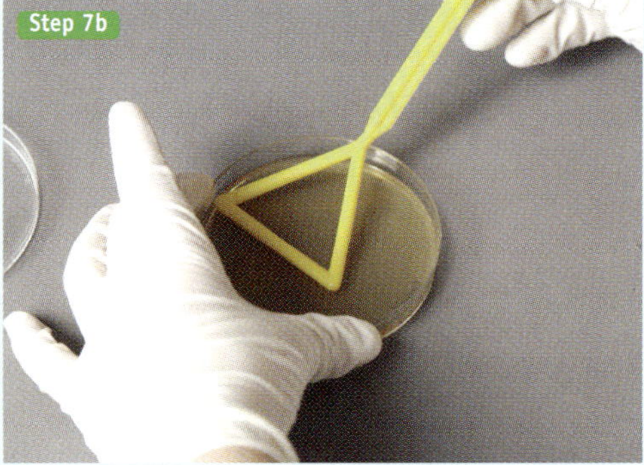

Step 8 After spreading your cells, allow the plates to sit upright at room temperature for 10 to 15 minutes, and then invert the plates and place them in the 37°C incubator O/N.

PROCEDURAL NOTE

For most transformation experiments, you generally should run two important controls using the same competent cells and protocol as you use for your experimental transformations. These controls are as follows:

    a. A negative control transformation with no DNA added to test for contamination of plates and solutions.

    b. A positive control transformation with uncut vector to confirm that the cells are competent to undergo transformation.

PROCEDURAL NOTE

Because the pDrive vector comes already linearized, it is not possible to do the second control in your experiment. Your instructor may do a positive control transformation using another plasmid that carries the ampicillin resistance gene (e.g., pUC19) to confirm the competency of the *E. coli* cells you are using for your

LAB PERIOD II.3.5.   "PICK" A RECOMBINANT COLONY FOR PLASMID PREPARATION

BEFORE THE start of this lab period and after the overnight incubation of your plates at 37°C, remove them from the 37°C incubator, and store them for several hours at 4°C. This will enhance the development of the blue color on your plates.

PROCEDURAL NOTE

The instructor can perform this step if the class schedule does not permit students to return to the lab at the appropriate time.

The blue color will indicate cells containing vectors with no insert in the β-galactosidase gene. The intact gene encodes a protein that can cleave the compound X-gal, giving rise to a "blue" colony. White colonies are those that have inserts (in this case, the cloned PCR products). You will need to come to the laboratory to pick one white bacterial colony made up of cells that contain a pDrive plasmid with a *Rvt* repeat inserted (if this is not possible, your instructor may pick a colony and start the overnight culture for you). To obtain sufficient plasmid DNA for the plasmid isolation, it is necessary to start a growing culture of *E. coli* containing the plasmid of interest the night before the plasmid preparation is to be performed. This will ensure that there will be sufficient numbers of cells harboring the plasmid for an efficient plasmid DNA isolation. This is done as follows:

Step 1 Remove the plates containing *E. coli* transformed with your recombinant pDrive/*Rvt* plasmid from 4°C, and look for *white* colonies (these are the transformant clones that contain inserts; the blue colonies contain no inserts). Examples of white and blue colonies are shown in here.

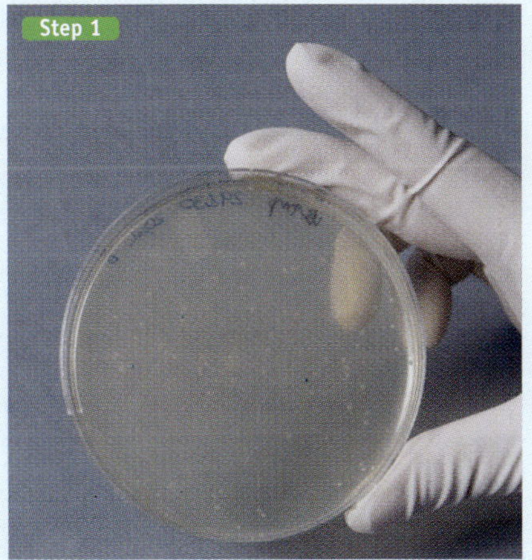

Step 2 Each student will locate a white colony on your transformation plate that is somewhat isolated from other colonies (to prevent picking cells from more than one colony).

Step 3 Using a sterile loop, pick a selected white colony as shown in here, and transfer it into a separate 15-ml conical tube containing 10 ml LBA medium (one colony per tube). Alternatively, use a 100-µl pipette tip to pick the colony, and transfer the tip into the culture tube.

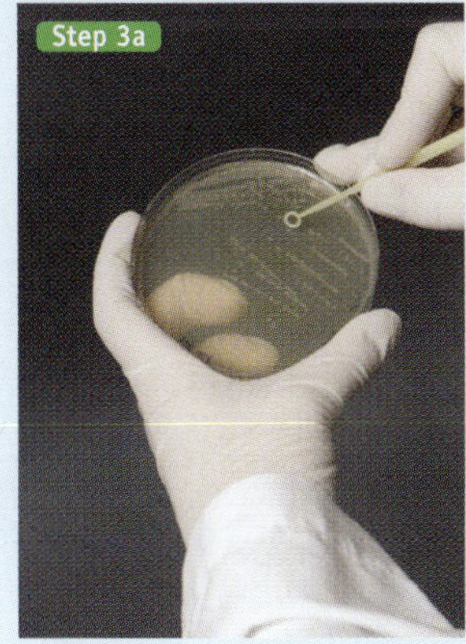

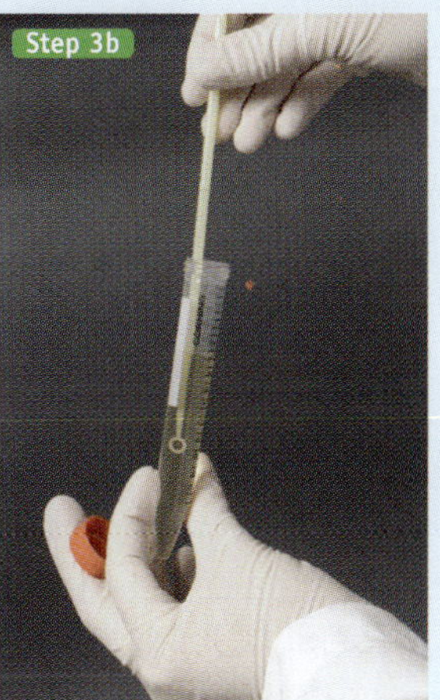

»»

LAB PERIOD II.3.5 CONT.

Step 4 Incubate the cultures at 37°C with vigorous shaking overnight.

PROCEDURAL NOTE

After the *E. coli* cells have grown overnight, you can preserve some or all of this *E. coli* culture by adding sterile glycerol to a final concentration of 50% and storing at –20°C. After a portion of the culture is saved in glycerol, it can be thawed and used at a later time. For this experiment, however, you should proceed immediately to the isolation of plasmid DNA from your culture.

LAB PERIOD II.3.6.   **PREPARE THE pDRIVE/*RVT* RECOMBINANT PLASMID**

IN THIS lab period, each student will isolate pure plasmid DNA from one pDrive/*Rvt* clone that you used to start your overnight culture. This method uses a silica-based column technology that allows rapid preparation of very pure plasmid DNA without ultracentrifugation or organic extractions. Yields from this "miniprep" should be in the 1- to 20-μg range. Other columns are available that can be used for "midipreps" (50 to 100 μg) or "maxipreps" (200 to 500 μg). Such columns are available from a number of companies. We use Qiaprep Miniprep kits that come from Qiagen.

Step 1 Centrifuge your plasmid culture tube at 2200x *g* for 10 minutes to pellet the cells.

Step 2 Pour off the supernatant into your waste beaker.

Step 3 Resuspend the cell pellet in 250 μl Qiagen buffer P1, and transfer to a new 1.5-ml microfuge tube. Buffer P1 contains RNase A to remove RNA selectively from the plasmid prep.

Step 4 To the cell mixture add 250 μl Qiagen buffer P2. Mix gently by inverting the tube five to six times. Do not vortex! Buffer P2 contains NaOH and SDS to break open the bacterial cells. This solution causes the cells to lyse, releasing the plasmid DNA from the cells.

Step 5 To the cell mixture add 350 μl Qiagen buffer N3. Mix immediately (but gently) by inverting the tube five to six times. The solution will now appear cloudy.

PROCEDURAL NOTE

This buffer contains guanidine hydrochloride, which neutralizes the buffer P2 and changes the binding conditions so that the DNA will bind to the silica in Step 7.

Step 6 Place your tube in the microfuge and spin at 16,000x *g* for 10 minutes to pellet cell debris. The *E. coli* chromosomal DNA is bound to the cell membrane/cell wall material and pellets with the debris. The plasmid DNA remains in the supernatant, however.

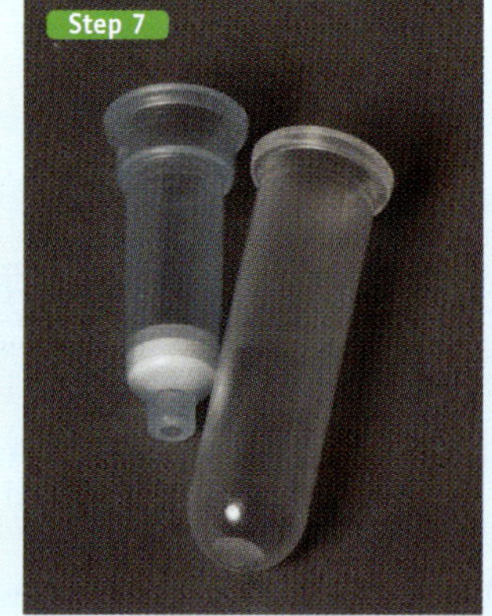

Step 7 Carefully transfer the supernatant onto a Qiaprep column. Place the column in its collection tube, and centrifuge the tube at 10,000x *g* (in a microcentrifuge) for 1 minute; discard the flow-through.

»»

LAB PERIOD II.3.6 CONT.

Step 8 To the column, add 750 µl Qiagen buffer PE. Centrifuge the column (with the same collection tube) at 10,000x g (in a microcentrifuge) for 1 minute, and discard the flow-through. The buffer PE helps to rinse contaminants from the column, leaving pure plasmid DNA in the column.

Step 9 Centrifuge the column again (with the same collection tube) at 10,000x g (in a microcentrifuge) for 1 minute to get rid of all the residual ethanol from the buffer PE.

Step 10 Transfer the column to a new 1.5-ml microfuge tube. Label the tube with your number and "pDrive/*Rvt* plasmid DNA."

Step 11 Add 50 µl of Qiagen buffer EB (10 mM Tris, pH 8.5) to the top of the column (in the center) and let the column sit for 1 minute at room temperature. Spin at 10,000x g in a microcentrifuge for 1 minute to recover the plasmid DNA in the bottom of the microfuge tube. Under these conditions, the plasmid DNA no longer stays bound to the silica and flows out of the column into the microfuge tube. Throw away the column and store the "pDrive/*Rvt* plasmid DNA" tube at 4°C.

LAB PERIOD II.3.7. **QUANTIFY THE pDRIVE/*RVT* DNA ON AN AGAROSE GEL**

YOU WILL electrophorese your isolated plasmid DNA on a 1% agarose gel.

Step 1 Pour a 1% agarose gel as described in detail in Lab Period I.1.4, Part B. Allow the gel to solidify for 20 minutes before pouring 1x TAE into the electrophoresis apparatus.

Step 2 Prepare your pDrive/*Rvt* plasmid DNA sample to determine the plasmid yield on the agarose gel: In a new 1.5-ml microfuge tube, mix 1 µl of your pDrive/*Rvt* plasmid DNA from Step 11 above with 49 µl ddH$_2$O (1:50 dilution). Mix thoroughly by pipetting up and down. Mix 5 µl of this dilution with 3 µl ddH$_2$O and 2 µl 5x BJ. This is your 1:50 diluted sample to load on the gel below. Label this tube "1:50 dil."

Step 3 In another new 1.5-ml microfuge tube, mix 10 µl of your pDrive/*Rvt* plasmid DNA from Step 11 above with 190-µl ddH$_2$O (1:20 dilution). Mix 5 µl of this dilution with 3 µl of ddH$_2$O and 2 µl 5x BJ. This is your 1:20 diluted sample to load on the gel below. Label this tube "1:20 dil."

Step 4 In a third new 1.5-ml tube, mix 2 µl of your pDrive/*Rvt* plasmid DNA from Step 11 above with 6 µl ddH$_2$O and 2 µl 5x BJ. This is your undiluted sample to load on the gel below. Label this tube "undil."

PROCEDURAL NOTE

Store the remainder of your undiluted plasmid DNA (labeled "pDrive/*Rvt* Plasmid DNA") at 4°C.

Step 5 You will also load several aliquots of your lambda BstEII DNA (which is at a concentration of 25 ng/µl in BJ). (Refer to Appendix III for a description of this molecular weight marker.) Heat 20 µl of the lambda markers at 65°C for 3 to 5 minutes, and place on ice before loading. Then load 2 µl (50 ng), 6 µl (150 ng), and 10 µl (250 ng) of this standard in three separate lanes on your gel as described here.

»»

LAB PERIOD II.3.7 CONT.

Step 6 Load the samples in the following order on your gel (lanes 1–5: partner 1; lanes 6–10: partner 2):

Lane 1:   10 µl lambda BstEII DNA (25 ng/µl = 250 ng)

Lane 2:   10 µl "undiluted plasmid" prepared in Step 4

Lane 3:   6 µl lambda BstEII DNA (25 ng/µl = 150 ng)

Lane 4:   10 µl 1:20 diluted plasmid prepared in Step 3

Lane 5:   10 µl 1:50 diluted plasmid prepared in Step 2

Lane 6   2 µl lambda BstEII DNA (25 ng/µl = 50 ng)

Lane 7:   Empty

Lane 8:   Empty

Lane 9:   Empty

Lane 10: Empty

Step 7 Electrophorese the gel at 60 V for 1.5 hours.

Step 8 Carefully stain your gel for 10 minutes in ethidium bromide DNA gel stain, followed by a 15-minute destain in a water bath (Lab Period I.1.4).

CAUTION!

Remember to wear gloves, lab coats, and safety glasses when using ethidium bromide!

PROCEDURAL NOTE

Appendix IV describes an alternative DNA stain, SYBRSAFE (Molecular Probes).

CAUTION!

Ethidium bromide is a known mutagen and carcinogen. Wear gloves, lab coats, and safety glasses! Handle gels carefully! Do not splash ethidium bromide! Always rinse the spatula in the water destain bath after it has been in contact with ethidium bromide! Always rinse the gel tray after sliding gels into the ethidium bromide bath. You may unknowingly contact the ethidium bromide! Be very careful not to drip the ethidium bromide anywhere!

Step 9 Place your gel on a UV transilluminator, and photograph the gel (or take a digital image). An example of the results expected on this gel is shown in **Figure 2-7**.

Step 10 Dispose of your gel in the appropriate waste container.

Step 11 Using the photograph of your gel and the reference chart in Appendix III (this chart shows the amount of DNA in each

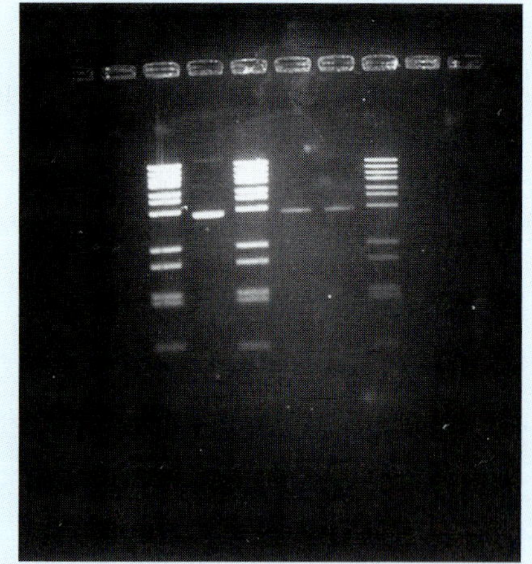

Fig 2-7 Expected appearance of gel.

LAB PERIOD II.3.7 CONT.

marker band), estimate the yield of your "pDrive/*Rvt* plasmid DNA" by comparing your plasmid bands with the marker bands. Calculate the concentration of plasmid DNA in ng per microliter in this preparation. Be sure to take into consideration the dilution factor of your DNA sample and how much you loaded in that gel lane. Compare your DNA, at an appropriate dilution, with a comparable band in one of the marker lanes. Because your plasmid DNA consists of both super-coiled plasmid and open circular plasmid, you will need to quantify both bands and add them together to get an accurate estimate of your plasmid DNA concentration.

Step 12 Write your calculated concentration on your gel photo and on the tube from Step 11, Lab Period II.3.6. You have now completed a plasmid isolation of your *Rvt* clone and you are ready to use this DNA to characterize further the cloned *Rvt* sequence as described in Modules II.4 and II.5.

Dideoxy Thermal Cycle Sequencing of a Cloned Gene

MODULE II.4

MODULE SUMMARY

PCR amplify and quantify the insert from the pDrive/Rvt plasmid. Determine the sequence of the insert by dideoxy thermal cycle sequencing followed by spin column purification. Load the DNA sequencing reactions into an automated DNA sequencer.

MODULE BACKGROUND

In the vast majority of organisms, genetic information is stored as DNA and is manifest as unique sequences of the four bases: adenine (A), guanine (G), cytosine (C), and thymine (T). In viruses, the genetic material can be DNA or RNA where the base uracil (U) replaces thymine (T). Long sequences of these bases called genes encode the specific amino acids to be incorporated into polypeptides during protein synthesis. In addition to genes that encode proteins, there are other genes that encode functional RNAs (e.g., rRNAs or tRNAs). Other DNA sequences that do not encode products can function as regulatory elements that control which genes are expressed in specific tissues or at specific developmental stages. DNA sequence also dictates where transcription of genes begins and ends, where introns will be spliced out of messenger RNAs, which regions of the genome will act as intergenic sequences to separate individual genes from one another and which regions encode functional RNA molecules that are not translated into proteins. Other DNA sequences function to form part of important chromosomal structures such as centromeres and telomeres. Therefore, after a region of DNA has been cloned, it is often desirable to determine the base sequence of that region to understand better the function of that sequence.

Multiple methods have been used to determine DNA sequences. The most commonly used method is called "dideoxy sequencing" or "chain termination sequencing." This method involves the enzymatic synthesis of a new strand of DNA that is complementary to the template strand to be sequenced. The unique feature of this method is that normal deoxyribonucleotides (dNTPs) are mixed with a small proportion of dideoxyribonucleotides (ddNTPs) to serve as the precursors for DNA synthesis. Dideoxynucleotides lack a hydroxyl group at the 3' carbon and thus cannot serve as a binding site for the addition

MODULE BACKGROUND, cont.

of the next nucleotide in the growing chain (**Figure 2-8**). Thus, as DNA synthesis progresses, the normal dNTPs become incorporated into the newly synthesized strand just as they do during normal DNA replication *in vivo*. At low frequency the ddNTPs can also be incorporated into the growing DNA chain on the basis of normal base complementarity. However, after a ddNTP is added to the newly synthesized DNA strand, no other nucleotides (dNTP or ddNTP) can be added beyond that site. Hence, these ddNTPs serve to terminate synthesis and are referred to as chain termination nucleotides. DNA sequencing can be performed using a variety of templates, including plasmid DNA, single-stranded DNA, PCR products, and even directly from bacterial colonies or phage plaques, although the quality is less from the latter two. The principle of the dideoxy, chain termination sequencing technique is schematically represented in Figure 2-8.

Two significant technical advances rendered DNA sequencing by the dideoxy chain termination method highly efficient. The first advance was the development of an "automated" DNA sequencing process. This automation was facilitated by (1) the development of unique fluorescent dyes bound to each of the four different ddNTPs so that each could be distinguished by the production of different emission spectra of light upon laser excitation and (2) the availability of automated DNA sequencing instruments that use gel or capillary electrophoresis to size-separate the newly synthesized DNA strands. These instruments can detect the dye emissions to identify which terminator nucleotide has been added to the end of each strand. Because of the use of four differentially labeled fluorescent dideoxy terminators, only a single reaction tube is required for this reaction. This single tube will contain all of the components necessary for the sequencing reaction to proceed, including the four dideoxy dye terminator nucleotides, a sequencing primer, the template DNA, a buffer to ensure proper reaction conditions, and a DNA polymerase.

The second significant advance was the implementation of "thermal cycle sequencing." This is essentially a standard dideoxy sequencing reaction that is cycled over and over again to generate additional product at each cycle. During each thermal cycle, a set of single-stranded DNA molecules is synthesized on the DNA template, each ending with a specific, labeled dideoxynucleotide. After the first cycle is complete,

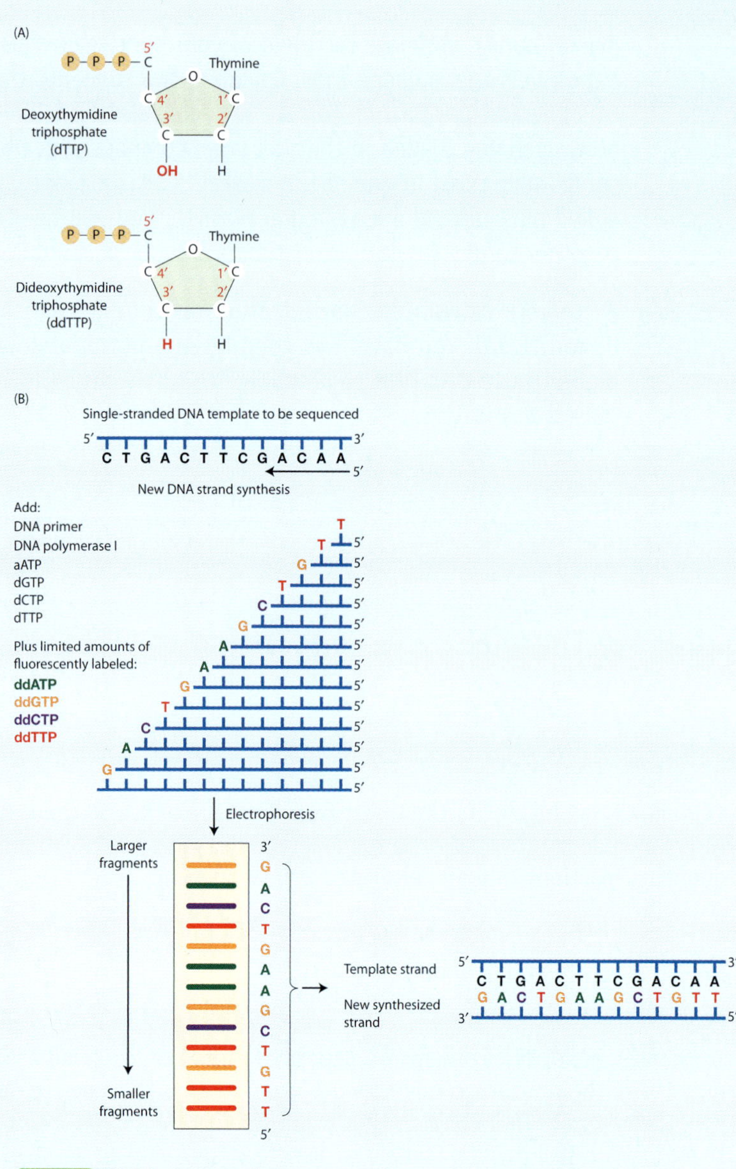

Fig 2-8 DNA sequencing.

MODULE BACKGROUND, cont.

the strands are heated to "melt" the newly synthesized strand away from the template strand. The temperature is then lowered to allow the annealing of another sequencing primer, and the reaction is run again. Because of the repeated cycles of denaturation, primer annealing and DNA synthesis, large numbers of newly synthesized single-stranded DNA molecules are produced by this process (**Figure 2-9**). When these products are separated by electrophoresis, the automated sequencing instrument "reads" the sequence by measuring the fluorescence of each synthesized molecule that passes the detector. After electrophoresis, a computer algorithm will convert the fluorescent patterns ("traces" or "chromatograms") into a linear DNA base sequence.

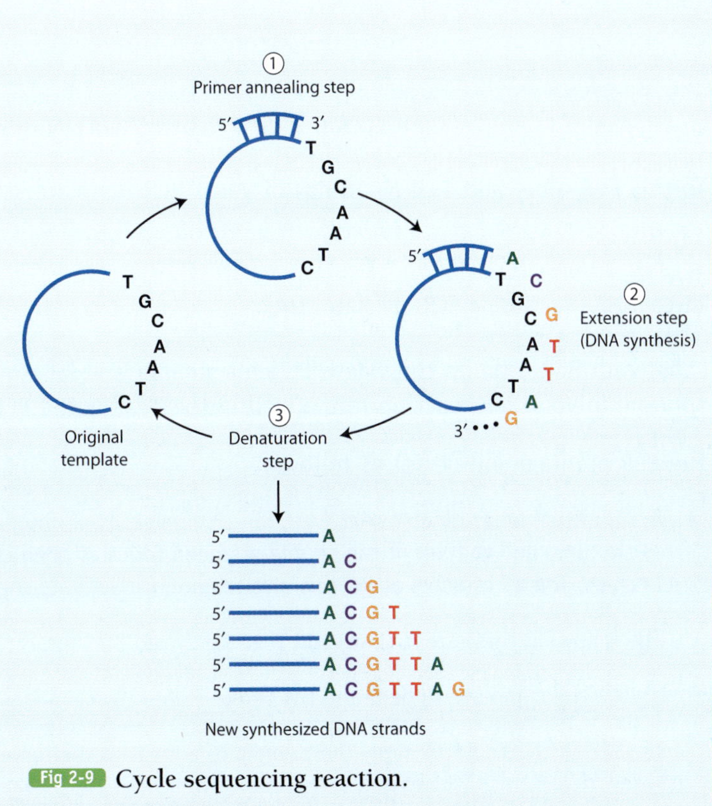

Fig 2-9 Cycle sequencing reaction.

    PCR AMPLIFY THE INSERT FROM THE pDRIVE/*RVT* PLASMID

Step 1 You will set up a PCR reaction to amplify the cloned insert from your pDrive/*Rvt* plasmid. The product of this reaction will serve as your template for DNA sequencing. Retrieve your tube of "pDrive/*Rvt* Plasmid DNA" from storage at 4°C.

Step 2 Using the concentration of your plasmid preparation determined in Lab Period II.3.7, determine the volume of that prep that will contain 1 µg (1000 ng) of your plasmid DNA. Transfer this amount of plasmid to a new 1.5-ml microcentrifuge tube. Add sufficient ddH$_2$O to bring the total volume to 100 µl. Label this tube with your number and "PCR template stock." The concentration of this template stock is 10 ng/µl.

Step 3 To a new 0.5-ml PCR tube, add the following components:

30.5 µl ddH$_2$O

8.0 µl dNTP mix

5.0 µl 10x PCR mix

3.0 µl MgCl$_2$ (25 mM)

1.0 µl M13 Reverse Primer

1.0 µl SP6 Primer

0.5 µl AmpliTaq™ Gold polymerase
____
49.0 µl Total volume

LAB PERIOD II.4.1 CONT.

PROCEDURAL NOTE

M13 Reverse Primer: 5' GGAAACAGCTATGACCATG 3' SP6 Primer: 5' CATTTAGGTGACACTATAG 3'

Step 4 To this PCR tube, add 1 µl (= 10 ng) of your "PCR template stock."

PROCEDURAL NOTE

You may discard your remaining PCR template stock.

Step 5 Label your PCR tube on the side with your number and "Plasmid PCR." Place the 0.5-ml tube containing your PCR reaction into a well in the designated PCR thermal cycler.

Step 6 The PCR program will be set as follows:

a. An initial denaturing step at 94°C for 12 minutes is performed to insure that the lambda phage are lysed and that all of the double-stranded DNA has been converted to single-stranded molecules. The 12 minutes at 94°C is also required to activate the *Taq* Gold DNA polymerase.

The following three steps (b–d) will then be cycled 35 times:

b. 94°C for 1 minute to denature the DNA.

c. 55°C for 1 minute to allow the primers to anneal to the template DNA.

d. 72°C for 3 minutes for DNA synthesis to extend from the primers (this is the temperature at which *Taq* DNA polymerase is most active).

e. A final extension step at 72°C for 7 minutes is added after the 35th cycle to insure that the extensions are complete.

The program will then maintain the tubes at 4°C until the tubes are removed and the instrument is turned off.

LAB PERIOD II.4.2.   **CONFIRM AND QUANTIFY THE PCR PRODUCT**

Step 1 Prepare a 1% agarose gel as described in Module I.1.4, Part B. Allow the gel to solidify for at least
20 minutes before adding 1x TAE buffer into the electrophoresis chamber.

Step 2 Retrieve your amplified DNA from the thermal cycler (or from its stored location at 4°C). Remove 8 µl of the reaction, and mix with 2 µl of 5x BJ. Label this tube with your number and "plasmid PCR sample."

Step 3 Retrieve your lambda BstEII marker in 5x BJ (25 ng/µl) from your 4°C box. (A description of the molecular weight marker is provided in Appendix III.) Remove 8 µl from this tube to a new tube labeled "Marker." Heat the marker tube at 65°C for 5 minutes. Place the marker tube on ice. Load your samples on your 1% agarose gel in the following order (a second group or individual may load their samples on the same gel):

»»

LAB PERIOD II.4.2 CONT.

Lane 1: Empty

Lane 2: 2 µl of your heated lambda BstEII marker (50 ng)

Lane 3: 4 µl of your heated lambda BstEII marker (100 ng)

Lane 4: 10 µl of your "plasmid PCR sample"

Lane 5: 10 µl of your "plasmid PCR sample" — second partner or group

Lane 6: 2 µl of your heated lambda BstEII marker (50 ng) — second partner or group

Lane 7: 4 µl of your heated lambda BstEII marker (100 ng) — second partner or group

Lane 8: Empty

Lane 9: Empty

Lane 10: Empty

Step 4 Run the gel at 65 V for 1.5 to 2 hours.

Step 5 Stain the gel in ethidium bromide DNA stain for 10 minutes, followed by a 15-minute destain, as described in Lab Period I.1.4.

CAUTION!

Ethidium bromide is a known mutagen and carcinogen. Wear gloves, lab coats, and safety glasses! Handle gels carefully! Do not splash ethidium bromide! Always rinse the spatula in the water destain bath after it has been in contact with ethidium bromide! Always rinse the gel tray after sliding gels into the ethidium bromide bath. You may unknowingly contact the ethidium bromide! Be very careful not to drip the ethidium bromide anywhere!

PROCEDURAL NOTE

Appendix IV describes an alternative DNA stain, SYBR SAFE (Molecular Probes).

Step 6 Photograph the gel. An example of the expected results is shown in **Figure 2-10**. Examine the PCR product generated from your plasmid. If a band is present, estimate its size (compare the size of the band containing your PCR product with the marker bands of known size using the marker chart for lambda BstEII) (Appendix III). What is the size of your PCR product? Does this correspond to the product size you expected? Remember that the PCR reaction will yield a product that includes approximately

CAUTION!

Always handle the gels carefully!

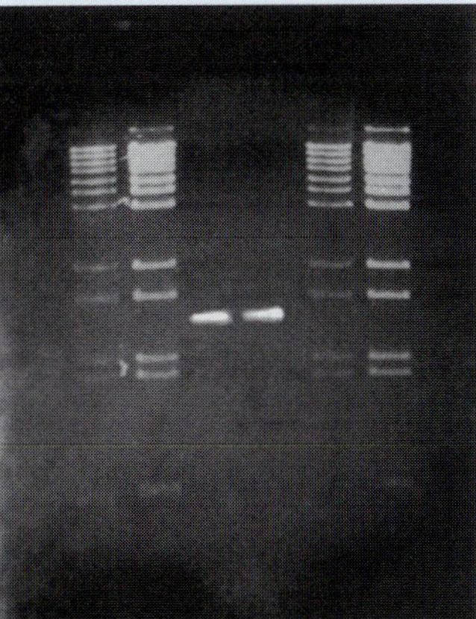

Fig 2-10 Expected appearance of gel.

LAB PERIOD II.4.2 CONT.

400 bp of vector sequence (approximately 210 bp from pDrive and approximately 190 bp from lambda ZAP Express) in addition to the cloned insert. If the product is not of the expected size, can you explain the size of the product you obtained?

Step 7 Determine the quantity of DNA in the band representing your plasmid PCR product by comparing the intensity of that band with the intensity of marker bands as described in detail in Appendix III. This amount divided by 8 will be the concentration of your "Plasmid PCR" product per µl because you loaded 8 µl of your PCR product on the gel. You will need this information to set up your sequencing reaction in Lab Period II.4.3. Write the concentration on the tube.

LAB PERIOD II.4.3.   **DIDEOXY THERMAL CYCLE SEQUENCING REACTION**

PROCEDURAL NOTES

The following protocol has been developed for the Applied Biosystems automated DNA sequencers (373, 377, 3100, 310, 3700 family). The protocols may differ depending on the DNA sequencer being used by your class. You will prepare a thermal cycle sequencing reaction using your "Plasmid PCR" DNA (from Lab Period II.4.2, Step 7) as a template.

Your instructor will provide you with the reaction premix. It contains a buffer, dNTP mix, the four labeled dideoxy nucleotides, and AmpliTaq FS DNA polymerase needed for the sequencing reaction.

Step 1 Each student will be given a new microfuge tube that contains 4.0 µl of a sequencing primer labeled "T3 Plasmid Seq."

PROCEDURAL NOTE

4.0 µl T3 Sequencing Primer = 3.2 pmol = 20 ng.

This T3 primer is one of the two primers that you used in the original PCR reaction to amplify the *Rvt* sequence from your lambda ZAP Express clone (Lab Period II.3.1) and therefore can also be used as a sequencing primer. Label this microfuge tube on the top and side with your number. This will be your DNA sequencing reaction tube.

Step 2 Centrifuge your primer tube briefly in the nanofuge to get all of the liquid to the bottom of the tube.

Step 3 Add 40 ng of your "Plasmid PCR" DNA to this tube, using no more than 8 µl of the DNA. You have calculated the concentration of the PCR product in II.4.2, Step 7. Using this calculation, determine the volume of your "Plasmid PCR" product that will provide 40 ng of template DNA. You may need to dilute your "Plasmid PCR" product to obtain 40 ng. Remember that the volume of this "Plasmid PCR" product must be no more than 8 µl. Verify this calculation with the instructor to make sure that you have properly calculated the amount of PCR product you need to add to your sequencing reactions. Add the amount of "Plasmid PCR" DNA needed to provide 40 ng of template to the tube containing the 4 µl T3 sequencing primer. Then add enough ddH$_2$O to raise the total volume to 12 µl (4 µl T3 primer + X µl "Plasmid PCR" [40 ng] + Y µl ddH$_2$O = 12 µl total volume).

»»

LAB PERIOD II.4.3 CONT.

Step 4 Add 8 µl of the Reaction Premix to your DNA sequencing reaction tube to bring the total reaction volume to 20 µl. Keep the tube cover closed as much as possible to prevent evaporation, and keep the tube on ice until you place it into the thermal cycler.

Step 5 Place your DNA sequencing reaction tube into the thermal cycler as directed by your instructor.

Step 6 The thermal cycling program will be set as follows:

   a. 96°C for 30 seconds to denature the double-stranded DNA.

   b. 50°C for 15 seconds to allow primers to anneal to the template DNA.

   c. 60°C for 4 minutes so DNA synthesis can extend from the primers.

The three steps (a–c) will be cycled 25 times. This thermal cycling reaction will take 2 to 3 hours to complete. The instrument will then hold the samples at 4°C until the samples are removed and stored at 4°C until the next lab period.

PROCEDURAL NOTE

For purposes of determining the sequence of your genomic clone of the *Rvt* gene, you actually have several options for the DNA sequencing template. These include the PCR product derived directly from the lambda vector (Lab Period II.3.1), the pDrive plasmid carrying the subcloned *Rvt* insert, or the PCR product derived from the pDrive plasmid, as described above. Each of these includes a representation of the cloned *Rvt* sequence flanked by several useful unique primer sites that can be used for sequencing. Indeed, you might choose to sequence both the lambda PCR product and either the subclone in the pDrive plasmid or the plasmid PCR product to confirm that you properly subcloned the insert from your original lambda clone. Verify that the primer you will use for sequencing has only one unique site in the PCR product that you will be sequencing (examine the maps of the pBK-CMV plasmid and the pDrive plasmid). Remember that pBK-CMV is the plasmid sequence contained within the lambda ZAP Express vector. Why can you not use a T7 primer to sequence the "Plasmid PCR" product in this case?

LAB PERIOD II.4.4.   **PURIFY AND LOAD THE DNA SEQUENCING REACTIONS**

PART A Spin Column Purify the DNA Sequencing Reactions

You will purify your thermal cycle sequencing reactions using the Performa Gel Filtration Cartridges from Edge Bioscience. These are gel exclusion columns: your sequenced DNA will move rapidly through the gel in the column, whereas the unincorporated dideoxy dyes used in the sequencing reaction will pass through very slowly. Because the unincorporated dideoxy dyes do not migrate as normal nucleotides (because of their charge) but migrate at positions within the DNA sequence, they can interfere with accurate base calling by the automated DNA sequencer. Thus, these columns are crucial for removing these dye terminators.

Step 1 You will receive one column and two tube(s) for this procedure. Label both the column and the tubes with your number. Label one tube "wash" and the other tube "Seq Product." Place the column in the "wash" tube, and spin in a variable speed microfuge at 720x *g* for exactly 2 minutes to remove the storage buffer.

Step 2 Remove the column from the "wash" tube, and insert it into the "Seq Product" tube.

»»»

LAB PERIOD II.4.4 CONT.

Step 3 Add the sequencing reaction sample (20 µl) to the top of the spin column (look at your column from the side). Apply the sample to the top of the column in the middle as shown in here.

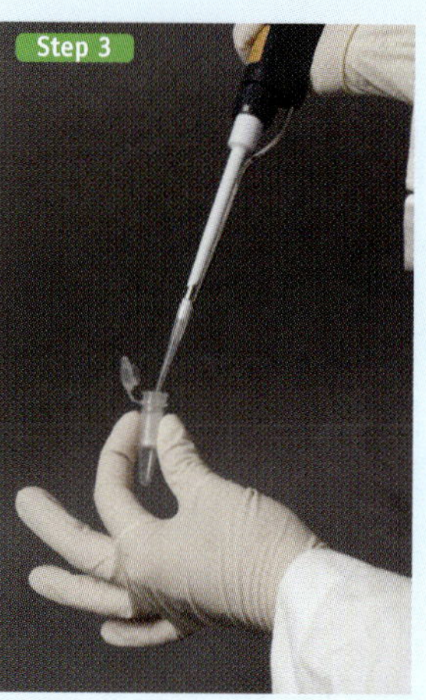

PROCEDURAL NOTE

Do not touch the sample to the side of the column, and do not poke through the spin column matrix with your pipette tip!

Step 4 Spin the column in the collection tube in a microfuge at 720x g for 2 minutes.

Step 5 Discard the column and close the cap on the collection tube.

Step 6 The collection tube labeled "Seq Product" now contains your purified DNA sequencing reaction that will be used to load into the automated DNA sequencer gel or capillary (depending on the model of the sequencer.

Step 7 Before loading the samples, they must be evaporated and then resuspended in a small volume of "loading buffer." Place your samples in a "spin vacuum concentrator," and dry the samples to completion. (Do no overdry them or else they will be hard to resuspend in loading dye.) After the samples are dry, resuspend them in the appropriate solution and correct volume before loading them in the DNA sequencer. The loading buffer differs depending on the DNA sequencing instrument being used but is usually water or formamide (a denaturant) and may include a non-fluorescent dye to aid in loading the samples. Your instructor will inform you as to which loading solution will be used. For example, for the 373 and 377 Applied Biosystems instruments, you will resuspend the samples in 3 µl of formamide, containing EDTA and Dextran Blue. For the 3100, 310, or 3700 family of automated DNA sequencers, you will use formamide or water.

PART B Load the "Seq Product" Samples into an Automated DNA Sequencer

Step 1 Load (or watch as the instrument loads) your "Seq Product" sample. The exact instrument that you will use will depend on what is available at your institution. Some automated sequencers "autoload" the reactions, whereas others require manual loading of samples. Accordingly, your instructor will provide you with further details on this step.

Step 2 The automated sequencer will electrophorese your samples and the sequence data will be collected automatically during the night. Sequence data will be returned to you the next day or at the next lab period.

PROCEDURAL NOTE

You can send your DNA sequence to a variety of companies that specialize in DNA sequencing if you don't have access to an automated DNA sequencer.

Bioinformatic Analysis of the DNA Sequence Data

MODULE II.5

MODULE SUMMARY

Analysis and quality assessment of the DNA sequence data. BLAST search to confirm the identity of the DNA sequence data. Search for an open reading frame (ORF) in your DNA sequence data. Perform a similarity search between two sequences. Electronically map restriction sites in your cloned *Rvt* sequence.

MODULE BACKGROUND

The DNA sequence results from Module II.4 should reflect the approach used to perform the recombinant DNA cloning of a portion of the *Rvt* gene. Recall that in this project, a mouse genomic EcoRI fragment was cloned (ligated) into the lambda ZAP Express vector (from Stratagene) and clones containing the *Rvt* gene were selected by hybridization with a labeled probe to the gene sequence. Confirmation that the clone is correct (i.e., that it is indeed the clone you were searching for) is one of many uses for DNA sequence analysis. This confirmation can often be determined by comparison of your sequence to known sequences that have been previously submitted to the database. We will use the BLAST computer suite of programs at the National Center for Biotechnology Information (NCBI) at the National Library of Medicine at the National Institute of Health (NIH). BLAST stands for "Basic Local Alignment Search Tool." A comparison of sequences with database entries can suggest, but not directly prove, the identity of unknown sequences.

PROCEDURAL NOTE

In this exercise, you are not *submitting* database entries, but rather *comparing* your DNA sequence to known, previously submitted, database sequences.

LAB PERIOD II.5.1.   PRELIMINARY ANALYSIS OF THE DNA SEQUENCE DATA

PART A Quality Assessment of Your DNA Sequence Data

You will examine your PCR product sequence data that was collected by the automated DNA sequencer to assess the quality of the data. Your DNA sequence data will be returned to you in two forms. One form will be a printout of the DNA sequence "trace" or "chromatogram" of your DNA sample. An example of a high-quality trace is shown in **Figure 2-11**, and examples of less desirable results are shown in **Figure 2-12a, b**, and **c**. The second form will be a computer file with a corresponding "textfile" of the base sequence that you will use to edit and to compare with sequences in the database via the Internet. An example of a "textfile" is shown in **Figure 2-13**.

Step 1 You first need to "edit" your sequence. There are several freeware editing programs available that allow you to edit your sequence trace files. For the Macintosh, the EditView program can be downloaded from the Applied Biosystems website at
http://www.appliedbiosystems.com/support/software/dnaseq/installs.cfm.

For the PC, the Chromas program can be downloaded from
http://www.technelysium.com.au/chromas.html.

>>>

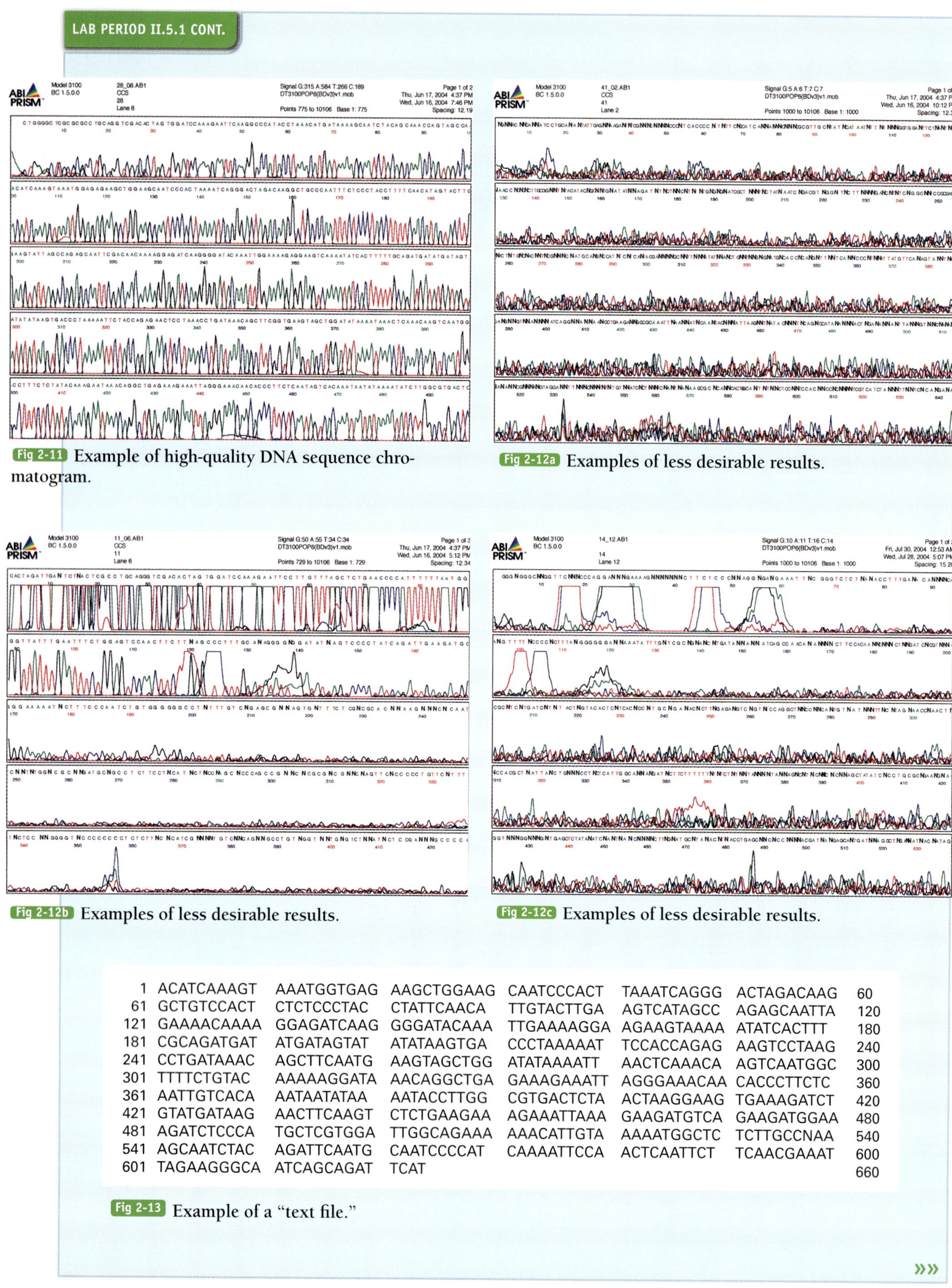
Fig 2-11 Example of high-quality DNA sequence chromatogram.

Fig 2-12a Examples of less desirable results.

Fig 2-12b Examples of less desirable results.

Fig 2-12c Examples of less desirable results.

1 ACATCAAAGT	AAATGGTGAG	AAGCTGGAAG	CAATCCCACT	TAAATCAGGG	ACTAGACAAG	60
61 GCTGTCCACT	CTCTCCCTAC	CTATTCAACA	TTGTACTTGA	AGTCATAGCC	AGAGCAATTA	120
121 GAAAACAAAA	GGAGATCAAG	GGGATACAAA	TTGAAAAGGA	AGAAGTAAAA	ATATCACTTT	180
181 CGCAGATGAT	ATGATAGTAT	ATATAAGTGA	CCCTAAAAAT	TCCACCAGAG	AAGTCCTAAG	240
241 CCTGATAAAC	AGCTTCAATG	AAGTAGCTGG	ATATAAAATT	AACTCAAACA	AGTCAATGGC	300
301 TTTTCTGTAC	AAAAAGGATA	AACAGGCTGA	GAAAGAAATT	AGGGAAACAA	CACCCTTCTC	360
361 AATTGTCACA	AATAATATAA	AATACCTTGG	CGTGACTCTA	ACTAAGGAAG	TGAAAGATCT	420
421 GTATGATAAG	AACTTCAAGT	CTCTGAAGAA	AGAAATTAAA	GAAGATGTCA	GAAGATGGAA	480
481 AGATCTCCCA	TGCTCGTGGA	TTGGCAGAAA	AAACATTGTA	AAAATGGCTC	TCTTGCCNAA	540
541 AGCAATCTAC	AGATTCAATG	CAATCCCCAT	CAAAATTCCA	ACTCAATTCT	TCAACGAAAT	600
601 TAGAAGGGCA	ATCAGCAGAT	TCAT				660

Fig 2-13 Example of a "text file."

PROCEDURAL NOTE

There are numerous other computer programs that enable chromatogram viewing and editing as part of larger software packages. Suggested resources are listed in Appendix VI.

Open your sequence file using the correct program (discussed in Step 1), and view your sequence results. Normally the first 10 to 15 nucleotides of sequence may be "messy" (low quality). Examine the peak heights and "cadence" of the peaks. The farther you read in the sequence, the less "tight" the peak cadence becomes and the more apt the base calling software is to make a mistake, especially in runs of the same nucleotide (i.e., at a particular position, are there three or four "A" nucleotides in a row?). At some point, you may decide that the base calling is no longer as accurate as you would like. Note this position. Be aware that you can edit the sequence or decide to eliminate low-quality data from further consideration.

PROCEDURAL NOTES

Editing the chromatogram also edits the text file of the sequence and vice versa.

If you do edit the sequence, save the file under a different name or version number. It is generally a good idea to keep the original sequence file for future use (in case you need to refer back the original data).

Step 2 Determine the length of high-quality DNA sequence data obtained. Record this information for each student in the class, and compare your results with those of the other students. Confirm that the sequence data have been correctly entered as a sequence file in the computer. Did the automated sequencer fail to identify certain bases that you can identify by examining the trace data? If so, correct these bases in the sequence file on the computer by editing.

PART B Further Bioinformatic Analysis of the DNA Sequence Data

Each student will continue to examine his or her own DNA sequence data derived from Module II.4.

Step 1 Your sequence should begin with "vector" sequence from lambda ZAP Express because you used a DNA sequencing primer that hybridized at the T3 sequence position in the vector at a position that begins 50 bases before the EcoRI cloning site (see Fig. 2-2 for the sequence and location of the T3 primer). Find this vector sequence in your DNA sequence readout. This vector sequence is not needed for the bioinformatics analyses you are about to perform on your gene sequence and, thus, should be removed (excluded) from the final sequence that will be used for your database analyses.

Step 2 Using the sequence editing feature of either the EditView program (if you are using a Macintosh) or the Chromas program (if you are using a PC), remove the lambda ZAP Express vector sequence from your DNA sequence textfile. Your sequence should now begin with an EcoRI site (5'-GAATTC-3'). To do this, compare your sequence to the vector sequence of the lambda ZAP Express vector, provided in Figure 2-2 . (This is the same sequence as found in the polylinker of the pBK-CMV vector.)

Begin reading your sequence at the leftmost (5' end) of your sequence trace. Compare your sequence to the above sequence of the vector until you find an EcoRI site (GAATTC). This is the cloning site into which you ligated your 1.4-kb EcoRI *Rvt* gene fragment from mouse genomic DNA. All sequence to the left of this site can be deleted (it is vector sequence). If you cannot locate the EcoRI site, it could be that the software base caller has made a mistake at that point (you should verify the sequence base calling from the chromatogram data). Exam-

»»

LAB PERIOD II.5.1 CONT.

ine your sequence at the point where it first deviates from that of the vector sequence, as this is where your insert sequence should begin. It is also remotely possible that your clone has a deletion of some type. If, in fact, your clone is difficult to interpret (verify this with the instructor), use the sequencing results from another member of the class.

PROCEDURAL NOTE

Because of the methods you used to generate your genomic library and the subsequent subcloning of your gene into the plasmid, there are two possible orientations of the cloned *Rvt* insert sequence relative to the T3 sequencing primer site. This is because the original EcoRI fragment that was cloned into the lambda Zap Express vector had identical EcoRI 5'-overhang sequences on both ends. As a result, this insert may have ligated into the lambda vector in either of two orientations (designated "A" or "B," as shown in **Figure 2-14**).

Thus, approximately half of the class should have cloned their insert in one orientation (relative to the T3 sequencing primer site), and the remainder of the class should have cloned their insert in the opposite orientation. Inserts that were cloned in one of the orientations will have a different sequence immediately downstream from the EcoR1 cloning site when compared with those that were cloned in the opposite orientation (we refer to the two possible orientations as the "A" orientation and the "B" orientation). Because the cloned *Rvt* insert is typically 1.4 to 1.5 kb in length and because a single sequencing run typically yields only about 700 to 800 nucleotides of sequence data, inserts cloned in opposite orientations may or may not yield sufficient sequence data to enable the class to find an overlap between the two orientations. To assist you in determining the orientation of your *Rvt* sequence, the figures below show the pDrive *Rvt* sequence beginning with the EcoRI cloning site. They also show the first 250 bp of expected *Rvt* gene sequence for each of the two orientations (**Figures 2-15a** and **b**).

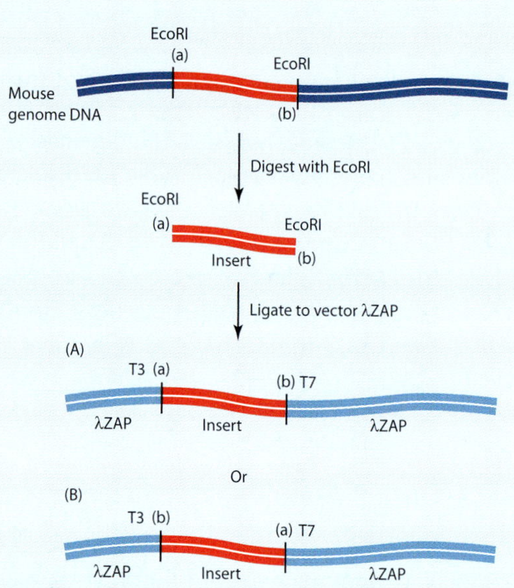

Fig 2-14 Two orientations, A and B, of cloned *Rvt* gene sequence.

PROCEDURAL NOTE

The majority of the class will have one of the two sequences shown in Figure 2-15. However, there is a great deal of sequence variation found among the hundreds of thousands of *Rvt* gene repeats in the mouse genome (about 5% variation between any two of the repeats). Therefore, your sequence may not be exactly the same as shown in Figure 2-15.

Step 3 Determine whether your sequence was derived from a clone in the "A" or "B" orientation, and record this information. Record the orientations of cloned *Rvt* genes for the entire class on the blackboard, and determine the proportion of each. How close to a 50:50 ratio of orientations was your class?

»»

LAB PERIOD II.5.1 CONT.

```
  1  GAATTCTTTG  TTCAGTTCTG  AGCCCCATTT  TTTAAGGGGG  TTATTTGATT  TTCTGAGGTC   60
 61  CACCTTCTTG  AGTTCTTTAT  ATATGTTGGA  TATTAGTCCC  CTATCTGATT  TAGGATAGGT  120
121  AAACATCCTT  TCCCAGTCTC  TTGGTGGTCT  TTTTGTCTTA  TAGACAGTGT  CTTTTGCCTT  180
181  GCAGAAACTT  TGGAGTTTCA  TTAGGTCCCA  TTTGTCAATT  CTCGATCTTA  CAGCACAAGC  240
241  CATTGCTGTT                                                              300
```
(a)
```
  1  GAATTCAAGG  CTCATACGTA  AACATGATAA  AAGCAATCTA  CAGCAAACCA  ATAGCCAACA   60
 61  TCAAAGTAAA  TGGTGAGAAG  CTGGAAGCAA  TCCCACTTAA  ATCAGGGACT  AGACAAGGCT  120
121  GTCCACTCTC  TCCCTACCTA  TTCAACATTG  TACTTGAAGT  CATAGCCAGA  GCAATTAGAA  180
181  AACAAAGGA  GATCAAGGGG  ATACAAATTG  AAAAGGAAGA  AGTAAAAATA  TCACTTTCGC  240
241  AGATGATATG                                                              300
```
(b)

Fig 2-15 The first 250 bp of expected *Rvt* gene sequence for each orientation, A (a) and B (b).

PROCEDURAL NOTE

You may find that your sequence does not match either of those shown in Figure 2-14. Remember, however, that you screened for positive clones containing the *Rvt* gene by DNA hybridization in Module II.3. Thus, the most likely explanation for not getting a match is that you have cloned a *variant* of the *Rvt* gene that contains an EcoRI site at a position other than the one expected. If this is the case, you may have cloned an unusual fragment of the *Rvt* gene. Because there are hundreds of thousands of copies of the *Rvt* gene in the mouse genome and because the different copies can vary in their DNA sequences, it is known that some of the repeat copies carry such a novel EcoRI site. We have found this occurs in about 1% of the *Rvt* sequences we have cloned. You will examine this possibility in Part C below.

Two other, less likely outcomes of this experiment are that you have cloned a fragment of the mouse genome *other* than the *Rvt* gene or that you lost all or most of your cloned insert sequence at some point during the cloning or subcloning process. If you have no cloned insert, your textfile will contain only Lambda Zap express vector sequence with no, or very few nucleotides, inserted at the EcoRI cloning site. You will also examine this possibility in Part E below.

Step 4 Ensure that the sequence data you will use to search the database is of sufficiently high quality to yield a valid database match. Examine your trace file along its entire length. Retain only that portion of the sequence for which you feel confident that the base sequence derived by the computer is an accurate interpretation of the trace data. You should retain only that portion of the textfile that corresponds to this valid region in the trace file and delete any remaining portions of the textfile.

PROCEDURAL NOTE

Without prior experience, this is a somewhat subjective process; however, your instructor will assist you with this. It is important that you edit your textfile so that it includes only sequence data that you feel are of high quality and that it no longer includes any vector sequence or any portions of the sequence you determine not to be of sufficiently high quality to ensure the accuracy of further data analyses.

PART C BLAST Search to Confirm the Source of the DNA Sequence Data

Step 1 Each student will analyze his or her own DNA sequence. Access the following public website: http://www.ncbi.nlm.nih.gov/BLAST/.

»»

LAB PERIOD II.5.1 CONT.

PROCEDURAL NOTES

BLAST (Basic Local Alignment Search Tool) is a free, public database suite of search programs that is produced and maintained by the National Center for Biotechnology Information (NCBI), which is a subsidiary of the National Library of Medicine at the National Institutes of Health (NIH). BLAST is a program that searches many databases (sequences, publications, structures, etc.), but you will access GENBANK, a repository of DNA sequences.

A full description of these programs and databases is provided at the NCBI website and includes a tutorial that is extremely useful.

Step 2

From the NCBI web page accessed via the URL listed above, select "Translated query vs. protein database (blastx)" under the "Translated" heading. This program translates your DNA sequence into protein sequence in all six possible translation reading frames (three forward and three reverse). You can submit and analyze your own nucleotide sequence by following these steps:

a. Copy your final, edited textfile sequence and paste it into the window labeled "search window." Set the default values so that the database is listed as "nr" (nonredundant), and the genetic code is listed as "standard," as shown in **Figure 2-16**.

b. Click "BLAST." A format window will then appear. Click "format," and a page should appear that provides the results of this "BLAST search," although it may take a minute or too

Step 3

When your data are returned, it consists of several sections. The top section will look similar to that shown in **Figure 2-17**. This is the "distribution of hits": a color-coded diagram of the sections of your query sequence (the sequence that you submitted) that have matches to sequences already in the database. The color-coding reflects the degree of match, with red being the most significant match and black being the least significant match.

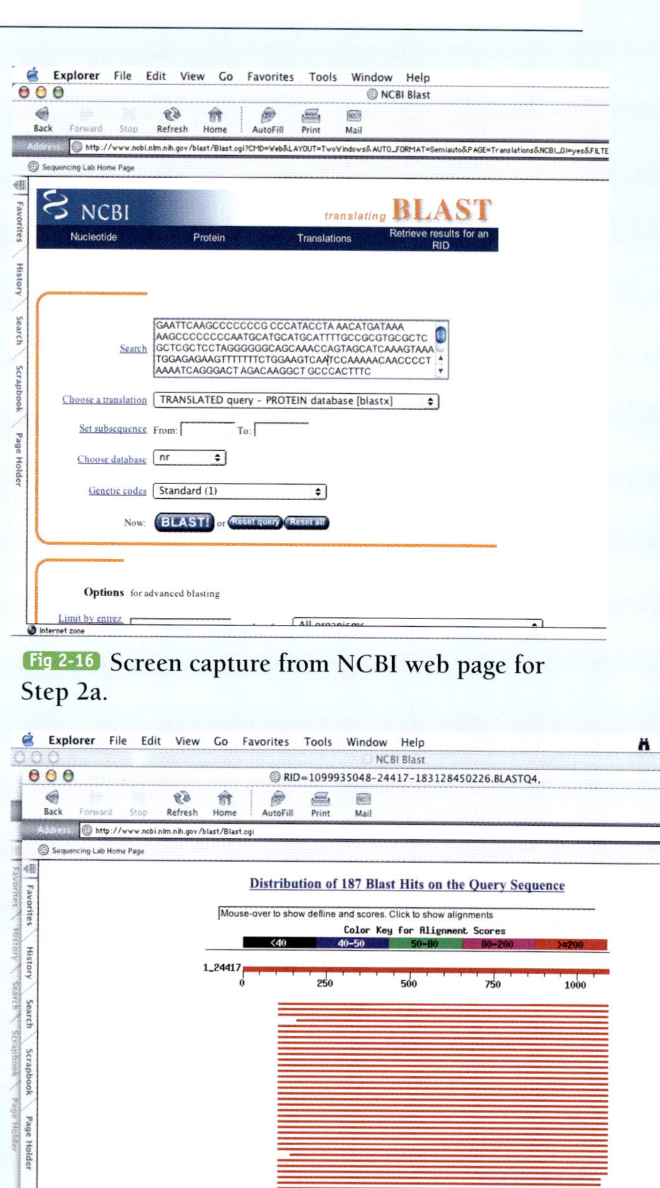

Fig 2-16 Screen capture from NCBI web page for Step 2a.

Fig 2-17 Screen capture from NCBI web page for Step 3.

The second section should look similar to **Figure 2-18**. This is a list of sequences in the database producing alignments (i.e., that match your query sequence). The score and probability values are also listed. The probability value is the probability that your query sequence matches the target sequence by chance; thus, the lower the e value, the more significant the "hit." Generally, e-5 or lower is considered significant.

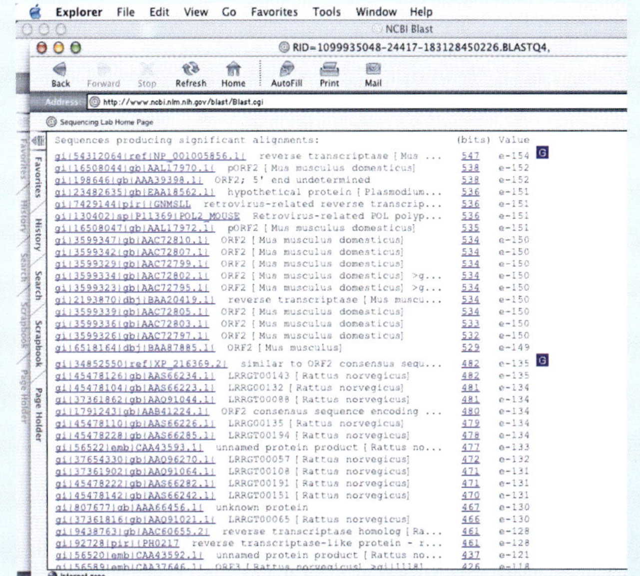

Fig 2-18 Screen capture listing sequences in database producing alignments.

PROCEDURAL NOTE

It is difficult to suggest a unique score or probability value that is always indicative of a significant database match. The probability depends on the size of the database and the length of the query sequence, among other parameters. A description of these parameters can be found at the BLAST website. For this analysis, a score of e^{-5} or lower is considered significant.

Step 4

The third section of the results shows the exact alignments of your query sequence to those matches found in the database. In this case, you will see amino acid (protein sequence) alignments that should look similar to **Figure 2-19**. These figures enable you to examine and evaluate the exact alignment of your query sequence to the various matches found in the database. This alignment can aid you in deciding if your query sequence is the expected Rvt gene sequence.

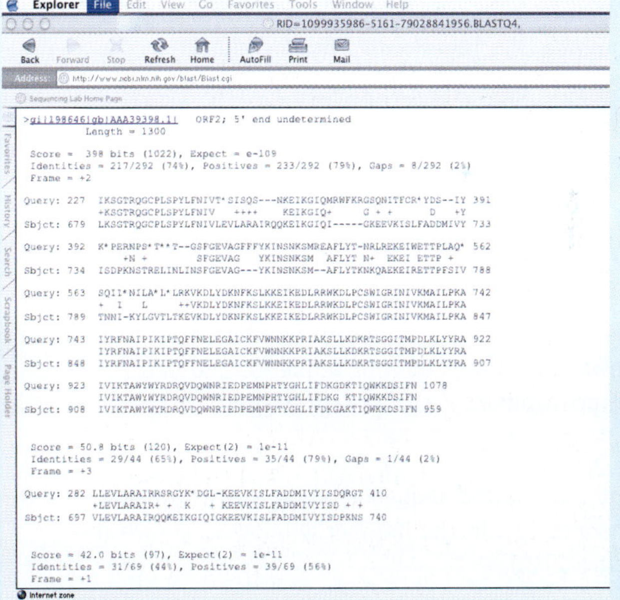

Fig 2-19 Screen capture of amino acid alignments.

PROCEDURAL NOTE

Each of the "windows" presented from the NCBI provides opportunities to connect to other pieces of information found elsewhere in the database, such as the publication data on the sequence from the database that matches your query sequence. Again, a full description of these capabilities is provided at the BLAST website.

Step 5 Based on the previous analysis, determine whether your sequence is what you expected. If you did not find one of the two *Rvt* orientations ("A" or "B" in Part B above) but still found a

»»

LAB PERIOD II.5.1 CONT.

match to the *Rvt* gene based on the BLAST search, it might suggest that your clone is a deletion product or a rearrangement of the *Rvt* gene. You can examine this further in Part E below.

PROCEDURAL NOTE

Be aware that the sequence that matches your query sequence is given a descriptor by the researcher submitting that sequence. If your *Rvt* gene query sequence finds a match in a sequence entitled "mouse immunoglobin region," this suggests that a copy of the *Rvt* sequence is located in that fragment of DNA submitted to the database. It does not necessarily mean that your query sequence is an immunoglobin gene. If you find such a database hit with your *Rvt* gene sequence query, is that expected? Keep in mind that the *Rvt* sequence is part of the LINE-1 repeat family that makes up 20% of the mouse genome! Also, remember that these sequences are scattered throughout the mouse genome and are often found in regions with other genes.

PART D Search for an Open Reading Frame in your DNA Sequence Data

Your EcoRI clone was derived from a mouse genomic DNA fragment and was not, *a priori*, expected to be a full length coding region for the *Rvt* gene. It is possible to identify exactly which part of this sequence is the *Rvt* gene by identifying that region of the clone that encodes the *Rvt* protein. One method that can be used to perform this analysis is to identify open reading frames (ORFs) in the clone sequence that potentially encodes the *Rvt* gene. BLAST analysis can then be used to confirm that assumption.

Step 1

Each student will search his or her sequence for potential ORFs. Access the following website: http://www.ncbi.nlm.nih.gov/gorf/gorf.html.

This site contains a computer program at NCBI that allows you to look at your sequence and determine ORF positions.

Step 2

Copy your edited sequence, and paste it into the sequence box in the browser window as shown in **Figure 2-20**.

Step 3

Click "OrfFind." Your sequence data will then be automatically searched for ORF sequences, and the data will be returned to you as a graphic similar to that shown in **Figure 2-21**.

The colored bars show the positions of ORFs, some of which may be "biologically relevant" — that is, they may encode actual proteins. Other potential ORFs may

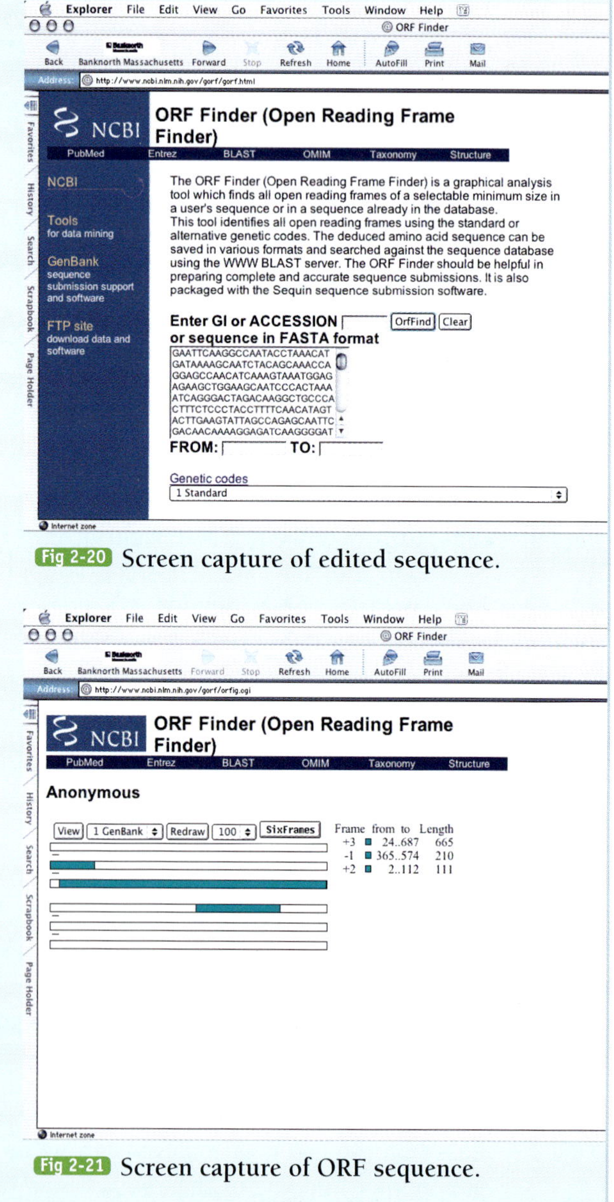

Fig 2-20 Screen capture of edited sequence.

Fig 2-21 Screen capture of ORF sequence.

simply reflect random short regions with translation start and stop codons that do not actually encode a protein. Search for matches to your sequence in the database.

Step 4 Each student will compare his or her own ORF sequence to the database. Click on the longest colored ORF bar in the 5' → 3' direction from your sequence data (displayed left to right on one of the top 3 lines in the display). This will lead to a window where your DNA sequence and protein sequence are indicated. The positions of the start and stop codons are also shown (**Figure 2-22**).

Step 5 Next, select *both* the "BLASTP" and "nr" database options, and click "BLAST." This launches the BLAST program that searches the GenBank protein database (maintained at NCBI) to look for matches to the protein sequence encoded by the ORF you have selected from your sequence.

Step 6 Now, click "Format." The displayed results provide information about the identification of the potential protein encoded by the selected ORF. When the data are returned, it will look similar to that shown in **Figure 2-23**.

This window provides you with information on (1) the database matches to your submitted sequence (query) and (2) the probability scores indicating the likelihood that each match is valid. In this case, probabilities of e^{-5} or less can be considered significant. The probability values reflect the likelihood that the results observed are due to chance. The lower the probability, the less likely the results observed are due to chance. For a full description of the analysis of probability values and scores, refer to the BLAST information page.

Step 7 Can you find any matches with your DNA sequence or the translated protein sequence? Remember that functional copies of the highly repeated *Rvt* gene encode the enzyme reverse transcriptase. Also, remember that very few copies of the *Rvt* repeat found in the mouse genome are actually functional.

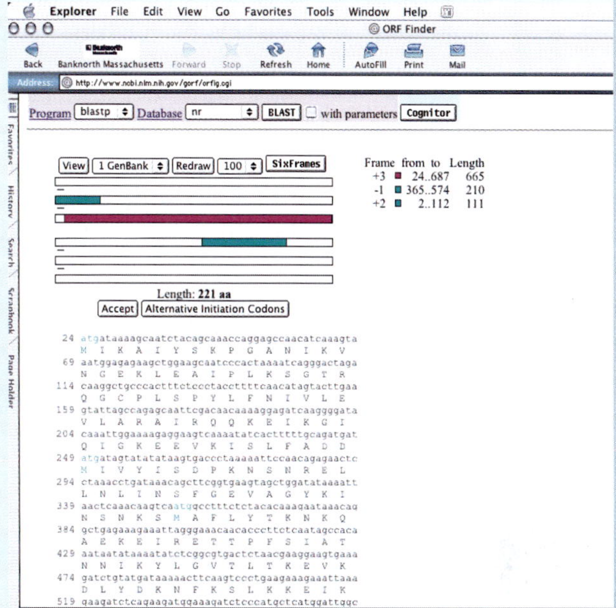

Fig 2-22 Screen capture of start and stop codons positions.

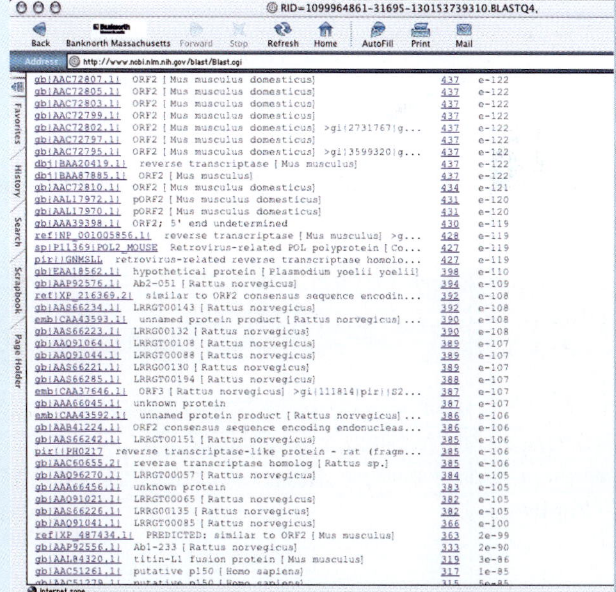

Fig 2-23 Screen capture of identification of potential protein encoded by selected ORF.

LAB PERIOD II.5.1 CONT.

Step 8 Remember that your sequence was derived from *mouse* genomic DNA. However, the BLAST search will provide you with potential matches to sequences from many different species. As you scroll down, can you find similarities between your sequence and sequences from other organisms? What do these similarities mean?

PART E Perform a Similarity Search between Two Sequences

It might be of interest to compare the sequence of your *Rvt* clone with your lab partner's sequence or with the sequence obtained by another student in the class. To do this, you can use another NCBI program that compares two different sequences with one another.

Step 1 Access the following website: http://www.ncbi.nlm.nih.gov/blast/.

Under the "Special" set of programs, click on "Align two sequences (bl2seq)."

Step 2 Each student will perform this exercise independently, but you will need to import your partner's (or another classmate's) sequence to compare with your own sequence. To do this, obtain a copy of your partner's sequence textfile on a CD or other transportable medium so that you can transfer it to your computer. Enter your own sequence textfile and the other sequence textfile in the two boxes provided: do this by sequentially selecting each sequence, copying each sequence, and then pasting each into the appropriate box. Then click "align" at the bottom of the window.

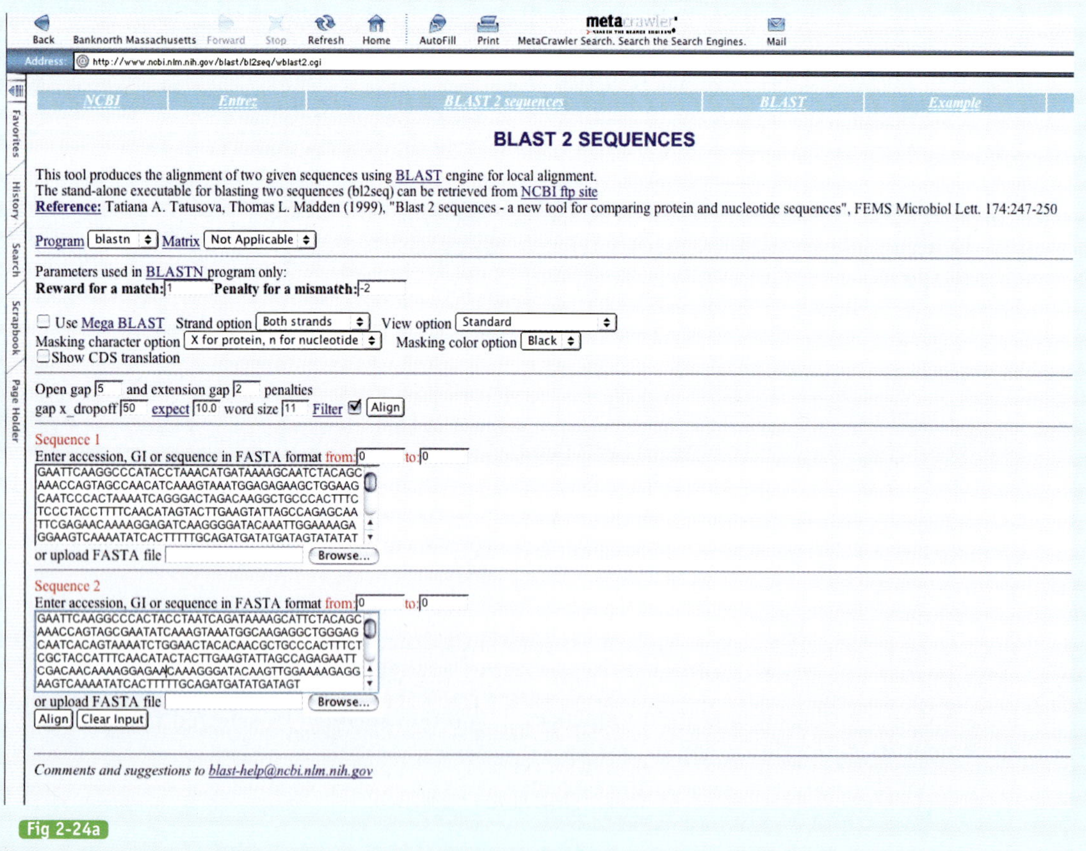

Fig 2-24a

LAB PERIOD II.5.1 CONT.

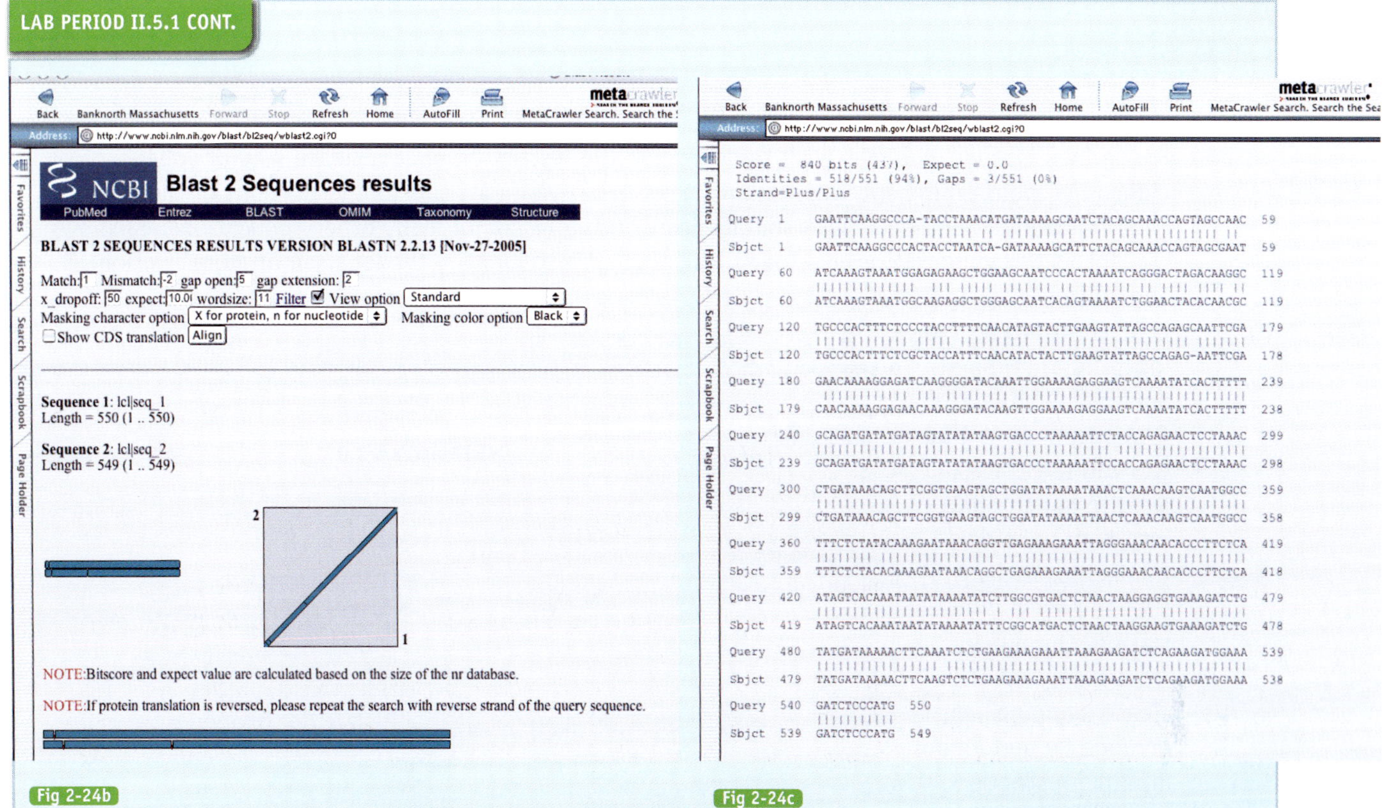

Fig 2-24b   Fig 2-24c

Step 3 The displayed results (**Figure 2-24a, b, c**) present a graphic of a pair-wise alignment of the two sequences, along with a probability estimate that the two sequences are from the same gene. Below this graphic is an alignment of the two complete sequences. How similar are your two sequences? If they are different, in what way(s) are they different? For example, are the sequences listed in the same or opposite orientations? Do the two sequences differ greatly, suggesting that they represent different genes (i.e., they are not both *Rvt* sequences), or do the two sequences show only minor differences (i.e., suggesting they are both *Rvt* sequences)? If the differences are minor, do they occur only at single nucleotides, or are there short stretches of sequence that differ between the two clones? Do the differences represent base substitutions or insertions or deletions between the two sequences? How do you think these differences arose?

PROCEDURAL NOTES

You may wish to compare your sequence to that of a database version of the *Rvt* gene, especially if you did not have an "A" or "B" orientation in Part B above. You can examine this further by comparing your sequence to the complete *Rvt* gene sequence provided at the following URL: http://www.ncbi.nlm.nih.gov/ blast/ (BLASTN Access = "gi|33342446").

This is the sequence of one version of this highly repeated *Rvt* gene from mouse. Sequences of other copies of this repeat may vary). Which part of the *Rvt* gene did you sequence? Does any part of your *Rvt* gene sequence align with this database version of the *Rvt* gene? Based on the analysis of your cloned *Rvt* gene from Parts A to E above, what do you conclude about the structure of your clone and the clones of your classmates?

PART F Electronically Map Restriction Sites in Your Cloned *Rvt* Sequence

Step 1 For various reasons, you may wish to determine the restriction endonuclease map of your cloned *Rvt* mouse fragment. Restriction sites can be useful for subcloning portions of the gene fragment, for doing *in vitro* mutagenesis on the insert, or many other applications. Each student will examine his or her cloned DNA sequence to identify the location of all restriction endonuclease sites. It is often of interest to locate restriction endonuclease sites in a DNA sequence so that you can use these sites to cut the DNA into defined, ordered fragments. Without DNA sequence information, the process of "restriction endonuclease mapping" must be done by performing multiple restriction digests followed by careful analysis of the results to determine the relative order and positions of the different restriction sites. However, after you have the DNA sequence of your cloned gene, you can use available computer software to analyze this sequence to locate ("map") all possible restriction sites very quickly and simply.

Step 2 A number of computer programs can be used to locate restriction sites in DNA sequence data. We recommend the "freeware" program available at the New England Biolabs website. Access the following website: http://www.neb.com.

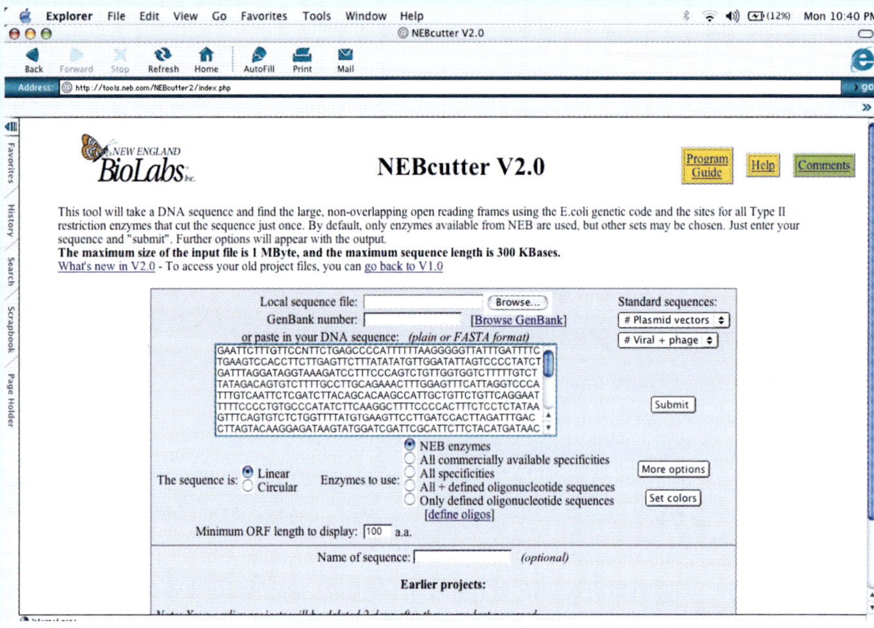

Fig 2-25 "NEB cutter" screen capture. (Copyright © New England Biolabs, Inc. Reprinted with permission.)

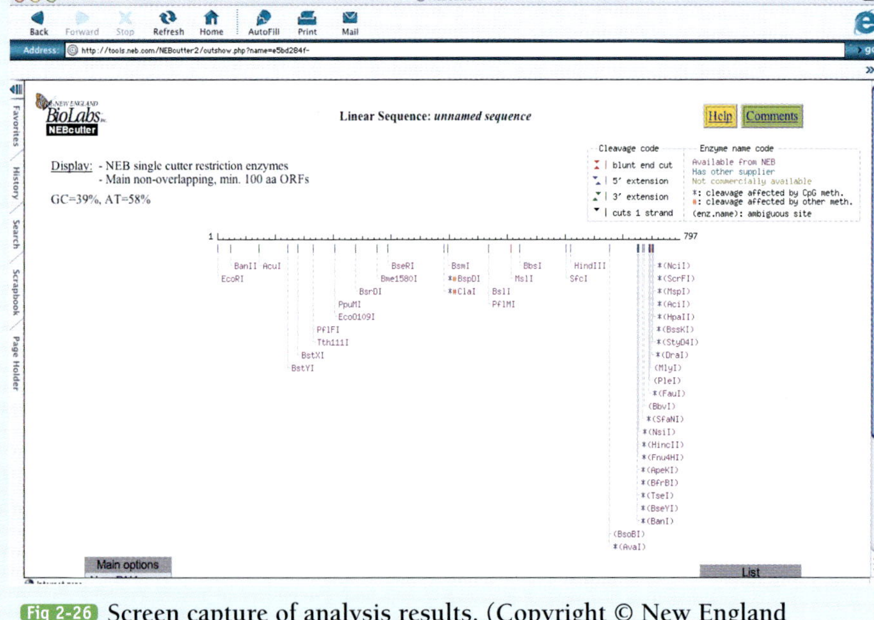

Fig 2-26 Screen capture of analysis results. (Copyright © New England Biolabs, Inc. Reprinted with permission.)

»»»

LAB PERIOD II.5.1 CONT.

Step 3 Click "NEBcutter." This will bring up the "NEBcutter" window (**Figure 2-25**). Copy and paste your DNA sequence into this window. There are many options for customizing this analysis, but for your initial analysis, we recommend that you leave all settings in their default positions and proceed by clicking "Submit."

The results of this analysis will then be displayed as a graphic similar to the one shown in **Figure 2-26**.

You can zoom in to select individual subregions of your cloned sequence to analyze in greater detail, or you can limit your search to a specific subset of enzymes. You can select enzymes sensitive to methylation or those with differing lengths of recognition specificity (i.e., those that recognize sites that are 4, 5, 6, 7, or 8 nucleotide pairs in length). Examine the frequency and distribution of sites for four-cutter enzymes as compared with sites for six-cutter enzymes. How do these compare? Suppose that you would like to selectively recover a fragment carrying only that part of the sequence with the *Rvt* ORF. Which restriction enzyme(s) could you use to yield this fragment? Suppose that you would like to confirm this "electronic restriction map" by performing a set of diagnostic restriction digests on your cloned fragment. What enzymes might you use for this purpose?

PROCEDURAL NOTE

The bioinformatics tools described above serve as only an introduction to the vast array of computer programs available to the researcher. Appendix VI lists sources of many of these programs.

MODULE MATERIALS AND REAGENTS (PER PAIR OF STUDENTS)

MODULE II.1

EcoRI restriction endonuclease (4 µl) (20 U/µl) (New England Biolabs)

EcoRI 10x reaction buffer (10 µl) (New England Biolabs)

Mouse genomic DNA (from Project 1 or supplied, 250 ng/40 µl; 2 tubes) (American Bioanalytical)

Uncut mouse genomic DNA control in BJ, 150 ng (15 µl) (American Bioanalytical)

Buffered phenol (50 µl) (American Bioanalytical)

Chisam (50 µl) (24:1v/v chloroform:isoamylalcohol) (American Bioanalytical)

1 g Agarose (35 ml)

1x TAE (500 ml). See Appendix II.

5x Blue Juice loading dye (BJ) (50 µl). See Appendix II.

Lambda BstEII markers (6 µl, 25 ng/µl). See Appendix II.

Lambda HindIII markers (6 µl, 25 ng/µl). See Appendix II.

T4 DNA ligase (1 µl) (4 U/µl) (Stratagene)

10x DNA ligase buffer, no ATP (1 µl) (Stratagene)

ATP, 10 mM (1 µl) (Stratagene)

Lambda ZAP Express vector (Stratagene, EcoRI predigested and dephosphorylated) (1 µg)

Gigapack Gold™ packaging extract (one) (Stratagene)

LBMM medium. See Appendix II.

10 mg/ml tetracycline. See Appendix II.

LB+ tetracycline agar plates (1). See Appendix II.

NZY agar plates (2). See Appendix II.

NZY Top agarose (6 ml). See Appendix II.

XL1-Blue MRF' cells (Stratagene) (200 µl prepared cells)

10 mM $MgSO_4$ (250 ml). See Appendix II.

SM buffer (10 µl). See Appendix II.

Chloroform (20 ml) (American Bioanalytical)

0.5 M IPTG (30 µl) (American Bioanalytical)

250 mg/ml X-gal in dimethyl formamide (100 µl) (American Bioanalytical)

6 ml NZY top agarose. See Appendix II.

21% dimethyl sulfoxide (DMSO) (if storing the library) (American Bioanalytical)

Phase Lock Light™ tubes (1) (Eppendorf)

Millipore type VSWP 04500, type VS dialysis membranes (1)

Agarose gel apparatus and power supply

Ethidium bromide DNA stain (200 ml)

UV transilluminator, gel photography equipment

Dry ice

Speed-vacuum concentrator

Heat block

Water bath

Nanofuge

Table top centrifuge

Access to fume hood

Microcentrifuge

Petri dish (1)

250-ml beaker (1)

Safety glasses

Plastic trays for staining and destaining gels

MODULE II.2

XL-1 Blue MRF' strain (Stratagene) (for the class)

LBMM medium (50 ml). See Appendix II.

Tetracycline (10 mg/ml stock). See Appendix II.

LB+ tetracycline agar plates (1). See Appendix II.

10 mM $MgSO_4$ (250 ml). See Appendix II.

Lambda ZAP phage library (produced in Module II.1 or supplied) (American Bioanalytical)

0.5 M IPTG (30 µl). See Appendix II.

X-gal (250 mg/ml), 100 µl. See Appendix II.

NZY agar plates (2). See Appendix II.

NZY top agarose (6 ml). See Appendix II.

Plaque lift denaturation solution (50 ml). See Appendix II.

Plaque lift neutralization solution (50 ml). See Appendix II.

2x SSC (50 ml). See Appendix II.

1 g Agarose. See Appendix II.

1x TAE buffer (500 ml). See Appendix II.

Agarose gel apparatus, power supply

UV transilluminator

Gel photography apparatus

Plaque Wash Buffer (50 ml). See Appendix II.

Probe Amp Hybridization Buffer (6.5 ml). See Appendix II.

1x SSC, 0.1% SDS (200 ml). See Appendix II.

0.5x SSC, 0.1% SDS (200 ml). See Appendix II.

ECL Buffer (300 ml). See Appendix II.

Antibody Blocking Solution (20 ml). See Appendix II.

Antibody Solution (20 ml). See Appendix II.

1% Tween (20 ml). See Appendix II.

ECL detection solution I (10 ml) (GE Healthcare)

ECL detection solution II (10 ml) (GE Heatlhcare)

SM buffer (200 µl)

5x Blue Juice loading dye (BJ) (2 µl). See Appendix II.

100-base-pair ladder in BJ (10 µl). See Appendix II.

Ethidium bromide stain. See Appendix II.

Plastic wrap or plastic development folder

X-ray cassette

GBX developer (Kodak)

GBX fixer (Kodak)

Darkroom access

X-ray film (Kodak BioMax™)

Nylon filters (Nytran+)(45 mm) (Schleicher & Schuell)

Whatman 3MM™ absorbant paper

UV cross-linker or baking oven

Plastic hybridization bags

Marking pen

Stick pins

PCR thermocycler

Tabletop centrifuge

Water bath

Lid locks

Heat block

Pencil

Stain and destain trays

The following items are duplicates of Module 1.2 and need not be reordered:

Fluorescein PCR labeling kit (Roche Diagnostics)

pBluescript plasmid with *Rvt* insert (4 ng/29.5 µl) (1) (American Bioanalytical)

Rvt Repeat Forward Primer 10 pm/µl (1 µl) (American Bioanalytical)

Rvt Repeat Reverse Primer 10 pm/µl (1 µl) (American Bioanalytical)

MODULE MATERIALS AND REAGENTS (PER PAIR OF STUDENTS)

MODULE II.3

Rvt phage stocks (from Lab Period II.2.6 or supplied by the instructor) (4) (American Bioanalytical

AmpliTaq Gold™ PCR kit (4 reactions) (Applied Biosystems)

T3 primer (10 pmol/µl) (4 µl) (American Bioanalytical)

T7 primer (10 pmol/µl) 4 µl (American Bioanalytical)

Qiaquick PCR purification Kit (Qiagen)
Buffer PB (1000 µl)
Qiaquick columns (4)
Buffer PE (3 ml)
1x TE, pH 8.0 (120 µl)

2 g Agarose

600 ml 1x TAE Buffer. See Appendix II.

Ethidium bromide DNA stain (200 ml). See Appendix II.

5x Blue Juice loading dye (BJ) (50 µl). See Appendix II.

λBstEII marker (25 ng/µl) (25 µl). See Appendix II.

Qiagen pDrive cloning kit (Qiagen)
2x Ligation master mix (10 ml)
pDrive plasmid (25 ng) (2)
Qiagen EZ competent cells (100 ml)

LB ampicillin plates, with Xgal and IPTG (2). See Appendix II.

SOC media (250 µl). See Appendix II.

L broth plus ampicillin (15 ml) (2)

Qiagen Miniprep Kit (Qiagen)
Qiaprep column (1)
Buffer P1 (500 µl)
Buffer P2 (500 µl)
Buffer N3 (700 µl)
Buffer PE (1.5 ml)
Buffer EB (100 µl)

Agarose gel apparatus and power supply

Ethidium bromide. See Appendix II.

Staining tray and destaining tray

UV transilluminator and camera apparatus

Sterile loops (2)

MODULE II.4

AmpliTaq™ Gold PCR reagent kit (Applied Biosystems) (two reactions)

M13 Forward Primer (10 pmol/µl) (2 µl) (American Bioanalytical)

SP6 Primer (10 pmol/µl) (2 µl) (American Bioanalytical)

Plasmid DNA with *Rvt* PCR product insert (10 ng/µl) (2) (from Lab Period II.4.1)

1 g Agarose (50 ml). See Appendix II.

1x TAE buffer (300 ml). See Appendix II.

T3 sequencing primer (5 ng/µl) (8 µl) (American Bioanalytical)

DNA sequencing reagents (sequencer dependent) (2) (Applied Biosystems)

Performa Gel Filtration Columns (2) (Edge Biosciences)

DNA sequencer loading dye, formamide, or water (sequencer dependent) (Applied Biosystems)

Ethidium bromide gel stain (200 ml). See Appendix II.

Gel staining and destaining trays

Gel electrophoresis apparatus and power supply

UV transilluminator

Photography equipment

Lambda BstEII marker (25 ng/µl) (16 µl). See Appendix II.

5x Blue Juice loading dye (BJ) (8 µl). See Appendix II.

Speed-vacuum concentrator

MODULE II.5

DNA sequence results

Computer with Internet access

DNA sequence viewing programs

EditView:
http://www.appliedbiosystems.com/support/software/dnaseq/installs.cfm

Chromas:
http://www.technelysium.com.au/chromas.html

Database searches:
http://www.ncbi.nlm.nih.gov/BLAST/

Open reading frame (ORF) analysis:
http://www.ncbi.nlm.nih.gov/gorf/gorf.html

Restriction Enzyme Site Analysis (NEB Cutter): http://tools.neb.com/NEBcutter2/index.php

Study Questions for Project II

Module II.1

1. In Lab Period II.1.1, is it important that the 10x EcoRI buffer be used instead of some other restriction enzyme buffer? Explain your answer.

2. In Lab Period II.1.2, Step 3, what is the purpose of adding the 5x BJ to each of your DNA samples before loading them on the agarose gel?

3. In Lab Period II.1.3, Step 8, did your EcoRI digest successfully cut your mouse genomic DNA? How can you tell?

4. In Lab Period II.1.3, Step 5, what is the physical explanation for how the gel material separates the aqueous phase from the organic phase?

5. In Lab Period II.1.4, Step 2, what is the recommended molar ratio of EcoRI cut mouse DNA to the lambda ZAP Express vector DNA?

6. In Lab Period II.1.4, Step 4, why is the ligation reaction done at a cold temperature (14°C) instead of the optimal temperature for DNA ligase (37°C)?

7. In Lab Period II.1.5, Step 5, what is the purpose of adding the chloroform to the genomic library tube?

8. In Lab Period II.1.6, why is it important to infect the correct *E. coli* strain (XL-1 Blue MRF') with the packaged phage library? Do you think other strains of *E. coli* could be used? Why or why not?

9. In Lab Period II.1.7, what is the titer of your genomic library in pfu/μl? What is the total titer of your library? What is the percentage of recombinant plaques in your library? What is the range of titers and percentage of recombinants seen in your class? What are the average titer and the average percentage of recombinants in your class?

Module II.2

1. In Lab Period II.2.1, what is the reason for replating the genomic library given that the library was already plated in Lab Period II.1.6?

2. In Lab Period II.2.2, why do the plaques need to be transferred to nylon membranes?

3. In Lab Period II.2.2, Step 7, what is the purpose of the denaturation solution? What is the purpose of the neutralization solution?

4. In Lab Period II.2.3, you labeled the mouse *Rvt* repeated gene sequence for hybridization to the plaque lifts. The *Rvt* sequence that you labeled was cloned previously by the authors into a plasmid vector (pBluescript). If this gene sequence has already been cloned, why is it of interest to clone additional copies in this experiment?

5. In Lab Period II.2.4, why does the 1.4-kb PCR product of the *Rvt* gene fluoresce after exposure to UV light even before staining the gel with ethidium bromide (or SYBR SAFE)?

6. With regard to the gel discussed in the previous question, did you obtain a PCR product at 1.4 kb that you know is labeled? How do you know its labeled?

7. In Lab Period II.2.4, why is the hybridization done at 65°C? Is this higher or lower than you would expect to use for an oligonucleotide probe in this hybridization? Explain your answer.

8. In Lab Period II.2.5, what is the purpose of the posthybridization washes done in Part A?

9. In Lab Period II.2.5, Part B, describe the purpose of the blocking solution, the antibody solution, and the ECL solutions. How do they result in a clean signal on the filters that exposes the X-ray film?

10. In Lab Period II.2.6, were you able to identify spots on your lumigram that aligned with plaques on your plates? How many positive clones did you obtain on your plates? What was the number of positive plaques relative to the total number of clear plaques? Is this more or less than you expected? Explain your answers.

Module II.3

1. In Lab Period II.3.1, why did you use the T3 and T7 primers to amplify your cloned DNA insert from the lambda vector? Where do the two primers hybridize?

2. In Lab Period II.3.2, Part A, what is the purpose of the purification of your PCR product using the spin columns?

3. In Lab Period II.3.2, Part B, what is the size PCR product that you expect to see on the gel? Explain why you expect this size. By comparing to the marker DNA, did you obtain the size PCR product that you predicted? If not, can you explain the size of the PCR product you obtained?

4. In Lab Period II.3.3, the plasmid used for subcloning comes in a linear form with a single U overhang on each 3' end. How is this useful in cloning your PCR products? Draw a diagram to explain your answer. Be sure to note the 5' and 3' ends of all of your DNA strands.

5. In Lab Period II.3.3, why is an insert to vector ratio of up to 10:1 acceptable in this particular ligation reaction?

6. In Lab Period II.3.4, why is ampicillin included in the medium in the Petri dishes on which you plate your competent cells?

7. In Lab Period II.3.4, why are IPTG and X-gal also added to the medium?

8. In Lab Period II.3.5, why is ampicillin included in the liquid medium for growing the *E. coli* culture?

9. In Lab Period II.3.6, why is it important to lyse the cells gently?

10. In Lab Period II.3.7, why is there more than one band seen in the plasmid lanes? The darkest band is the one that migrates farthest down the gel. What is the conformation of the plasmid DNA found in this band? What are the conformations of the plasmid DNA found in the bands that run slower in the gel?

11. What is your estimate for the concentration of your plasmid DNA? What is the total amount of plasmid DNA that you isolated?

Module II.4

1. In Lab Period II.4.1, why are the M13 Reverse and SP6 primers used for the PCR? Where do these primers hybridize to amplify your cloned insert for sequencing?

2. In Lab Period II.4.2, what is the approximate size of the PCR product that you obtained? Is this the size that you expected? If yes, explain why this is the expected size. If not, propose an explanation for the size of the product that you obtained.

3. In Lab Period II.4.2, what is the concentration of your PCR product that you estimated from the gel?

4. In Lab Period II.4.3, why is the T3 primer used for sequencing your PCR product? Are there other primers that could have been used? How would the sequencing results differ with these other primers? Why can't you use the T7 primer to sequence your PCR product?

5. In Lab Period II.4.4, why is it important to use the purification columns to remove the unused dye terminators? What do you think your sequence data would look like if the dye terminators were not removed?

Module II.5

Follow all of the directions and answer all of the questions included in the text for this module.

III GENE EXPRESSION ANALYSIS: RNA ISOLATION, NORTHERN BLOT, AND REVERSE TRANSCRIPTASE-POLYMERASE CHAIN REACTION

PROJECT SUMMARY

In the series of experiments described in **Module III.1**, you will isolate total RNA from mouse liver and then fractionate that RNA to enrich for poly A+ mRNA. Genes that encode proteins are transcribed to produce mRNA that typically becomes polyadenylated (poly A+ tail) at the 3'-end. Fractionation of total RNA to yield poly A+ mRNA yields a sample of RNA that is highly enriched for transcripts from protein-encoding genes. You will then run aliquots of both your total and poly A+ RNA preparations on a denaturing agarose gel to assess the integrity and purity of each. You will also run an aliquot of mouse spleen poly A+ on the gel.

In the experiments described in **Module III.2**, you will use the Northern blot technique to analyze your total and poly A+ RNA preparations to detect transcripts from a specific gene (*Ttr*) that is expressed in the liver. You will also be able to compare expression of the *Ttr* gene between liver and spleen tissues. To do this, you will transfer the electrophoresed RNA samples to a blotting membrane and hybridize that membrane with a biotin-labeled probe specific to the *Ttr* gene. This will allow you to compare the relative abundance of the *Ttr* transcript in total and poly A+ RNA samples from mouse liver and in poly A+ RNA from spleen.

The experiments described in **Module III.3** will allow you to use a PCR-based method to detect the same liver-specific transcript from the *Ttr* gene in a very small amount of total RNA. To do this, you will first convert a small amount of total RNA to "complementary DNA" (cDNA) using reverse transcriptase (RT). You will use an oligo dT primer for this reaction so that you will selectively convert poly A+ mRNA into cDNA. You will then use primers specific to the *Ttr* gene to amplify a 150-bp fragment of this gene. This process is called reverse transcriptase polymerase chain reaction or RT-PCR. To detect the presence of the *Ttr* transcript, you will visualize the PCR amplification product on an agarose gel to confirm that a product of the correct size was produced.

PROJECT BACKGROUND

All of the cells of an individual organism contain the same DNA. Different types of cells develop as the result of differential gene expression. Because different genes are expressed in different tissues or organs, there will be a different set of mRNA transcripts in each. Gene expression can be studied by assaying the encoded mRNA by a variety of means. In this experiment, you will use two different methods to detect mRNA from the mouse transthyretin gene (*Ttr*), which is expressed at very high levels in the liver and at much lower levels in certain other tissues. The methods you will use are Northern blot analysis and RNA- or RT-PCR. To do this, you will purify mRNA from mouse liver and then specifically detect the mRNA encoded by the *Ttr* gene. This same mRNA will also provide the starting material for cDNA synthesis in Project IV. There are three primary types of RNAs found in most eukaryotic cells: transfer RNA (tRNA), ribosomal RNA (rRNA), and messenger RNA (mRNA). When these three types of RNA are isolated together, they are called "total RNA." Genes that encode proteins produce mRNAs that are subsequently translated to yield the protein. Only 1% to 2% of the total RNA in a cell is mRNA. Approximately 90% of the total RNA is comprised of rRNA, and 8% to 9% is tRNA and other small RNAs.

Typically, about 10,000 different protein-encoding genes are expressed in any one mammalian cell type. Thus, any particular mRNA will usually represent only a very small proportion of all of the mRNAs in that cell, and the mRNAs represent only a small proportion of the total RNA. It is therefore often

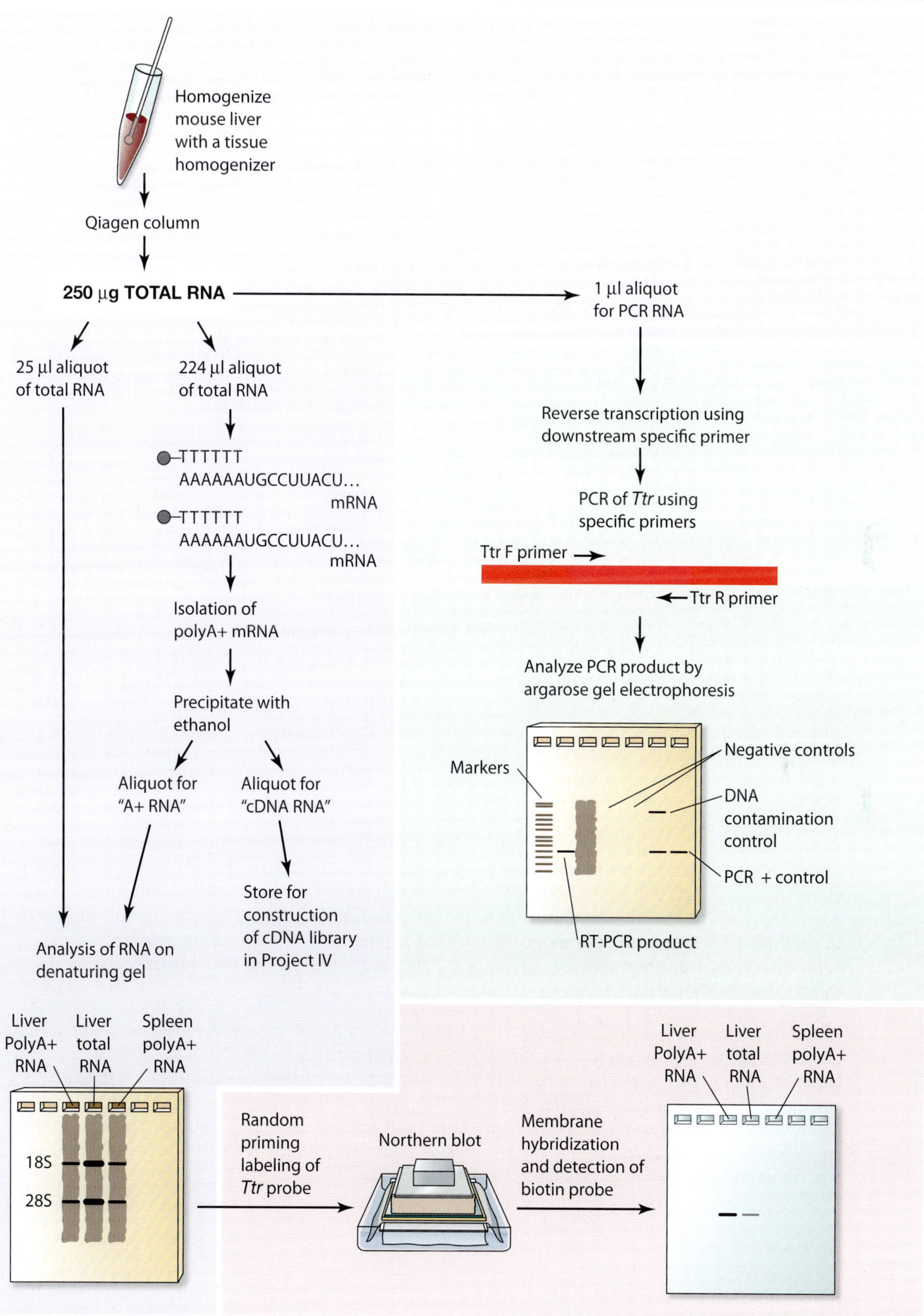

Project III flow diagram.

useful to first fractionate mRNA away from the other types of RNA and then to examine directly the population of mRNA from the cell type or tissue of interest to detect a specific mRNA.

For this project, you will first recover total RNA from mouse liver and then fractionate that to selectively enrich for mRNA (Module III.1). To do this, you will take advantage of a unique feature of mRNA molecules. Most mRNAs become "polyadenylated" shortly after they are synthesized. Polyadenylation involves the addition of a string of approximately 50 to 200 adenine ribonucleotides onto the 3′ end of each mRNA molecule. This produces what is known as a "poly A tail" on each mRNA. Neither rRNA nor tRNA is polyadenylated, and thus, it is possible to enrich selectively for mRNA by hybridizing total RNA to oligo dT molecules that are bound to beads. These oligo dT chains are single-stranded molecules of 12 to 15 thymidine nucleotides to which the poly A tail can hybridize. When total RNA is hybridized to these oligo dT beads under favorable conditions, the mRNA will selectively bind to the beads because of hybridization of the mRNA poly A+ tails to the oligo dT molecules attached to the beads. The rRNA and tRNA do not bind to the beads and can be washed away. The conditions can then be changed to elute the purified poly A+ mRNA from the beads to be used for subsequent analyses.

You will then subject samples of both total and poly A+ RNA from mouse liver to agarose gel electrophoresis to separate different transcripts according to molecular size. You will also run an aliquot of spleen poly A+ mRNA (provided to you by your instructors) on this same gel. It is not necessary to digest the RNA samples with endonucleases before performing electrophoresis as was the case for the genomic DNA you prepared and analyzed in Project I. This is because RNA transcripts are synthesized as discrete molecules that are typically 500 to 3000 nucleotides in length, with most of the mRNAs averaging 1000 to 1500 nucleotides in length.

An important difference between RNA and DNA is that RNA is typically single stranded, whereas DNA is typically double stranded. The bases in single-stranded RNA molecules are normally unpaired and can therefore hybridize to other complementary bases at other locations in the same RNA molecule (think of the stem–loop structure that is so familiar in tRNA molecules). Such internal hybridization in RNA creates a molecular structure that alters the migration of the RNA molecules in the agarose gel such that they do not separate strictly on the basis of size. To avoid this confounding effect, the agarose gel is run under "denaturing conditions" that inhibit the formation of secondary structures so that the RNA molecules will migrate solely on the basis of their linear, molecular size.

To determine whether the transcript (mRNA) from a particular gene is among the transcripts represented in a particular sample of RNA, the electrophoresed RNA can be transferred to a blotting membrane (Northern blot) and hybridized with a gene-specific probe (Northern hybridization). Detection of bound probe will indicate that a transcript from that gene is present in the sample and will allow you to determine the size of that transcript. This method is often used to compare the relative levels of expression of a gene in two different tissues (e.g., liver and spleen). This is the experiment that you will perform in Module III.2.

PCR can also be used to detect the presence of a particular transcript in an RNA sample from any particular tissue or cell type. To do this, an aliquot of RNA is first converted into cDNA using reverse transcriptase. The cDNA is then subjected to PCR with primers specific to the gene of interest to determine whether that particular transcript was present in the sample of RNA. Because of the sensitivity of PCR, this approach can be used on total RNA from a very small amount of starting material. This has provided an extremely valuable technique for analysis of gene expression because it allows you to examine genes that are expressed at low levels from small numbers of cells.

Isolation and Analysis of Total RNA from Mouse Liver

MODULE SUMMARY

Isolation of total RNA from mouse liver, fractionation of poly A+ RNA from the total RNA, and electrophoresis and visualization of both types of liver RNA plus poly A+ RNA from spleen on a denaturing agarose gel.

BACKGROUND

In this module, you will be provided with mouse liver tissue that was dissected from a euthanized mouse and stored frozen before this experiment. Liver tissue was dissected from the mice, cut into 1 g pieces, and flash frozen in an RNase-free 50-ml conical tube containing liquid nitrogen. From this tissue, you will prepare total RNA using the RNeasy Maxi Kit from Qiagen. This method uses silica to isolate RNA greater than 200 nucleotides in length. Small RNA molecules, including 5.8S rRNA (160 nt), 5S rRNA (120 nt), and tRNA (70–90 nt), will not be efficiently isolated using this method. However, larger RNAs, including most mRNAs and larger rRNAs, including 28S and 18S rRNA, will be efficiently recovered. You will then use another kit, the Ambion Micro Poly(A) Purist kit, to fractionate poly A+ mRNA from polyA(–) RNA in this sample. This will yield an RNA sample that is highly enriched for poly A+ mRNA. Finally, to assess the integrity of the RNA samples you have recovered and to obtain an indication of the extent of enrichment you have achieved for poly A+ mRNA, you will run aliquots of your total and poly A+ RNA samples on a denaturing agarose gel. This gel will also provide the starting point for the Northern blot experiment described in Module III.2. You will also run an aliquot of poly A+ mRNA isolated from spleen that will be provided to you.

PROCEDURAL NOTES

An alternative method for isolating total RNA using ultracentrifugation is given in Appendix IV.

Unlike DNA, RNA is a very labile molecule and is subject to degradation by RNases that are ubiquitously present. You must take several precautions to minimize the potential for degradation, including wearing gloves at all times when handling tubes containing RNA (to avoid contaminating your samples with RNases that may be on your fingers). You also must use solutions that have been made with water treated with diethyl pyrocarbonate (DEPC) to inactivate contaminating RNases. (A protocol for making DEPC-treated water is provided in Appendix IV.) You should also use special pipette tips and microfuge tubes that have been dedicated to use with RNA-containing solutions (this reduces the chance that these materials will become contaminated by RNases during use with other macromolecules). Finally, you should keep your tubes that contain RNA, and those that contain solutions that will come in contact with your RNA, on ice unless specifically instructed to do otherwise.

LAB PERIOD III.1.1.   PREPARATION OF TOTAL RNA FROM FROZEN MOUSE LIVER TISSUE

THE FOLLOWING protocol will be done in pairs. If purified total RNA is provided, proceed directly to Lab Period III.1.2.

Step 1 To each 1-g piece of frozen mouse liver tissue in a 50-mL conical polypropylene tube, add 15 ml of Qiagen buffer RLT (containing β2- mercaptoethanol), and homogenize the tissue for 1 minute using a tissue homogenizer set on high speed. (An ultrasonic tissue homogenizer such as a Polytron works best for RNA isolation.)

PROCEDURAL NOTES

For tissue culture cells, you can first pellet the cells by centrifugation and then proceed with buffer RLT addition.

Although the composition of the buffers is proprietary, the company is required for safety reasons to divulge that this buffer contains guanidinium isothiocyanate (a potent RNase inhibitor).

Step 2 Centrifuge the conical tube at 3000 to 5000x g for 10 minutes at room temperature. Carefully pipette the solution (being careful not to transfer the fatty layer on top of the solution or the pellet at the bottom of the tube) into a new RNase-free 50-ml conical tube.

Step 3 Estimate the volume of RNA solution that you recovered by placing a similar volume of water in a separate tube and measuring the volume. Add an equal volume of 70% ethanol (made with DEPC ddH$_2$O) to the tube containing your RNA solution, and mix immediately by shaking vigorously.

PROCEDURAL NOTE

Because RNA molecules are typically of relatively low molecular weight, you do not need to worry about shearing them with vigorous shaking.

Step 4 Immediately add 15 ml of the mouse liver solution from Step 3 onto a Qiagen RNA column (silica). Close the tube gently, and centrifuge at 3000 to 5000x g for 5 minutes. Discard the flow-through, and then add the remaining solution to the column. Close the tube gently, and centrifuge again at 3000 to 5000x g for 5 minutes. Discard the flow-through.

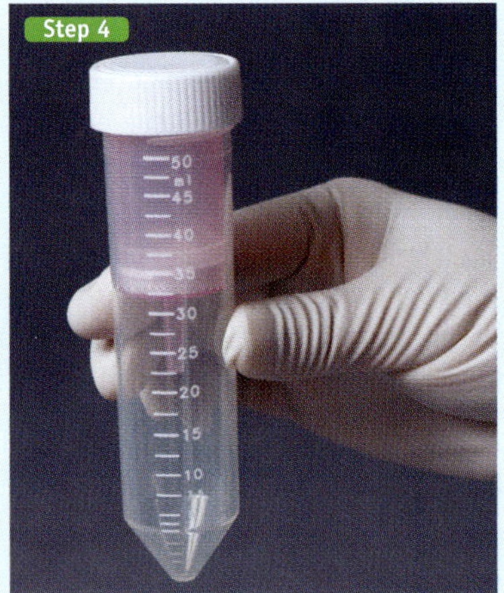

PROCEDURAL NOTE

Total RNA binds to the silica in the column in the presence of ethanol. The maximum volume that can be added to the column is 15 ml.

Step 5 Pipette 15 ml of buffer RW1 onto the column, and centrifuge the tube at 3000 to 5000x g for 5 minutes. Discard the flow-through.

»»

LAB PERIOD III.1.1 CONT.

Step 6 Pipette 10 ml of Qiagen Wash buffer RPE (supplied as a concentrate to which the user adds four volumes of 100% ethanol to obtain the working solution) onto the column, and close the tube gently. Centrifuge at 3000 to 5000x g for 2 minutes. Discard the flow-through.

Step 7 Pipette another 10 ml of wash buffer RPE onto the column, and close the tube gently. Centrifuge at 3000 to 5000x g for 10 minutes to get rid of all of the residual RPE buffer. The ethanol in this buffer can interfere with downstream reactions, and thus, the centrifuge time is critical. Discard the flow-through.

Step 8 To elute the purified RNA, transfer the RNA column to a new RNase-free sterile 50-ml conical tube (supplied by Qiagen). Add 250 µl of RNase-free ddH_2O (also supplied by Qiagen) to the center of the column membrane, and close the tube gently. Centrifuge at 3000 to 5000x g for 3 minutes. *Do not discard the flow-through! This is your RNA!*

PROCEDURAL NOTE

In the absence of ethanol, the RNA comes off the silica column.

Step 9 Add another 250 µl of RNase-free ddH_2O to the center of the column membrane. Close the tube gently, and centrifuge at 3000 to 5000x g for 3 minutes. Save this flow-through; it contains additional RNA! Combine this with the flow-through you recovered from Step 8 in a single 1.5-ml microfuge tube. The total volume should be ~500 µl. Label this tube with your number and "Flow-Through RNA."

Step 10 Pipette 100 µl of your flow-through RNA sample into a new 1.5-ml microfuge tube. Add 900 µl DEPC-H_2O (= 1:10 dilution). Label this tube with your number and "spec RNA." Read the OD_{260} using an ultraviolet (UV) spectrophotometer to determine the concentration and total quantity of RNA. Spectrophotometric readings should be taken at wavelengths of 260 and 280 nm to quantify the amount of RNA and to help determine its purity. An $OD_{260} = 1$ (1.0-cm path length) corresponds to approximately 40 µg/ml of single-stranded RNA. To determine the quantity of RNA in your sample, multiply the OD_{260} reading x 40 µg/ml x 10 (for the dilution) = µg/ml total RNA recovered. Divide this number by 1000 to get the con-

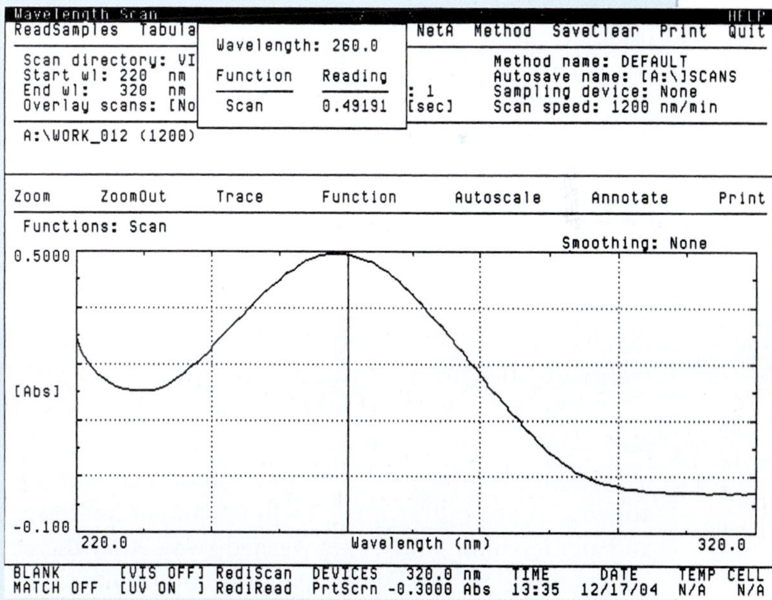

Fig 3-1 Spectrophotometer output of pure RNA.

centration of your sample in µg/µl. The ratio between the readings at 260 and 280 nm gives an estimate of RNA purity. An OD_{260}/OD_{280} of about 1.6 to 2.0 indicates a pure preparation. A continuous scan between OD_{220} to OD_{310} provides an even better estimate of nucleic acid purity. An example of a scan of pure RNA is shown in **Figure 3-1**.

»»

LAB PERIOD III.1.1 CONT.

Step 11 If your sample is at a concentration of >1 µg/µl, add sufficient RNase-free ddH$_2$O to your sample to adjust the concentration to 1 µg/µl. If your original sample is at a concentration of <1µg/µl, reduce the volume by precipitating your RNA with ethanol and resuspending in a smaller volume of RNase-free ddH$_2$O to yield a final concentration of ~1 µg/µl.

PROCEDURAL NOTES

If it is necessary to precipitate the total RNA in your remaining 400-µl sample in the tube labeled "Flow-Through RNA," add 40 µl of 3-M NaOAc (pH 7.0). Mix briefly, and then add 1 ml of ice-cold 100% EtOH. Let this sit on ice for at least 10 minutes.

This sample can be stored indefinitely at –70°C. Recover the precipitated RNA by spinning the tube at maximum speed in a microfuge for 60 minutes. When the spin is complete, use a flame-drawn Pasteur pipette to remove and discard the supernatant, and then wash the RNA pellet with 500 µl of 70% EtOH (made up in DEPC-treated ddH$_2$O). Spin again in the microfuge at top speed for 15 minutes. Refer to the figures in Step 4 in Lab Period I.1.6 for making the flame-drawn pipette and the figures in Step 5 of the same lab period that illustrates the procedure for removing the ethanol from the pellet with the flame-drawn pipette. Remove the supernatant. Air dry the pellet, and resuspend it in a sufficient volume of DEPC-treated ddH$_2$O to yield a final concentration of 1 µg/µl (for instructions on the flame-drawn Pasteur pipettes, refer to the figures in Step 4, Lab Period I.1.6.).

DEPC water is water treated with diethyl pyrocarbonate (DEPC), a potent RNAse inhibitor. Refer to Appendix IV for the procedure for preparation of DEPC-treated water.

Step 12 Pipette 250 µl (= 250 µg) of your total RNA into a new 1.5-ml microfuge tube. Label this tube with your number and "Liver RNA." Store this tube at –70°C. You can discard the remaining total RNA from your preparation, or give it to any of your classmates who may need additional material.

LAB PERIOD III.1.2.   **FRACTIONATION OF POLY A+ mRNA FROM TOTAL RNA**

BECAUSE WE are interested in detecting a particular mRNA (encoded by the *Ttr* gene) and because that will be only one of thousands of different mRNAs found in liver cells and because mRNA makes up only about 1% to 2% of the total cellular RNA, it is very helpful to enrich specifically for mRNA before proceeding with your analysis. Because mRNA transcripts in eukaryotes are typically polyadenylated, this enrichment can be accomplished by suspending, under conditions that favor hybridization, your total RNA in a solution containing cellulose to which oligo dT molecules are attached. The poly A+ mRNA will remain bound to the oligo dT cellulose while the poly A– rRNA and any remaining tRNA are washed away. After the poly A– RNAs are washed away, you can elute your purified mRNA by changing the conditions so that they no longer favor hybridization between the poly A tails on the mRNA and the oligo dT molecules. You will then use the purified mRNA that you recover for subsequent analyses.

PROCEDURAL NOTE

If you did not perform the purification of total RNA from mouse liver, as described in Lab Period III.1.1, you will be provided with a tube of mouse total liver RNA (about 400 µg in a 250-µl volume) in a tube labeled "Liver RNA," as in Step 12 above. Label this tube with your number, and store on dry ice or at –70°C until use.

»»

LAB PERIOD III.1.2 CONT.

PROCEDURAL NOTES

For this procedure, you will use the Ambion Micro Poly(A) Purist kit to fractionate poly A+ RNA from total mouse liver RNA. You will use one of these columns to fractionate poly A+ RNA from your preparation of mouse liver total RNA.

Wear gloves, and use the specially labeled RNA tubes and pipette tips for *all* procedures in this experiment. Keep your RNA sample on ice with the top of the tube closed at all times unless instructed otherwise!

Step 1 Label one 1.5-ml microfuge tube "Total RNA" and one 0.5-ml microfuge tube "PCR RNA." Label both tubes with your number. Pipette 250 μl of your mouse liver total RNA sample (from Step 12 above in the tube you labeled "Liver RNA") into a new 1.5-ml microfuge tube. Pipette 25 μl into the 1.5-ml tube labeled "Total RNA-N." To this, add 2.5 μl of 3 M NaOAc (pH 7.0). Mix briefly, and precipitate by adding 62.5 μl of ice-cold 100% EtOH. You will run this "Total RNA-N" on a denaturing Northern gel in the next lab period. Store this tube at –70°C.

Step 2 Pipette another 1 μl of your Liver RNA sample (from your original "Liver RNA" sample) into the 0.5-ml tube labeled "PCR RNA." Store this tube at –70°C. This will be your starting material for the RT-PCR experiment described in Module III.3.

Step 3 To the remaining 224 μl Liver RNA sample, add 26 ml of DEPC-treated ddH$_2$O and 250 μl of Ambion 2x Binding Solution (contains the oligo-dT cellulose resin).

Step 4 Mix by inversion to resuspend the resin thoroughly. If necessary, clumps can be broken up by pipetting up and down. Label this tube with your number and "RNA/oligo dT."

Step 5 Incubate the RNA/oligo dT mixture at 70°C for 5 minutes to denature the RNA. Tape your tube on its side to your rack, and place on a rocker for 30 to 60 minutes at room temperature. During this time, the poly A+ mRNA will hybridize to the oligo-dT cellulose.

Step 6 Pellet the oligo-dT cellulose at 3000x *g* for 3 minutes at room temperature. Remove the supernatant with a pipette, and store it in a new 1.5-ml microfuge tube labeled with your number and "RNA sup."

Step 7 Label your tube of elution buffer (the Ambion RNA Storage Buffer) with your group number, and put it at 70°C to prewarm the solution.

Step 8 Resuspend the oligo-dT cellulose pellet in 500 μl of Ambion wash solution 1 by vortexing briefly. Transfer the cellulose into an Ambion RNA Spin Column (provided by your instructor).

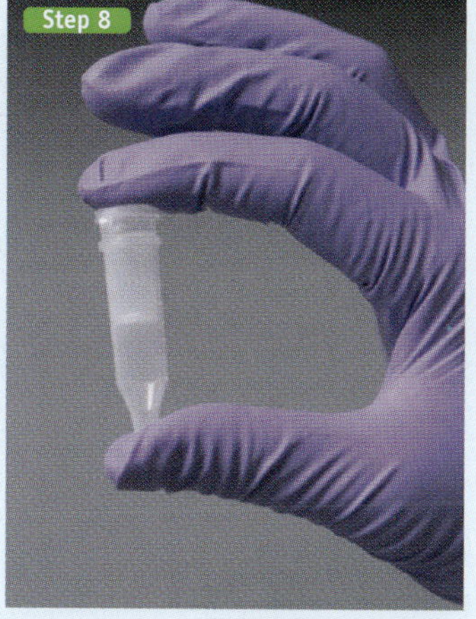

Step 9 Centrifuge the column at 3000x *g* for 3 minutes at room temperature, and discard the flow-through. Place the column back into the same tube.

Step 10 Add 500 μl of Ambion wash solution 1 to the oligo-dT cellulose. Close the tube, and vortex briefly to mix well.

Step 11 Repeat Step 9.

»»

LAB PERIOD III.1.2 CONT.

Step 12   Add 500 µl of Ambion wash solution 2 to the oligo-dT cellulose. Close the tube, and vortex briefly to mix well. Centrifuge again as in Step 9 and discard the flow-through.

Step 13   Repeat Step 12 with 500 µl of fresh Ambion wash solution 2.

Step 14   Label the special white 1.5-ml microfuge tube that comes in the Ambion kit with your number and "A+ RNA." Place the column into this microfuge tube.

Step 15   Add 100 µl of warm (70°C) elution buffer (the Ambion RNA storage buffer) to the Ambion oligo-dT cellulose. Close the tube, and vortex briefly to mix thoroughly. *Immediately* centrifuge at 4900x g for 2 minutes.

Step 16   After the centrifugation, the poly A+ RNA is in the bottom of the tube. *Do not discard!* Add 100 µl of additional warm elution buffer to the column. Vortex briefly, and spin again as in Step 15.

Step 17   You will now have a 200-µl volume of mouse liver poly A+ RNA in the bottom of the tube. Discard the column.

Step 18   To this tube, add 2 µl of Ambion glycogen carrier, 20 µl of Ambion 5-M ammonium acetate, and 560 µl of ice-cold 100% EtOH. Mix thoroughly by pipetting up and down.

PROCEDURAL NOTE

The glycogen helps small quantities of nucleic acid to precipitate efficiently. It acts as a carrier.

Step 19   Transfer 145 µl of A+ RNA from this tube to a new 1.5-ml microfuge tube. Label this tube with your number and "A+ RNA-N." You will run this A+ RNA sample on the denaturing Northern gel in the next lab period.

Step 20   Transfer the remaining 637 µl into another new 1.5-ml microfuge tube labeled with your number and "cDNA RNA." This will be your starting material for the cDNA library experiment described in Project IV.

Step 21   Store both tubes at –70°C.

At this point you should have four tubes stored at –70°C: (1) "Total RNA-N" and (2) "A+ RNA-N," both of which will be run on the denaturing Northern gel in the next lab period, (3) "PCR RNA," which will be used in the RT-PCR experiment described in Module III.3, and (4) "cDNA RNA," which will be used for the cDNA experiment described in Project IV.

LAB PERIOD III.1.3.   RECOVERY OF PRECIPITATED RNA SAMPLES

YOUR "Total RNA-N" and "A+ RNA-N" samples that have been stored at –70°C will be returned to you on ice. You will prepare these to be run on a denaturing agarose gel. This is called a "Northern gel."

Step 1   Recover the RNA in each sample by centrifuging your two RNA ethanol precipitations for the Northern gel ("Total RNA-N" and "A+RNA-N") in a microfuge at top speed for 60 minutes at 4°C.

»»

LAB PERIOD III.1.3 CONT.

PROCEDURAL NOTE

A full-hour spin is recommended to maximize recovery of the precipitated RNA.

Step 2 When the spin is complete, remove the supernatants with a flame-drawn Pasteur pipette. Refer to the figures in Step 4 in Lab Period I.1.6 for making the flame drawn pipettes and the figure in Step 5 of the same lab period that illustrates the procedure for removing the ethanol from the pellet with the flame-drawn pipette. You will use two of these pipettes. Using tape, label them "T" and "A+." Save these for use again in Step 4.

Step 3 Add 1-ml cold 70% ethanol (made with DEPC water and reserved for RNA work only) to wash the pellets in each of your two RNA tubes. Mix by vortexing for 30 seconds, and then centrifuge for 15 minutes in a microfuge at maximum speed at 4°C.

Step 4 Remove the ethanol supernatant from each tube with the appropriate flame-drawn Pasteur pipette.

Step 5 Spin the tubes for 30 seconds in your nanofuge, and use the same flame-drawn pipettes to remove any residual ethanol (be sure to avoid disrupting the RNA pellet!).

Step 6 Allow the tubes containing the RNA pellets to air dry with the caps open on ice for 10 minutes.

Step 7 Add 15 µl of Northern loading dye (this is *not* the 5x Blue Juice loading dye [5x BJ] used for DNA samples on agarose gels) to each tube, but *do not* pipette up and down. These tubes will be collected by the instructor and stored for you at –70°C until you need them for your Northern gel.

LAB PERIOD III.1.4.   **ELECTROPHORESIS AND VISUALIZATION OF RNA ON A DENATURING AGAROSE GEL**

YOU WILL run your RNA samples on a denaturing agarose Northern gel. This will allow you to visualize your RNA samples and will prepare your samples for the Northern blot experiment described in Module III.2.

CAUTION!

The denaturing agent in this gel will be formaldehyde. Formaldehyde is a noxious substance that emits fumes that are toxic if directly inhaled. Therefore, you should prepare your gel in a fume hood and run the gel in the fume hood.

Step 1 In a fume hood, to 100 ml of melted 1% agarose, add 5.4 ml of 37% formaldehyde (commercially available formaldehyde comes as a 37% solution). Mix gently, and pour about half of the agarose into the gel tray (depending on the gel electrophoresis apparatus, two gels can typically be poured per bottle of agarose). After pouring, allow the gel to set for at least 1 hour in the fume hood.

PROCEDURAL NOTE

For Northern gels, use 1% agarose made up in 1x MOPS/EDTA buffer using DEPC-H$_2$O.

LAB PERIOD III.1.4 CONT.

Step 2   In the hood, cover the gel with MOPS/EDTA buffer to just above the level of the wells (leave the comb in place).

Step 3   Set up the electrophoresis apparatus at your bench as described in Lab Period I.1.4, Part B, Steps 3 to 6. Carefully remove the comb. The formaldehyde in the gel will leach into the wells, and this can interfere with sample loading. Therefore, before loading the gel, be sure to flush any excess formaldehyde from the sample wells by pipetting the electrophoresis running buffer in and out of the wells just before use. Note that the running buffer for Northern gels is 1x MOPS/EDTA.

Step 4   Retrieve your two samples of RNA that you prepared previously (total RNA-N and A+ RNA-N) and stored at –70°C. Pipette each of these samples up and down to complete the resuspension of the pellets in Northern gel loading dye. You will also be given a 15-µl sample of spleen poly A+ RNA in Northern loading dye.

CAUTION!

Wear gloves, lab coats, and safety glasses. Handle these gels with extreme care! They are slippery. Support the entire gel with the spatula to keep it from breaking.

Ethidium bromide is a known mutagen and carcinogen. Wear gloves, lab coats, and safety glasses! Handle gels carefully! Do not splash ethidium bromide! Always rinse the spatula in the water destain bath after it has been in contact with ethidium bromide! Always rinse the gel tray after sliding gels into the ethidium bromide bath. You may unknowingly contact the ethidium bromide! Be very careful not to drip the ethidium bromide anywhere!

Step 5   Denature your three 15-µl RNA samples by heating them to 65°C for 2 minutes. Spin briefly in your nanofuge. Place on ice immediately for 1 minute (or until you are ready to load the gel).

Step 6   Flush the wells with gel buffer just before loading. Load the three RNA samples in this order, but skip a lane between each:

    a. Liver poly A+ RNA (15 µl in Northern loading dye; approximately 2–4 µg).

    b. Liver total RNA (15 µl in Northern loading dye, approximately 25 µg).

    c. Spleen poly A+ RNA (15 µl in Northern loading dye; approximately 2–4 µg).

PROCEDURAL NOTE

Passing total RNA through a digodT column one time enriches for mRNA but does not completely purify mRNA.

Step 7   Electrophorese the gel at 50 V (constant voltage) at room temperature for 2 hours. The bromphenol blue dye should be about two thirds of the way down the gel before stopping the electrophoresis.

Step 8   Wear gloves while handling your gel. After electrophoresis, disconnect the leads from the power supply. Remove the lid from the gel box, and cut a corner of the gel in an asymmetric manner to orient it.

LAB PERIOD III.1.4 CONT.

Step 9   Carefully stain the gel for 8 to 10 minutes in an ethidium bromide bath using a spatula to transfer the gel from the electrophoresis box to the gel stain container. (The instructor will demonstrate this.) Using the spatula, transfer the gel to a water bath (to destain the gel) for a minimum of 15 minutes (30 to 60 minutes is better if you have time).

PROCEDURAL NOTE

Appendix IV describes an alternative DNA stain, SYBR Safe (Molecular Probes).

Step 10   Place your gel on a UV transilluminator, and place a fluorescent ruler along the top of the gel to mark where the wells are located. Photograph the gel (or take a digital image with a gel documentation system). Your instructor will demonstrate this. Your gel image should look similar to that of **Figure 3-2**. You should be able to see the 18S (1894 ribonucleotides) and the 28S (4718 ribonucleotides) rRNA bands in the total RNA lane. The 28S band should be about double the intensity of the 18S band. The 18S and 28S rRNA bands should be much less prominent in the A+ RNA lane. This is an indication of the extent of enrichment you achieved in fractionating the A+ RNA on the oligo dT column. Remember that your total RNA sample contains much more RNA than your A+ RNA sample (about 250 times as much).

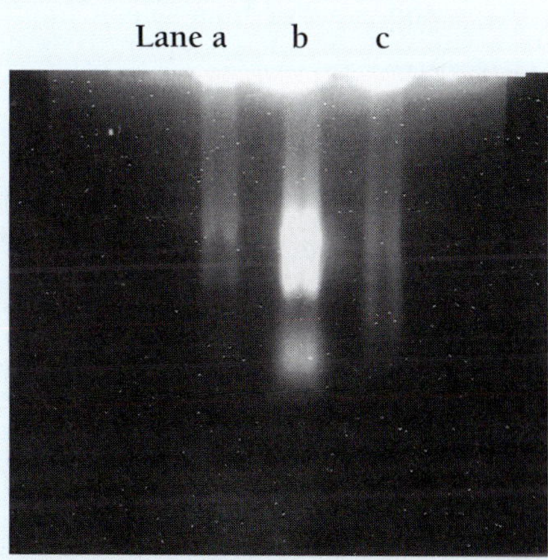

Fig 3-2   Expected RNA gel image.

PROCEDURAL NOTE

Save this gel! Do not discard it. You may use it to transfer the RNA from this gel onto a membrane in the Northern blot transfer described in Module III.2. If you are performing the Northern blot experiment, proceed directly to Module III.2, Lab Period III.2.1.

Northern Blot Experiment to Detect a Specific Gene Transcript

MODULE III.2

MODULE SUMMARY

Northern blot transfer of electrophoresed RNA. Random-priming reaction to nonradioactively label a *Ttr* DNA probe, followed by Northern hybridization and detection of the bound probe to indicate the presence and size of the *Ttr* transcript in liver and spleen tissue.

MODULE BACKGROUND

Because only a fraction of the genes in the genome are expressed in any one cell type or tissue, it is necessary to examine RNA samples from specific tissues to determine whether a specific gene is expressed in that tissue. The classic method for this type of analysis is the Northern blot hybridization

BACKGROUND, continued

technique. This technique is similar to the Southern blot hybridization technique described in Project I, except that samples of RNA rather than DNA are run in a gel and subsequently transferred to a hybridization membrane. In this experiment, you will start with the Northern gel you ran in Module III.1 and transfer the electrophoresed, size-separated RNA samples to a hybridization membrane by "Northern transfer," which is a capillary transfer process that is exactly analogous to Southern transfer. You will then label a probe that is specific to the *Ttr* gene and hybridize this probe to your Northern blot to detect the presence of RNA transcripts from this gene. You will detect the *Ttr* mRNA using a chemiluminescent method to identify the biotinylated *Ttr* probe (New England Biolabs Phototope kit). The results will indicate the relative abundance of the *Ttr* mRNA in total and poly A+ RNA samples from liver and poly A+ RNH from spleen. You will also be able to estimate the size of the *Ttr* transcript.

LAB PERIOD III.2.1.   PROCESSING THE NORTHERN GEL AND SETTING UP THE NORTHERN BLOT TRANSFER

LAB PERIOD III.2.1 is an extension of Lab Period III.1.4 and should be performed on the *same* day.

Step 1 Cut a piece of nylon membrane (Immobilon; Millipore) to the exact size of the Northern gel. Write your group numbers and "RNA side" across the top of the membrane. Use a #2 pencil, not a pen to label the filters! Prewet the membrane in DEPC-H_2O for 5 minutes and then in 10x SSC for 5 minutes. These are minimum times — leave the membrane in the 10x SSC until you are ready to use it.

PROCEDURAL NOTE

Remember to handle the membrane with gloves or forceps only! Oils from your fingers are easily transferred to these membranes and will prevent the transfer and binding of nucleic acids to the membranes.

Step 2 To set up the Northern transfer, begin by cutting three sheets of Whatman 3MM paper to the exact dimensions of the gel (use your gel tray as a guide). Then cut two sheets of Whatman 3MM paper to the same width as your gel but increase the length by 3 to 4 inches (these will serve as your wick). Also, cut paper towels to the same dimensions as your gel. You will need a stack at least 3 inches thick.

PROCEDURAL NOTE

Refer to Module I.2 (Lab Period I.2.1) for figures of how to set up a Southern transfer (Figures 1-5a and b and figures in Steps 11–13, 15–17). The Northern blot transfer is set up in an identical manner.

Step 3 Place your agarose gel tray upside down in a plastic container. Add enough 10x SSC so that the level of the liquid is halfway up the side of the gel tray.

Step 4 Saturate the wick of filter papers in 10x SSC and align them on the gel tray. Smooth out any air bubbles.

Step 5 Place the gel *face down* on the saturated filter papers. Be sure that no air bubbles are trapped between the gel and wick.

Step 6 Place the wetted nylon membrane on the gel ("RNA side" down!). Do not let the membrane hang over the gel and come in contact with the wick or else you will "short circuit" the transfer. Again, be sure that no air bubbles are trapped between the gel and the membrane.

»»

LAB PERIOD III.2.1 CONT.

Step 7 Wet one precut piece of filter paper in 10x SSC, and place it on top of the membrane. You can roll a pipette over the surface of the filter paper to eliminate any air bubbles. Add the two dry pieces of filter paper to the stack.

Step 8 Cover the filter paper with the stack of precut paper towels at least 3 inches thick. Be sure that the filter papers and paper towels do not directly contact the gel or the wick.

Step 9 Place the top of your gel box on top of the paper towels to provide a rigid support.

Step 10 Cover the whole apparatus with plastic wrap, and place a blot weight of about 200 to 500 g on the gel box top. Be sure that the entire apparatus is stable!

Step 11 The transfer is done with 10x SSC and is usually complete in 8 to 12 hours depending on the thickness of the gel. You should perform the transfer overnight.

PROCEDURAL NOTE

It will be necessary to disassemble and process the Northern blot the following day.

LAB PERIOD III.2.2.   DISASSEMBLE THE NORTHERN TRANSFER; UV CROSS-LINK/BAKE THE FILTER

Step 1 Remove all of the paper above the membrane. *Before* removing the membrane from the gel, be sure to cut one corner of the membrane to orient the blot (record which corner that you cut). Also, mark the location of the wells using your comb for alignment. Be sure your number and "RNA side" are already written on the RNA side of the membrane.

Step 2 Next, remove the membrane from the gel, and place it into 6x SSC in a plastic container (use forceps and gloves!). Rinse for 2 to 5 minutes.

Step 3 Check your gel on the UV light box to confirm that all or most of the RNA has been transferred. Dispose of the gel in an appropriate waste container.

Step 4 Lay the membrane *RNA side up* on a clean sheet of filter paper, and allow the membrane to air dry.

PROCEDURAL NOTE

If necessary, Steps 5 and 6 can be performed during Lab Period III.2.3. If so, store the membranes at room temperature during the interim.

Step 5 UV cross-link the membranes in a UV cross-linking instrument (if available) according to instructions from the manufacturer.

Step 6 Place your membrane in between two sheets of filter paper, and bake at 80°C for 1 hour (if no UV cross-linker is available to do Step 5, bake instead for 2 hours). The baking and UV cross-linking will fix the transferred RNA onto the membrane. After baking the membranes, mark the location of the wells on the RNA side of the membranes using a #2 pencil. Store the membrane at 4°C in a plastic bag until ready for prehybridization.

LAB PERIOD III.2.3.   RANDOM PRIMING REACTION TO LABEL NONRADIOACTIVELY THE *TTR* DNA PROBE

YOU WILL do the random priming reaction using the NEBlot Phototope Labeling Kit from New England Biolabs. You will label a DNA template with biotinylated nucleotides. Your instructor will provide the DNA template. This template is a purified 580-bp *Ttr* cDNA PCR product produced from mouse liver RNA by RT-PCR. Random priming is a common technique for labeling double-stranded DNA probes of more than 200 bp in length. The DNA template is heated to denature the double-stranded DNA, and then a set of short oligonucleotides of random sequence, each 8 nt long, is mixed with the DNA template along with appropriate buffers, DNA polymerase, and dNTPs (including biotin-conjugated dATP). The primers, which are also labeled with conjugated biotin, will anneal to complementary sequences in the DNA template and will then be extended by the DNA polymerase. This reaction yields copies of the original template molecules that include the biotin-conjugated dATP and are therefore nonradioactively labeled.

Step 1   You will be given a microfuge tube containing 34 µl of the mouse *Ttr* cDNA clone (25 ng). Label this tube with your number and "Northern Probe."

Step 2   To denature this DNA before labeling, heat the tube at 95°C for 5 minutes (use a locking ring to secure the cap). *Place immediately on ice* for 5 minutes to cool quickly to prevent the DNA from reannealing. Spin 5 seconds in a nanofuge to bring the solution to the bottom of the tube.

Step 3   To this tube, add the following:

> 10 µl 5x Labeling Mix (buffer and primers)
>
> 5 µl dNTP Mix (1-mM each dNTP)
>
> 1 µl Klenow DNA Polymerase (5 units)
> ___________________________________
>
> 50 µl total reaction volume (mix by stirring)

PROCEDURAL NOTE

In this labeling reaction, the random octamers (random primers) *and* the dATP are biotinylated.

Step 4   Incubate the labeling reaction at 37°C overnight.

LAB PERIOD III.2.4.   STOP THE LABELING REACTION AND PRECIPITATE THE *TTR* PROBE

Step 1   Retrieve your tube of biotinylated *Ttr* probe from the 37°C water bath.

Step 2   Add 5 µl of 0.2 M EDTA, pH 8.0, to stop the labeling reaction.

Step 3   Next add the following:

> 5 µl 4 M LiCl
>
> 100 µl 100% ice-cold EtOH

>>>>

LAB PERIOD III.2.4 CONT.

PROCEDURAL NOTE

This protocol using LiCl and ethanol effectively precipitates the labeled DNA but not the biotinylated primers. It is very important to use this protocol to remove the labeled primers because otherwise the random biotinylated primers will hybridize nonspecifically to all RNA on the Northern blot.

Step 4 Mix the solution by vortexing and store at –70°C for 30 minutes or longer to precipitate the labeled DNA.

Step 5 Centrifuge the tube at 4°C for 30 minutes at 16,000x g in a microfuge to pellet the DNA. Be sure to orient the microfuge tube so that the hinge is oriented directly away from the center of the rotor (see the figure in Step 3, Lab Period I.1.6). If the tube is properly oriented during the spin, the pellet will be at the bottom of the tube directly below the hinge.

Step 6 Using a flame-drawn Pasteur pipette, remove the EtOH supernatant, and discard into your waste container (be careful not to disturb the pellet!) (see figure in Step 5, Lab Period I.1.6). Save the pipette to use again in Step 9.

> **CAUTION!**
>
> Flame-drawn pipettes are extremely sharp! Handle with care!

Step 7 Add 500 µl of ice-cold 70% EtOH, and invert the tube gently several times to wash the pellet.

Step 8 Centrifuge at 16,000x g at 4°C for 10 minutes.

Step 9 Using the same flame-drawn pipette as in Step 6, remove the EtOH supernatant, and discard into your waste beaker. Invert the tube on Kimwipes, and air dry the pellet for 10 minutes at room temperature.

Step 10 Resuspend the pellet in 20 µl of 1x TE buffer by pipetting up and down for about 30 seconds.

Step 11 Store the *Ttr* biotinylated probe at –20°C until you are ready to do the Northern hybridization.

LAB PERIOD III.2.5.   **NORTHERN BLOT PREHYBRIDIZATION**

PROCEDURAL NOTE

Refer to figures in Step 1, Lab Period I.2.5, for general methods of the Northern blot prehybridization and hybridization protocol. These methods are identical to the methods used for Southern blots.

Step 1 Place one membrane in a hybridization bag or two membranes back-to-back (RNA side facing out) in a hybridization bag. (The hybridization bag should have a spout, as shown in the first figure of Step 1, Lab Period I.2.5.)

Step 2 Add 6 ml of Northern prehybridization solution to the bag (second figure of Step 1, Lab Period I.2.5). Carefully remove most of the air bubbles, and seal the bag using a heat sealer (third figure of Step 1, Lab Period I.2.5).

> **CAUTION!**
>
> The Northern prehybridization solution contains 50% formamide. Formamide is hazardous! Wear gloves!

»»»

LAB PERIOD III.2.5 CONT.

PROCEDURAL NOTE

The Northern prehybridization solution contains Denhardt's solution and salmon sperm DNA that serve as blocking agents that bind to "sticky" spots on the membranes. This prevents the DNA probe from binding to these random sticky regions and then dramatically reduces the background signal on the film following detection.

Step 3 Prehybridize the Northern blots at 55°C with gentle rocking for at least 3 to 6 hours (for a laboratory course, an overnight prehybridization is sometimes convenient).

LAB PERIOD III.2.6.   **NORTHERN BLOT HYBRIDIZATION**

Step 1 Each group should retrieve their *Ttr* probe (a 1.5-ml tube that you stored at −20°C and labeled "Northern Probe"). Thaw at room temperature.

Step 2 To a tube containing 60 µl of Northern hybridization solution, add 20 µl of your Northern probe. Denature the probe by heating in a temp block at 95°C for 10 minutes (use a lid lock to prevent your tube from popping open). Place on ice immediately for 5 minutes to cool the reaction rapidly to prevent the labeled DNA from reannealing. Spin your tube briefly in a nanofuge to collect all of the solution at the bottom of the tube, and place back on ice.

Step 3 Retrieve your Northern membrane that is prehybridizing in the 55°C incubator. Use scissors to open the bag, and pour out the Northern prehybridization solution into a waste beaker (cut just a small corner of the spout of the bag).

Step 4 Pipette 6 ml of Northern hybridization solution into the hybridization bag.

CAUTION!

The Northern prehybridization solution contains 50% formamide. Formamide is hazardous! Wear gloves!

Step 5 Pipette all of your denatured probe into the hybridization bag containing your membrane. Do not pipette the probe directly onto the membrane.

Step 6 Remove air bubbles from the hybridization bag and use the heat sealer to reseal the bag. The membrane will be hybridized overnight at 55°C with gentle rocking.

PROCEDURAL NOTE

55°C with 50% formamide is the equivalent of 85°C without the 50% formamide in the Northern hybridization solution.

LAB PERIOD III.2.7.   NORTHERN BLOT DETECTION OF THE HYBRIDIZATION PROBE

PROCEDURAL NOTE

All Phototope wash solution recipes are listed in Appendix II.

Step 1 After the overnight hybridization, cut open the hybridization bag with scissors and transfer the Northern membrane to a plastic container with 500 ml of Northern wash solution I at room temperature. Reseal and discard the hybridization bag.

PROCEDURAL NOTE

All of the membranes from an entire class can be washed together in a single plastic container.

Step 2 Transfer all of the membranes to a new container with 500 ml of the same wash solution. Incubate the filters at 55°C for 5 minutes with shaking.

PROCEDURAL NOTE

Be sure to use a thermometer to check the temperature of the wash solutions!

Step 3 Discard this wash solution and repeat with 500 ml of fresh Northern wash solution I for 5 minutes at 55°C with rocking.

Step 4 Discard this wash solution and wash with 500 ml of Northern wash solution II in another plastic container for 5 minutes at 55°C with rocking. This is a more stringent wash due to the reduced salt concentration.

Step 5 Repeat Step 4 with 500 ml of fresh Northern wash solution II.

Step 6 Discard the wash solution and rinse the membrane briefly in 500 ml of Northern wash solution I at room temperature.

Step 7 Place your Northern membrane into a hybridization bag with a spout as described above (see the first figure of Step 1, Lab Period I.2.5), and seal the sides of the bag.

Step 8 Through the spout, add 6 ml of Phototope blocking solution. Remove the bubbles (refer to the second figure of Step 1, Lab Period I.2.5), and seal the spout with a dialysis clip. Refer to Lab Period I.3.6 for a complete description of the Phototope blocking solution.

PROCEDURAL NOTE

The dialysis clips are useful for sealing and resealing the bags through the several wash and detections steps, as described below.

Step 9 Place your bag on a rocking platform at room temperature. The membrane should be in the Phototope blocking solution for at least 5 minutes.

Step 10 Discard the Phototope blocking solution into the plastic waste container at your bench.

»»

LAB PERIOD III.2.7 CONT.

Step 11   Add 4 ml of Streptavidin solution to your membrane. Remove the bubbles, and reseal the spout with a dialysis clip. Incubate the membrane in this solution for 5 minutes on the rocking platform. Discard the Streptavidin solution into your plastic waste container.

Step 12   Add 30 ml of Phototope detection wash solution I. Reseal and incubate for 5 minutes on the rocking platform. Discard the solution into the plastic waste container at your bench.

Step 13   Repeat the 5-minute wash with 30 ml of fresh Phototope detection wash solution I (on the rocking platform). Discard the solution into your plastic waste container at your bench.

Step 14   Add 4 ml of alkaline phosphatase solution to the bag. Reseal and incubate for 5 minutes on the rocking platform. Discard this solution into your plastic waste container.

Step 15   Add 30 ml of Phototope blocking solution to the membrane. Reseal and incubate for 5 minutes on the rocking platform. Discard the solution into your plastic waste container.

Step 16   Add 30 ml of Phototope detection wash solution II. Reseal and incubate for 5 minutes on the rocking platform. Discard the solution into your plastic waste container at your bench.

Step 17   Repeat this 5-minute wash with 30 ml of fresh Phototope wash solution II. *Do not* discard this final wash solution until you are ready to add the Phototope detection mix in the dark room.

PROCEDURAL NOTE

This last 5-minute wash is a minimum time, and you can leave your membrane in the wash as long as necessary.

Step 18   When you are ready to add the Phototope detection mix, discard the last wash solution; add 4 ml of Phototope detection mix to your membrane (the Phototope detection mix is 3.84 ml ddH$_2$O + 160 µl 25x Star Diluent + 4 µl Detection Reagent). Reseal and incubate in detection mix for 5 minutes on the rocking platform. Discard the detection mix into the plastic waste container at your bench, and carefully remove any remaining bubbles from the bag.

Step 19   Heat-seal the bag over the spout. Wash the outside of the bag in tap water, and dry with a paper towel.

Step 20   In the dark room with red safe lights on, place the blot (still sealed in the hybridization bag) "RNA side" up in an X-ray film cassette. Place a sheet of autoradiography film (Kodak BioMax) over the blot (see figure in Step 14, Lab Period I.2.7). Orientation marks can be made on the film and the bag so that they can later be oriented properly. Close the cassette and expose the film for 1 to 5 minutes. It is recommended to first do a short exposure for 1 to 5 minutes and then a longer exposure. Put a new piece of film in the cassette immediately upon removal of the first film from the cassette in case you need a longer exposure.

Step 21   Remove the X-ray film from the cassette in the darkroom with the red safelights on. Develop in GBX (Kodak) developer for 2 minutes, rinse in tap water for 2 minutes. Fix in GBX fixer (Kodak) for 2 minutes, and finally, rinse for 2 additional minutes in tap water. If you have an automatic X-ray film developer, you can use that instead.

Step 22   Dry the film (air drying works fine or you can use a hair dryer to speed up the process), and then replace the film into the X-ray cassette with the Northern membranes. Use the alignment marks to be sure that they are properly oriented. Now use a marking pen to outline the

»»

LAB PERIOD III.2.7 CONT.

blot on the film. Trace the alignment lines from the filter onto the film, and write the group number next to the filter image on the film. Cut out the image of the filter from the film to tape into your lab book (this film is referred to as a lumigram because it results from a chemi-luminescent reaction!).

Step 22 Examine the lumigram and analyze your results! Your data should look similar to **Figure 3-3**. What can you conclude about the expression of the *Ttr* gene in liver and spleen tissue? What can you conclude about whether or not the *Ttr* transcript is polyadenylated? What is the approximate size of the *Ttr* transcript?

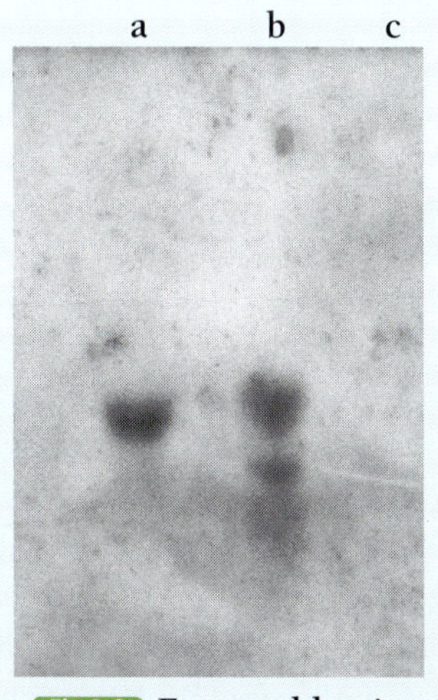

Fig 3-3 Expected lumigram image.

Reverse Transcriptase-Polymerase Chain Reaction (RT-PCR) Amplification of a Specific Gene Transcript

MODULE III.3

MODULE SUMMARY

Conversion of a small amount of RNA to cDNA using reverse trancriptase followed by PCR amplification of the *Ttr* gene sequence from this cDNA to indicate the presence of the *Ttr* transcript in the original RNA sample.

MODULE BACKGROUND

An alternative to standard Northern blot analysis for the detection of a specific RNA sequence is the reverse transcriptase polymerase chain reaction (RT-PCR) technique. This method is also referred to as RNA-PCR. With this technique, a sample of RNA is first subjected to reverse transcription to generate complementary DNA (cDNA) copies of the mRNA sequence. This cDNA is then used as a template for standard DNA PCR amplification. The products of this reaction are then run on an agarose minigel and stained to visualize the amplified fragment.

RT-PCR affords several advantages that complement or replace those of standard Northern blot analysis. First, RT-PCR is very sensitive so that only very small amounts of total RNA are required for the analysis (poly A+ mRNA is usually not required). Second, because only a small subfragment of the entire mRNA transcript is usually amplified, RNA that is somewhat degraded can still provide a reasonable template for RT-PCR. As with DNA PCR, however, the extreme sensitivity of the RT-PCR method renders this technique susceptible to false-positive results. Several controls should be run routinely to verify the validity of any positive result. For this experiment, you will use the 1-µl "PCR RNA" sample that you prepared from the total liver RNA in Module III.1. A cDNA copy of your RNA can be generated using either random primers, oligo-dT primers, or a primer specific to the target RNA. You will use the latter (a "downstream specific" primer).

BACKGROUND, continued

After one RT-PCR cycle to generate a cDNA copy of each *Ttr* mRNA transcript present in the sample, you will then run 43 DNA PCR cycles to amplify the desired fragment from the *Ttr* cDNA template. The primers you will use for the *Ttr* cDNA will amplify a fragment about 148 bp in length. The upstream primer is in exon 1, whereas the downstream primer is in exon 2 of the *Ttr* mRNA (Figure 1-38). Thus, these two primers flank intron 1 in genomic DNA. Therefore, only a cDNA product derived from the *Ttr* mRNA will provide a template that will yield the 148-bp PCR product. If contaminating genomic DNA were to act as a template for PCR with the same primers, it would yield a much larger product because of the presence of the intron (approximately 1100 bp) (see Fig. 1-8).

The PCR product amplified from the cDNA template will then be run on an agarose gel to facilitate direct visualization.

PROCEDURAL NOTES

For further confirmation of the identity of the RT-PCR product, you can run an aliquot of the RT-PCR product on a Southern blot gel and probe with an oligonucleotide probe specific to an internal sequence as described for the DNA PCR product in Project I. Because you may have already performed this procedure in Project I, you will not repeat it here.

A protocol for treatment of the initial RNA sample with DNase to eliminate any chance of DNA contamination is found in Appendix IV. This is especially useful if you are working with a gene that has no introns, as the products of PCR from DNA contamination and the RT-PCR products from mRNA would be indistinguishable by size.

LAB PERIOD III.3.1.   SET UP THE RT-PCR REACTION

Step 1 Retrieve your "PCR RNA" sample that has been stored at –70°C from Lab Period III.1.2. *Keep this tube on ice!*

Step 2 Pipette 49 µl of "RT-PCR Buffer" into your "PCR RNA" tube that already contains the 1 µl of your RNA sample. Your instructor will provide the "RT-PCR Buffer" components. All of the components except the gene specific primers and the RNA template come from the Applied Biosystems GeneAmp Gold RNA PCR Reagaent Kit.

Step 3 Mix gently by pipetting the solution up and down.

PROCEDURAL NOTES

The mix contains the primers, the AmpliTaq Gold polymerase, and all of the other required reagents. See Appendix II.

The *Ttr* Forward Primer (upstream specific) is 5' ACACAGATCCACAAGCTCCTGACAG 3'. The *Ttr* reverse primer (downstream specific) is 5' CGGACAGCATCCAGGACTTTGAC 3'.

Step 4 Place your "PCR RNA" tube into the thermal cycler. You will run one cycle to generate cDNA and then 43 cycles of standard PCR to generate the double-stranded DNA PCR product.

»»

LAB PERIOD III.3.1 CONT.

The following parameters will be used for this PCR:

a. 25°C for 10 minutes to anneal the *Ttr* reverse primer (downstream specific primer)

b. 42°C for 12 minutes (first-strand cDNA synthesis)

c. 95°C for 10 minutes (to denature the cDNA and RNA, to heat kill the RT enzyme, and to activate the *Taq* Gold DNA polymerase)

d. 43 cycles of 94°C for 20 seconds and 62°C for 1 minute (two-step PCR)

e. 72°C for 7 minutes (final extension)

f. Hold at 4°C

Your instructor will prepare the following RT-PCR controls (the second and third controls contain 750 ng of total RNA):

a. A "no template" control reaction in which 1 µl of ddH$_2$O will be added to the RT-PCR in place of the RNA template.

b. A "no RT" control reaction in which 1 µl of ddH$_2$O will be added to the RT-PCR in place of the reverse transcriptase enzyme.

c. A "DNA Contamination" control reaction spiked with 75 ng of genomic mouse DNA (this is done simply to show you what an RT-PCR result would look like if you had DNA contamination; what result do you expect with this control?).

d. A positive control should also normally be run. In this case, 2 ng of a plasmid containing a cDNA clone of the *Ttr* gene (which is identical to the *Ttr* mRNA) will be used.

LAB PERIOD III.3.2.   **ELECTROPHORESIS TO VISUALIZE THE RT-PCR PRODUCT ON AN AGAROSE GEL**

Step 1 Pour a 1.8% agarose gel as described in Lab Period I.1.4, Part B, except that you should dissolve 1.8 g of agarose in 100 ml of 1x TAE buffer to produce a 1.8% gel (refer to figures in Lab Period I.1.3, Part B, Steps 3–6 and 12).

Step 2 After the completion of the 43 PCR cycles, remove your "RNA PCR" tube from the thermal cycler and place it on ice.

PROCEDURAL NOTE

Your instructor may have already removed your tube from the thermal cycler and stored it at 4°C for you.

Step 3 Remove 8 µl from the "RNA PCR" tube, and place it into a new 1.5-ml microfuge tube labeled "RNA PCR Gel." Add 2 µl 5x BJ (loading dye), and mix by stirring with your pipette tip. Save the remainder of the "RNA PCR" tube at −20°C.

Step 4 Load the agarose gel with samples in the following order:

Lane 1: Empty

Lane 2: 10 µl 100-bp ladder in 5x BJ (50 ng/µl)

Lane 3: 10 µl from "RNA PCR Gel" tube (from Step 3 above)

»»

LAB PERIOD III.3.2 CONT.

PROCEDURAL NOTE

The 100-bp marker is used on this electrophoresis gel to provide known DNA sizes for comparison to your sample. A description of this marker is provided in Appendix III. Save your marker DNA at 4°C.

CAUTION!

Wear gloves, lab coats, and safety glasses. Handle these gels with extreme care! They are slippery. Support the entire gel with the spatula to keep it from breaking.

Ethidium bromide is a known mutagen and carcinogen. Wear gloves, lab coats, and safety glasses! Handle gels carefully! Do not splash ethidium bromide! Always rinse the spatula in the water destain bath after it has been in contact with ethidium bromide! Always rinse the gel tray after sliding gels into the ethidium bromide bath. You may unknowingly contact the ethidium bromide! Be very careful not to drip the ethidium bromide anywhere!

Step 5 Your instructor will provide you with aliquots of the four control reactions to run on your gel. These samples will already contain BJ loading dye and will have been prepared by your instructor to contain 8 µl of the RT-PCR control reaction plus 2 µl of 5x BJ dye. Load these control samples in the following order:

Lane 4:	10 µl "no template" control
Lane 5:	10 µl "no RT" control
Lane 6:	10 µl "DNA Contamination" control
Lane 7:	10 µl "PCR positive control"
Lanes 9 and 10:	Empty

Step 6 Run the gel at 70 V for 1 to 2 hours.

Step 7 Wear gloves while handling your gel. After electrophoresis, disconnect the leads from the power supply. Remove the lid from the gel box, and cut a corner of the gel in an asymmetric manner to orient it.

Step 8 Carefully stain the gel for 8 to 10 minutes in an ethidium bromide bath using a spatula to transfer the gel from the electrophoresis box to the gel stain container. Your instructor will demonstrate this. Using the spatula, transfer the gel to a water bath (to destain the gel) for 15 minutes minimum (30–60 minutes is better).

PROCEDURAL NOTE

Appendix IV describes an alternative DNA stain, SYBR Safe (Molecular Probes).

»»

Step 9 Examine your photograph. It should appear similar to **Figure 3-4**. Do you have a product in the RT-PCR lane? Is it the correct size? Is there a similar product in the positive control lane? Is there an absence of this product in the negative control lanes (lanes d and e)? Can you conclude that your data demonstrate that the *Ttr* gene is expressed in liver tissue?

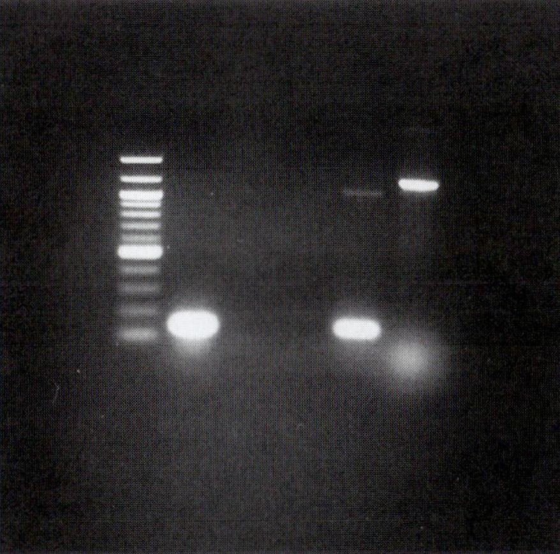

Fig 3-4 Expected RT-PCR gel image.

CAUTION!

Wear gloves, lab coats, and safety glasses. Handle these gels with extreme care! They are slippery. Support the entire gel with the spatula to keep it from breaking.

Ethidium bromide is a known mutagen and carcinogen. Wear gloves, lab coats, and safety glasses! Handle gels carefully! Do not splash ethidium bromide! Always rinse the spatula in the water destain bath after it has been in contact with ethidium bromide! Always rinse the gel tray after sliding gels into the ethidium bromide bath. You may unknowingly contact the ethidium bromide! Be very careful not to drip the ethidium bromide anywhere!

Handle these gels with extreme care! Support the entire gel with a spatula to keep it from breaking.

Be careful not to cause unwanted contact with the ethidium bromide DNA gel stain by splashing or spreading this solution unnecessarily — always rinse the spatula and gel tray with water after each use.

MODULE MATERIALS AND REAGENTS (PER PAIR OF STUDENTS)

MODULE III.1

Frozen mouse liver tissue (1 g) (American Bioanalytical)
Buffer RLT (containing β2-mercaptoethanol) (15 ml) (Qiagen)
70% ethanol (made with DEPC ddH$_2$O) (25 ml). See Appendix II.
RNA column (1) (Qiagen)
Buffer RW1 (15 ml) (Qiagen)
Wash Buffer RPE plus ethanol (20 ml) (Qiagen)
(4 ml RPE concentrate, 16 ml 100% ethanol) (Qiagen)
RNase-free ddH$_2$O (2.5 ml) (Qiagen)
3-M NaOAc (pH 7.0) (45 µl). See Appenidix II.
Ice-cold 100% EtOH (2 ml) (American Bioanalytical)
DEPC-treated ddH$_2$O (2 ml). See Appendix IV.

Mouse liver total RNA (optional) (250 µl containing 400 µg) (American Bioanalytical)
Micro Poly (A) Purist kit (Ambion)
250 µl of 2x Binding Solution (250 µl)
Wash Solution 1 (1 ml)
RNA spin column (1)
Wash Solution II (1 ml)
Warm (70°C) Elution Buffer (RNA Storage Buffer) (200 µl)
Glycogen carrier (2 µl)
5 M ammonium acetate (20 µl)
Northern gel loading dye (30 µl). See Appendix II.
Mouse spleen poly A+ RNA in Northern gel loading dye (0.1 µg in 15 µl) (American Bioanalytical)
1 g agarose (American Bioanalytical)

1x MOPS/EDTA Buffer (600 ml). See Appendix II.
37% formaldehyde (5.4 ml) (American Bioanalytical)
Ethidium bromide staining solution. See Appendix II.
Water destain bath
Gel photography set up
Ultrasonic tissue homogenizer (Polytron or similar instrument)
RNAse-free centrifuge tubes, tips, and pipettes. See Appendix II.
Gel electrophoresis apparatus, power supply
Flame-drawn Pasteur pipettes (2)
Nanofuge
Microfuge
Tabletop centrifuge

MODULE III.2

DEPC-treated H$_2$O (50 ml). See Appendix II.
10x SSC (600 ml). See Appendix II.
6x SSC (500 ml). See Appendix II.
Mouse *Ttr* DNA PCR Product (25 ng in 34 µl) (American Bioanalytical)
NEBlot Phototope™ Labeling kit (American Bioanalytical)
5x Labeling Mix (buffer and primers) (10 µl)
dNTP Mix (1 mM each dNTP) (5 µl)
Klenow DNA Polymerase (5 units) (1 µl)
0.2 M EDTA, pH 8.0 (5 µl). See Appendix II.
4 M LiCl (5 µl). See Appendix II.
Ice-cold 100 % EtOH (100 µl) (American Bioanalytical)
Ice-cold 70 % EtOH (500 µl) (American Bioanalytical)
1x TE (20 µl). See Appendix II.
Northern Prehybridization Solution (6 ml). See Appendix II.

Northern Hybridization Solution (6.1 ml). See Appendix II.
Northern Wash Solution I (1 liter). See Appendix II.
Northern Wash Solution II (1 liter). See Appendix II.
Phototope™ Blocking Solution (36 ml). See Appendix II.
Streptavidin Solution (4 ml). See Appendix II. Phototope™ Detection Wash Solution I (60 ml). See Appendix II.
Alkaline phosphatase solution (4 ml). See Appendix II.
Phototope™ Detection Wash Solution II (60 ml). See Appendix II.
Phototype Detection Solution (4 ml). See Appendix II.
Plastic containers
Plastic wrap

Blot weight of about 500 g
Nylon membrane (1) (Immobilon™; Millipore)
Whatman 3MM™ paper
Paper towels
UV cross-linker or baking oven
Lid locks
Flame-drawn pipette (1)
X-ray film (Kodak BioMax™)
Darkroom developer/fixer (Kodak GBX)
X-ray film cassette
Sealable hybridization bags with spouts (2)
UV transilluminator
Heat block
Marking pen
Dialysis clip
Rocking platform
Heat sealer (for plastic bags)
Vortex mixer

MODULE III.3

PCR RNA sample (from Lab Period III.1.2)
RT-PCR Buffer (49 µl) (GeneAmp™ Gold RNA PCR Reagant Kit, Applied Biosystems)
Ttr Forward Primer (American Bioanalytical)
Ttr Reverse Primer (American Bioanalytical)
Agarose (1.8g) (American Bioanalytical)
1x TAE Buffer (600 ml). See Appendix II.
100-bp marker in 5x BJ (10 µl). See Appendix III.

5x Blue Juice loading dye (5x BJ) (2 µl). See Appendix III.
No-template control for RT-PCR reaction (10 µl). See Appendix II.
No-RT control for RT-PCR reaction (10 µl). See Appendix II.
DNA contamination control for RT-PCR reaction (10 µl). See Appendix II.
PCR-positive control for RT-PCR reaction (10 µl). See Appendix II.

Agarose gel apparatus, power supply
Ethidium bromide solution
Staining/destaining trays
Photography Unit
Spatula
Mouse liver genomic DNA (75 ng in 1 µl) (American Bioanalytical)
cDNA clone of *Ttr* gene (2 ng in 1 µl) (American Bioanalytical)

Study Questions for Project III

Module III.1

1. In Lab Period III.1.1, what were your OD readings at 260 nm and at 280 nm? What is your calculated concentration of RNA in your sample? What is the estimated total amount of RNA that you recovered?

2. In Lab Period III.1.1, what was your 260/280 ratio? Is this an acceptable ratio for an RNA preparation? What is the likely problem if your 260/280 ratio is less than 1.9?

3. In Lab Period III.1.2, what is the purpose of the oligo-dT cellulose, and how does it enable the separation of mRNA (poly A+ RNA) from the poly A− RNA?

4. In Lab Period III.1.3, what is the reason for adding the Northern loading dye to the RNA sample but *not* pipetting up and down to mix?

5. In Lab Period III.1.4, why is MOPS buffer used instead of the TAE buffer used for DNA gels?

6. In Lab Period III.1.4, were you successful in enriching for poly A+ mRNA from the total RNA? Explain your answer.

7. In Lab Period III.1.4, were you able to visualize the two ribosomal RNA bands in the Liver Total RNA lane? Could you see these bands in the two poly A+ RNA samples? What does this indicate? Hint: Remember that the poly A+ isolation procedure *enriches* for poly A+ RNA.

8. Explain why fractionating poly A+ RNA is potentially beneficial before performing a Northern blot analysis to detect a specific mRNA.

9. What is your estimate for the range of molecular sizes of your RNA molecules after electrophoresis (use your two rRNA bands as markers)?

10. Explain why your undigested mouse genomic DNA molecules (in Project I) were of high molecular weight (> 20 kb), whereas your undigested RNA molecules were of significantly lower molecular weight.

Module III.2

1. In Lab Period III.2.1, why do you place the gel "face down" for the transfer?

2. In Lab Period III.2.2, what is the purpose of doing both the cross-linking and the baking steps?

3. In Lab Period III.2.3, why are the random primers labeled with biotin?

4. In Lab Period III.2.4, how does the addition of EDTA stop the labeling reaction?

5. In Lab Period III.2.5, what is the purpose of the Denhardt's solution in the prehybridization buffer?

6. In Lab Period III.2.6, what is the purpose of the formamide in the hybridization solution? How does this relate to the temperature of hybridization?

7. In Lab Period III.2.7, what is the purpose of adding the Streptavidin solution in Step 10?

8. In Lab Period III.2.7, what is the purpose of adding the alkaline phosphatase solution in Step 11?

9. Describe how the chemiluminescent reaction results in the emission of light that exposes the film.

10. Did you successfully transfer your RNA samples from your gel to the blotting membrane?

11. Were you successful in nonradioactively labeling the *Ttr* gene probe?

12. Did you specifically detect the *Ttr* sequence in any of your RNA samples on your Northern blot?

13. What is the molecular size of the *Ttr* RNA sequence you detected?

14. Did you observe a difference in the intensity of the band representing the *Ttr* sequence in the total and poly A+ samples of RNA on your Northern blot? If so, explain why you think this difference appeared.

15. What can you conclude about the expression of the *Ttr* gene in mouse liver and spleen tissue?

16. What can you conclude about whether or not the *Ttr* transcript is polyadenylated?

Module III.3

1. In Lab Period III.3.1, Step 3, why do you think more of the *Ttr* reverse primer is added than the *Ttr* forward primer?

2. In Lab Period III.3.1, Step 4.A, what is the purpose of the 25°C step done for 10 minutes?

3. In Lab Period III.3.1, Step 4.D, why does a two-step PCR work when usually PCR is done with three steps (denaturing, annealing, and synthesis)?

4. In Lab Period III.3.1, what is the purpose of the "no RT" control (i.e., what does this control show)?

5. In Lab Period III.3.2, what does your "DNA Contamination" control show?

6. Did you succeed in amplifying a PCR product representing *Ttr* RNA?

7. What was the size of the PCR product you obtained representing *Ttr* RNA? Is this the size you expected?

8. Is there a similar-sized product in the positive control lane?

9. Is there an absence of this product in the negative control lanes (lanes 5 and 6)?

10. Explain any of your results that are unexpected.

11. Why is the size of the RT-PCR product representing *Ttr* RNA different than the size of the band of *Ttr* RNA you detected on your Northern blot?

12. What evidence do you have that your RT-PCR product actually represents *Ttr* RNA?

13. Why is it necessary to use RT-PCR instead of regular "DNA-PCR" to detect a specific sequence in RNA?

IV

cDNA CLONING AND cDNA LIBRARY ANALYSIS

PROJECT SUMMARY

In the series of experiments described in **Module IV.1**, you will construct a unidirectional cDNA library from mouse liver poly A+ mRNA in the *Escherichia coli* bacteriophage lambda expression vector Uni-ZAP XR (Stratagene). You will synthesize double-stranded cDNA from mouse liver poly A+ mRNA, ligate the cDNA into the lambda vector, and then package and titer the cDNA library.

In the experiments described in **Module IV.2**, you will use the polymerase chain reaction (PCR) to determine the insert size of several clones picked at random from the cDNA library in order to estimate the average size of the cDNA inserts in your library. Finally, you will also use PCR to specifically detect the cDNA derived from the *Ttr* gene in your cDNA library. The *Ttr* gene is expressed in mouse liver and should therefore be represented in the cDNA library you construct from this tissue.

PROJECT BACKGROUND

Although it is relatively easy and straightforward to make a genomic library (Project II), it is often difficult to isolate a clone of a unique, single-copy gene from such a library. This is because a genomic library is very complex, with representation of all genes as well as all of the noncoding DNA sequences in the genome. Thus, clones of any single-copy gene are present at an extremely low frequency in a genomic library. The use of a cDNA ("copy" or "complementary" DNA) library affords several potential advantages for the initial isolation of a clone of a specific gene of interest. A cDNA library is made from mRNA and includes clones of *only* those sequences represented among the transcribed genes expressed in the tissue from which the mRNA was isolated. Thus, a cDNA library is a "snapshot" collection of those mRNAs that were present in the cells or tissue at the time the mRNA was recovered. For eukaryotic cells, a cDNA library is typically much less complex than a genomic library. Therefore, a cDNA library made from a tissue in which your gene of interest is known to be expressed will contain clones of that gene at a much higher frequency than would be found in a genomic library. This increases your chances of identifying a clone of your particular gene of interest in the library. Of course, it is critical that the cDNA library you use be made from mRNA isolated from a tissue in which your gene of interest is expressed to ensure that clones of this gene will be included in the library.

Once constructed, a cDNA library can be screened by hybridization with a nucleic acid probe in the same way you screened your genomic library in Project II. This requires an available nucleic acid probe with sufficient sequence similarity to your gene of interest to facilitate specific hybridization under stringent conditions. If such a probe is available, this is one of the most common methods for screening a cDNA library. Another common method is to use specific DNA sequence information for your gene of interest (or from a closely related gene) to design primers to PCR amplify your sequence of interest from the cDNA library (you will use this method in Module IV.2). However, such nucleic acid probes and/or DNA sequence information are not always available. Another significant advantage of cDNA libraries is that they can be screened in ways that do not require a nucleic acid probe or sequence information. By constructing the cDNA library in a specially designed "expression vector," the library can be plated on a bacterial lawn such that each inserted cDNA sequence will be transcribed into mRNA and then translated to produce the encoded protein. In this way, a specific clone can be detected on the basis of the product it encodes (e.g., using an antibody to detect an encoded protein) in addition to detecting it on the basis of its nucleic acid sequence alone.

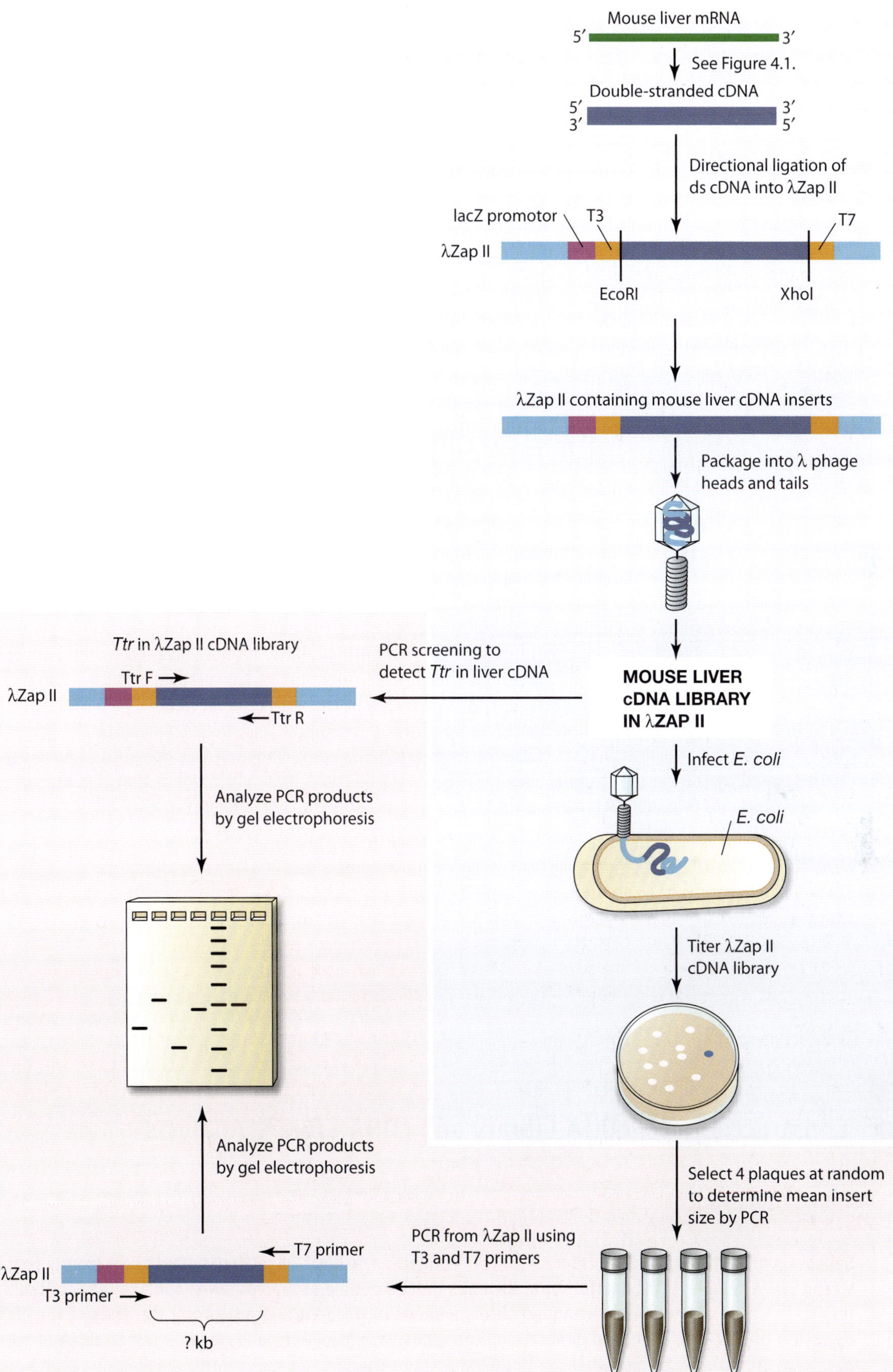

Project IV flow diagram.

BACKGROUND, continued

PROCEDURAL NOTE

This requires that the transcribed mRNA contains an appropriate open reading frame (ORF) so that it can be translated into the encoded protein by the host bacterial cells. This affords the opportunity to use an antibody probe or other method to specifically detect the encoded protein product to identify the specific cDNA clone containing your gene of interest in the library.

The construction of a cDNA library also differs significantly from that of a genomic DNA library. As noted previously, a cDNA library is derived from mRNA recovered from a particular tissue or cell type. The standard protocol for construction of a cDNA library involves a series of enzymatic reactions, each designed to carry out one part of the process required to copy mRNA into single-stranded (or "first-strand") cDNA, synthesize a complementary second strand of cDNA on each molecule to produce double-stranded cDNA, prepare the ends of the double-stranded cDNA molecules for ligation to the lambda phage vector arms, and ligate the cDNA molecules into the lambda vector arms, all of which are required to produce recombinant lambda genomes containing cDNA inserts that can then be packaged into functional, infectious lambda bacteriophage to produce the cDNA library. Particularly important considerations for the construction of a high-quality cDNA library include the following: (1) efficiently copying as much of the initial mRNA into cDNA as possible to produce the largest, most representative cDNA library possible and (2) producing "full-length" cDNA replicas of each original mRNA molecule. A large library is desirable because it will maximize the likelihood of inclusion of cDNA clones of all mRNAs in the initial sample, including those mRNAs that were present in particularly low abundance in the original mRNA sample. Full-length cDNAs are desirable because they will include all parts of the transcribed sequence of each gene, and this maximizes the information that can be obtained from each cDNA as well as the potential utility of each cDNA for subsequent studies. A concern about producing a large cDNA library is that the multiple step process typically involved in synthesis of double-strand cDNA requires recovery and purification of product after each step, and this leads to significant loss of material throughout this process. To minimize this problem, you will use the Uni-ZAP cDNA Synthesis Kit from Stratagene. An advantage of this kit is that several steps have been combined to minimize the number of recoveries and purifications required and, hence, maximize the potential yield. A second advantage is that all of the necessary reagents are provided in this one kit and have been quality tested by the manufacturer to ensure their compatibility. A final advantage is that this kit will allow you to produce double-stranded cDNA that can be unidirectionally cloned into a lambda expression vector (see below).

MODULE IV.1

Construction of a cDNA Library and cDNA Library Analysis

MODULE SUMMARY

Synthesis of first-strand cDNA from an mRNA template, synthesis of second-strand cDNA from a first-strand cDNA template, production of blunt ends on the double-stranded cDNA, addition of EcoRI adaptors to the ends of the double-stranded cDNA, removal of phosphate groups from the ends of the cDNA, digestion of cDNA with XhoI to generate unique ends, size selection, and purification of cDNA of more than 400 bp, unidirectional ligation of cDNA inserts to the lambda vector arms, packaging recombinant lambda genomes into infectious phage particles, and plating and titering the lambda cDNA library.

MODULE BACKGROUND

In this module, you will use an aliquot of mouse liver poly A+ mRNA that you prepared as part of Project III (or that will be provided to you) to generate a mouse liver cDNA library. The complete process that you will use to convert single-stranded poly A+ mRNA into double-stranded cDNA ready to be unidirectionally cloned into a lambda expression vector is shown as a 12-step process in **Figure 4–1**. The production of "full-length" double-stranded cDNA copies of mRNAs requires the complete, high-fidelity synthesis of a first-strand cDNA copy of the single-stranded mRNA and the complete, high-fidelity synthesis of a second-strand cDNA copy of the first-strand cDNA. In this module, first-strand cDNA will be synthesized from poly A+ mRNA using an "oligo-dT" primer (Fig. 4-1, step 1). This primer is a single-stranded oligonucleotide consisting of approximately 13 Ts that will hybridize with the poly A+ tail of the mRNA to initiate cDNA synthesis from the 3'-end of each mRNA. Reverse transcriptase (RT), which synthesizes DNA from an RNA template, is then used to copy the mRNA into first-strand cDNA. At this point, you will have a set of mRNA-cDNA "heteroduplex" molecules. To remove the mRNA strand from each of these and to synthesize simultaneously a second strand of cDNA that is complementary to each first strand, you will use a combination of enzymes including RNase H and DNA pol I (Fig. 4-1, step 2). The RNase H acts as an RNA-specific endonuclease to introduce "nicks" in the RNA strand at random locations. A nick is a cleaved phosphodiester bond that connects two adjacent nucleotides. Once this process begins, exposed 3'-ends of the RNA strand act as primers for DNA synthesis by DNA polymerase I. DNA polymerase I carries out the dual activity of degrading the RNA 5' to 3' while simultaneously synthesizing DNA 5' to 3'. In this way, all of the RNA is eventually digested and replaced with second-string cDNA complementary to the first-strand cDNA. However, because this second strand is synthesized in pieces, nicks remain in the phosphodiester backbone of this strand. T4 DNA ligase (which actually is not added until step 7 in Fig. 4-1) will act to seal these nicks, yielding a continuous second strand of cDNA. This will yield a set of double-stranded homoduplex cDNA molecules.

The sequence on one end of each cDNA molecule will consist of a string of Ts hybridized to a string of As; however, the sequence on the other end of each molecule will be unique, reflecting the different sequence at the 5'-end of each of the thousands of different mRNAs that acted as templates for cDNA synthesis. To facilitate ligation of each of these cDNAs into a single vector, you must modify the ends of these molecules so that they all share the same ends. To maximize the efficiency of ligation of cDNAs to the lambda vector arms, it is best to have complementary "sticky" ends (typically a four-base overhang) on each end of the molecules to be ligated. If orientation of the insert cDNA relative to the lambda vector arms is not a consideration, then the same sticky ends can be added to both ends of each cDNA. As noted above, however, for purposes of subsequent detection of specific cDNA clones based on the identity of the encoded protein product, it is desirable to clone the cDNAs into an "expression vector" such that the cDNAs can be transcribed and translated to produce the encoded protein. This is achieved by ligating the cDNAs to lambda vector arms, one of which has a promoter that can direct transcription of the cloned cDNA in *E. coli*. This will only produce the correct encoded protein if the cDNA is inserted in the proper orientation relative to this promoter such that the "sense" or "coding" strand of the cDNA will be transcribed rather than the "nonsense" or "noncoding" strand. To ensure proper orientation of the cDNA, it is preferable to perform "unidirectional cloning." This requires different sticky ends on each end of the cDNA molecules that will be compatible with complementary, unique sticky ends on each lambda vector arm.

To prepare the ends of your cDNAs for unidirectional ligation into the lambda vector arms, you will first "blunt" or "polish" the ends (remove any single-strand overhangs) to leave perfectly blunt ends on both ends of each double-stranded cDNA. This is done by treating with *Pfu* polymerase, which has 5'-3' synthesis activity and 3'-5' single-strand exonuclease activity (Fig. 4-1, step 3). Typically, during a multiple step synthetic process such as you are using in this project to produce cDNA from mRNA, it is necessary to purify and recover the product after each reaction. This presents a serious potential disadvantage in that a loss of material is usually incurred with each recovery and purification. A significant advantage of the kit that you are using to synthesize cDNA is that the manufacturer has deduced sequential, compatible buffer conditions for steps 1 to 3 in Figure 4-1 so that purification and recovery of product are not necessary prior to this point. After step 3, however, you will carry out an organic extraction and ethanol precipitation (Fig. 4-1, steps 4 and 5).

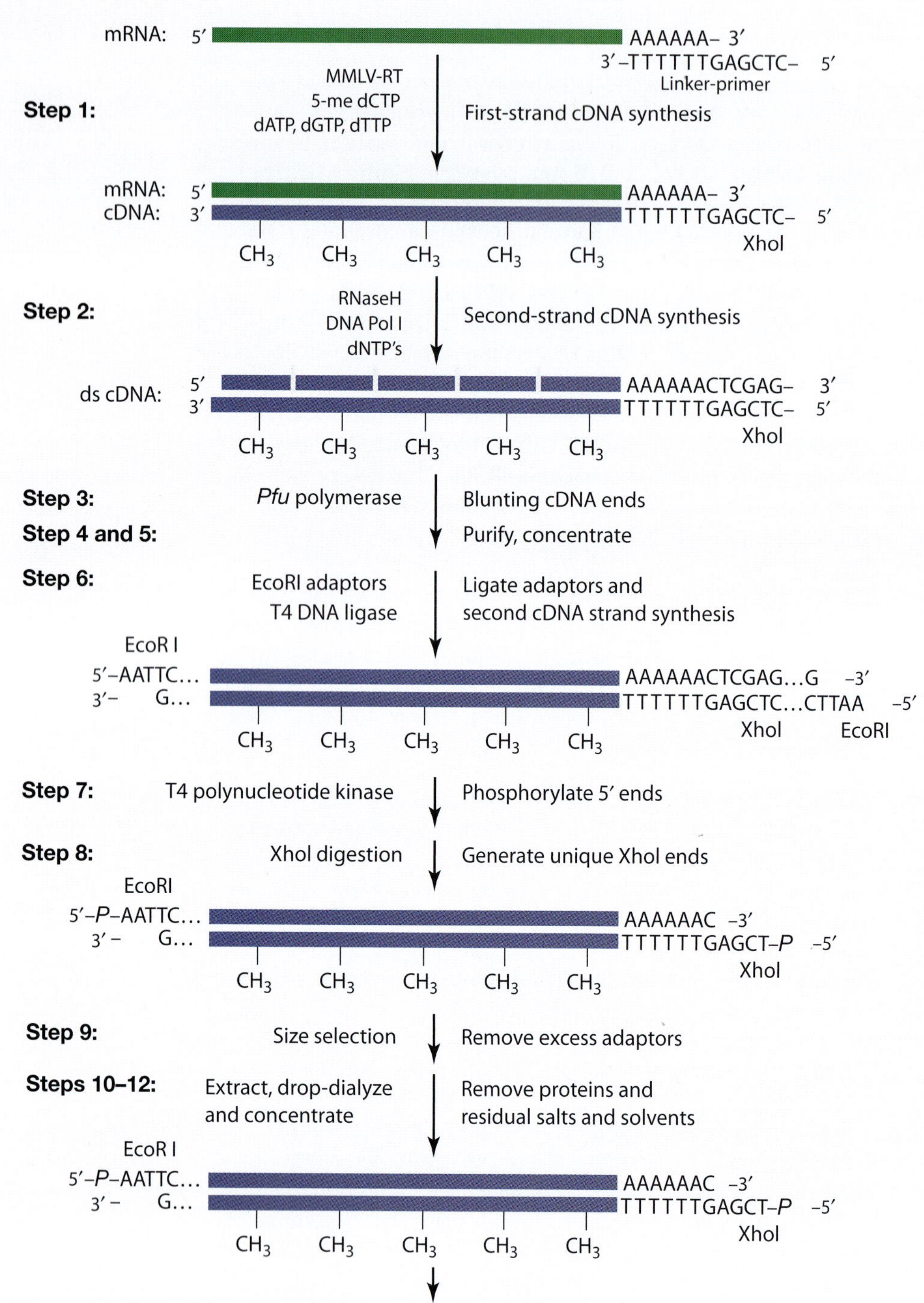

Fig 4-1 Multiple steps in the synthesis of cDNA.

Twelve different steps are shown for the synthesis of double-stranded cDNA from first-strand synthesis to ligation to the bacteriophage lambda cloning vector. Step 1: synthesis of first-strand cDNA from the mRNA template. Step 2: elimination of the mRNA with RNase H/DNA pol I and simultaneous synthesis of second-strand cDNA with DNA pol I. Step 3: blunting of ends of ds cDNA with *PFU* polymerase. Step 4: extraction of ds cDNA with organic solvents. Step 5: ethanol precipitation of ds cDNA. Step 6: ligation of EcoRI adaptors to the blunt ends of the ds cDNA. Step 7: phosphorylation of 5' ends with T4 polynucleotide kinase. Step 8: digestion of ds cDNA with XhoI to create a unique sticky end on the downstream end of each ds cDNA. Step 9: size selection of ds cDNA molecules to exclude molecules less than 400 bp in length. Step 10: extraction of ds cDNA with organic solvents. Step 11: drop dialysis of ds cDNA. Step 12: concentration of ds cDNA to 5 μl.

After the purification and recovery, you will resuspend your cDNA in fresh buffer and then ligate special "adaptor" molecules (also provided in the kit) to each end of your cDNA molecules (Fig. 4-1, step 6). Each adaptor molecule is a double-stranded DNA molecule with one blunt end that can be ligated to a blunt end of your cDNA, and one EcoRI sticky end (a 5'-overhang that is compatible with a similar 5'-overhang on one of the lambda vector arms). After this ligation is completed, you will add phosphate groups to the EcoRI adaptors on the ends of your cDNA molecules by treating with T4 polynucleotide kinase (Fig. 4-1, step 7) (this will allow the cDNAs to ligate to the vector arms during the vector-insert ligation step). At this point, your cDNAs will have "sticky" EcoRI 5'-overhangs on both ends. However, to facilitate unidirectional ligation of each cDNA into the lambda vector arms, you will digest your cDNAs with XhoI (Fig. 4-1, step 8). This will cleave a special "linker" sequence that was included on the 5'-end of the oligo dT primer you used for first-strand cDNA synthesis (Fig. 4-1, step 1) and was, therefore, incorporated onto one end of each cDNA molecule you produced. By cleaving this with XhoI, you will produce a unique XhoI-5'-overhang on one of each cDNA molecule while leaving the EcoRI-5'-overhang on the upstream end of each molecule. The upstream lambda vector arm (the one with the promoter) has an EcoRI-5'-overhang, while the downstream arm has an XhoI-5'-overhang. Thus, because of these unique ends, there is only one (unidirectional) orientation in which the insert cDNA can be ligated between the two lambda vector arms (**Figure 4-2**).

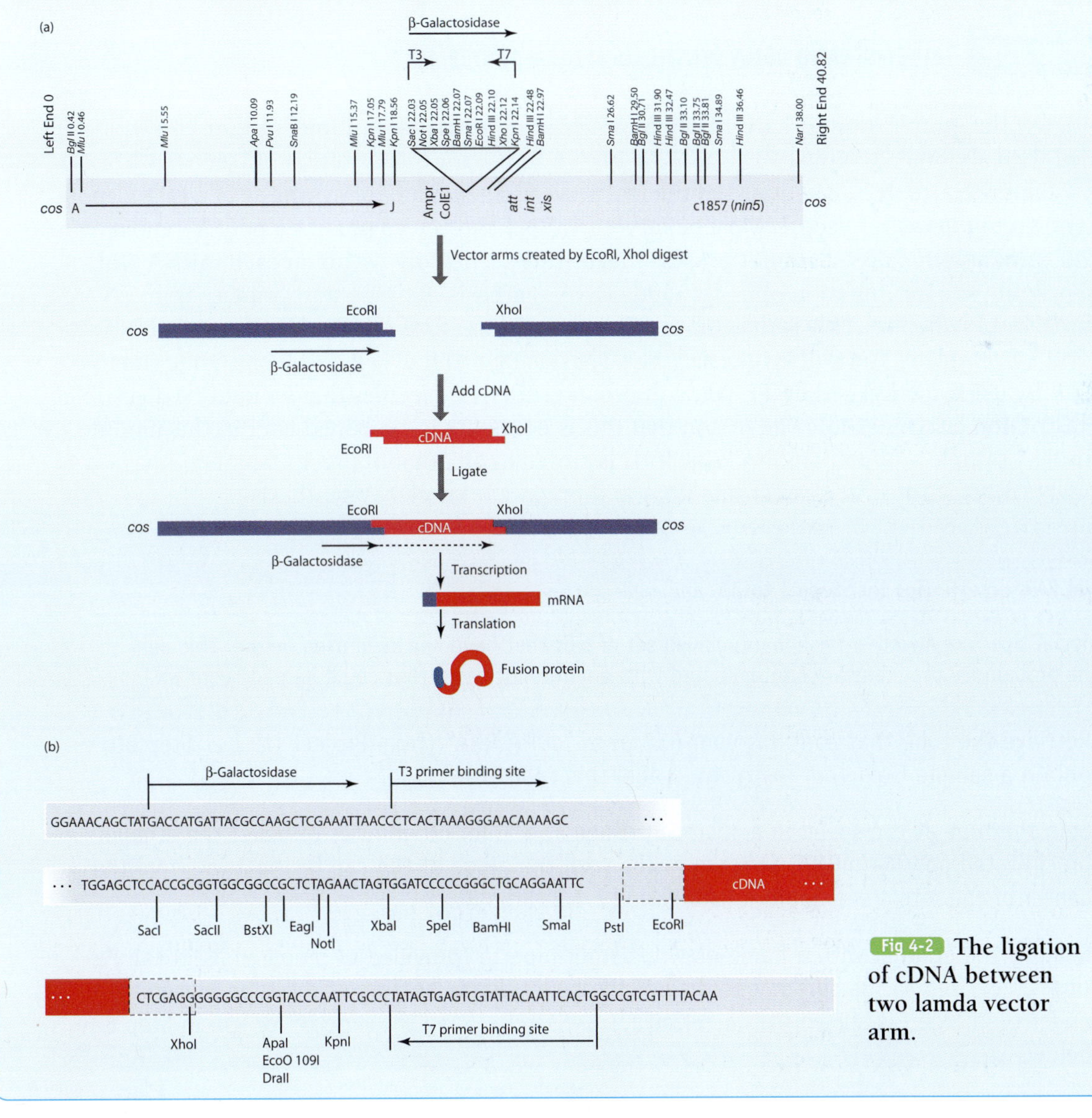

Fig 4-2 The ligation of cDNA between two lamda vector arm.

MODULE BACKGROUND, cont.

Before you perform this ligation, however, you will pass your population of newly synthesized cDNA molecules over a size-selection column to remove partial cDNAs and any excess adaptor molecules (Fig. 4-1, step 9) and then purify the cDNA by drop dialysis and concentrate it by speed-vacuum evaporation so that it is ready for ligation to the lambda vector arms (Fig. 4-1, steps 10–12). The ligation and packaging of the cDNA inserts into the lambda vector arms and the subsequent titering of your library will be done using the same methods that you used with your genomic DNA library in Project II.

PROCEDURAL NOTE

Synthesis of cDNA is a multistep process that begins with mRNA. Care is required throughout this process to protect against degradation of the mRNA template and/or loss of material during any of the steps of cDNA synthesis. With 12 steps in this process, as little as a 10% loss at each step would reduce your yield of cDNA to essentially nothing! Follow all instructions carefully. Keep reagents on ice. Use special tips and tubes, and wear gloves when instructed to do so.

LAB PERIOD IV.1.1.   FIRST-STRAND cDNA SYNTHESIS (FIG. 4-1, STEP 1)

YOU WILL use the sample of mouse liver poly A+ mRNA you prepared in Project III, or you will be provided with a sample of mouse liver poly A+ mRNA to use for this experiment (see Step 1 of Fig. 4-1). Each mRNA molecule in your sample will serve as a template for synthesis of a cDNA molecule, as shown in Figure 4-1, step 1. Most eukaryotic mRNA molecules possess a poly A+ tail at their 3' end (prokaryotic mRNAs do not possess this feature). The poly A+ tail in each mRNA will serve as a priming site for initiation of first-strand cDNA synthesis. Reverse transcriptase is an RNA-dependent DNA polymerase. This is the enzyme that you will use to synthesize first strand cDNA from the mRNA template. You will synthesize the cDNA using dATP, dGTP, dTTP, and 5-me (methylated) dCTP. By using methylated dCTP, you will produce cDNA molecules that will not be digestible with the restriction enzyme XhoI. The reason that this is necessary is explained later in this module.

The mRNA sample you prepared in Project III is stored as an ethanol precipitate in a tube you labeled "cDNA RNA." You will recover this RNA and prepare it for cDNA synthesis.

PROCEDURAL NOTES

Use special RNA pipette tips for Steps 1 to 20, and wear gloves!

These "special tips" are merely a freshly autoclaved set of tips that have not been used before. This decreases the probability that they are contaminated with any ribonucleases that could degrade your RNA.

Step 1 Retrieve the tube that contains your 637 µl of "cDNA RNA" (from Project III,1.2, Step 20) or obtain a sample from your instructor. Label this tube "#1."

Step 2 Spin the tube at top speed in a microcentrifuge at 4°C for 60 minutes to pellet the ethanol-precipitated poly A+ mRNA. Spin your tube with the hinge of the cap facing away from the center of the rotor (see figure in Lab Period I.1.6, Step 3).

Step 3 Remove the ethanol with a flame-drawn Pasteur pipette (provided or prepared as in Lab Period I.1.6, Step 4). Be sure not to disrupt the pellet at the bottom of the tube immediately under the hinge (as shown in Lab Period I.1.6, Step 5). Save the pipette for use in Step 6 below.

»»

LAB PERIOD IV.1.1 CONT.

Step 4   Wash the pellet by adding 500 µl 70% ice-cold ethanol (EtOH). Rock the tube *gently* back and forth for 15 seconds — this step is designed to wash, not disrupt, the pellet.

Step 5   Place the tube in the refrigerated microcentrifuge with the hinge pointing away from the center of the rotor. Spin at top speed for 5 minutes (this will ensure that the pellet is firmly in place on the bottom of the tube).

Step 6   Remove the ethanol with the same flame-drawn Pasteur pipette (from Step 3 above), making sure not to disturb the pellet, which will be under the hinge side of the microcentrifuge tube.

Step 7   After removing the contents, centrifuge the tube again for 30 seconds in the microcentrifuge. This will bring any residual ethanol to the bottom of the tube.

Step 8   Remove the residual ethanol with the same flame-drawn pipette (from Step 6 above).

Step 9   Place the *open* tube on ice for 5 minutes to air dry. During this time, any trace amounts of ethanol should evaporate.

Step 10   Label a new 1.5-ml microcentrifuge "tube #2." This will be your "first-strand synthesis tube." You will also be provided with a tube of methylated dNTP mix and a tube of linker-primer.

Step 11   Prepare the first-strand cDNA reaction mix by adding the following components to your #2 microfuge tube:

    5.0 µl 10x First-Strand buffer (Stratagene buffer 1)

    3.0 µl methylated dNTP mix

    2.0 µl linker-primer (0.4 µg/µl)

PROCEDURAL NOTES

The linker-primer sequence is 5'-(GA)$_{10}$ACTAGT**CTCGAG**(T)$_{13}$-3'. (The sequence in bold is an XhoI restriction enzyme recognition sequence that will be important later in this module.) Note that this sequence is shown in an abbreviated form in Figure 4-1.

At this point, a molecule of linker-primer will be hybridized to the poly A+ tail of each mRNA molecule and will be used to prime First-Strand cDNA Synthesis.

Step 12   Mix the contents of tube 2 by tapping the tube on the bench top or flicking the tube with your finger.

Step 13   Spin the tube for 5 seconds in your nanofuge to be sure that all of the liquid is at the bottom of the tube.

Step 14   Place this tube on ice.

Step 15   You will also be given a microfuge tube containing 1.0 µl RNase inhibitor (40 U/µl). Label this "tube #3."

Step 16   Add 37.5 µl of diethylpyrocarbonate (DEPC)-treated ddH$_2$O to tube 3 (for a description of the use of DEPC as an RNase inhibitor, see Project III).

»»»

LAB PERIOD IV.1.1 CONT.

Step 17 Pipette all of the solution in tube 3 into tube 1 (the tube containing your poly A+ mRNA pellet). Let tube 1 stand on ice for 5 minutes to soften the pellet. This will help prevent loss of the pellet caused by it sticking inside the pipette tip. After 5 minutes, resuspend the pellet in tube #1 by pipetting up and down for 1 minute.

Step 18 Add the 10 µl of first-strand reaction mix (from tube 2) to tube 1 (now containing your resuspended mRNA), and mix it with your pipette tip. Discard tube #2. Spin tube 1 for 5 seconds in your microcentrifuge.

Step 19 Place tube 1 in a rack on your bench, and let it stand at room temperature for 10 minutes. During this time the oligo d(T) linker-primer will anneal to the poly A+ tails of the mRNAs.

Step 20 After the 10-minute annealing period, add 1.5 µl of reverse transcriptase (50 units/µl). Mix briefly with your pipette tip, and then incubate your tube at 37°C for 1 hour. Synthesis of first-strand cDNA will occur during this time.

Step 21 When First-Strand Synthesis is complete, remove your tube from the 37°C water bath. Spin it briefly in your nanofuge, and place it at –20°C.

PROCEDURAL NOTE

If necessary, your instructor can do Step 21. If your schedule enables you to proceed directly to Second-Strand Synthesis, place the tube on ice while setting up the reagents for second-strand synthesis.

LAB PERIOD IV.1.2.   **SECOND-STRAND cDNA SYNTHESIS (FIG. 4-1, STEP 2)**

Step 1 Remove your First-Strand Synthesis tube (tube 1) from storage at –20°C and place it on ice.

Step 2 When the contents of the tube are defrosted, add the following Second-Strand Synthesis components:

PROCEDURAL NOTE

These components and your tube containing First-Strand cDNA should be kept on ice at all times!

> 20.0 µl 10x Second-Strand buffer (Stratagene buffer 2)
>
> 6.0 µl dNTP mix
>
> 116.0 µl DEPC–ddH$_2$O

Step 3 Mix by pipetting up and down, spin the tube for 5 seconds in the nanofuge, and then add the following components:

> 2.0 µl RNase H (1.5 U/µl)
>
> 11.0 µl DNA polymerase I (9 U/µl)

»»

LAB PERIOD IV.1.2 CONT.

PROCEDURAL NOTE

RNase H nicks the mRNA strand at various points and continues to slowly digest the mRNA to generate gaps in this strand. DNA polymerase I uses the 3'-ends of the remaining RNA as primers and fills in these gaps with DNA. DNA polymerase I can also remove the RNA with a 5' to 3' exonuclease activity and replace it with its 5' to 3' DNA synthesis activity.

Step 4   Mix briefly with your pipette tip. Spin the tube for 5 seconds in the microcentrifuge, and place at 16°C for 2.5 hours.

PROCEDURAL NOTE

Either a thermal cycler or a refrigerated water bath can be used to provide a constant temperature of 16°C. Removal of the original RNA template and synthesis of the second strand of cDNA will occur during this time. This reaction is conducted at 16°C to ensure that it occurs gradually to ensure that the mRNA strand will not be digested too quickly.

Step 5   After the incubation, place the tubes at –20°C. (If necessary, this can be done by your instructor.)

LAB PERIOD IV.1.3.   **PRODUCTION OF BLUNT ENDS ON THE DOUBLE-STRANDED cDNA AND PURIFICATION OF THE DOUBLE-STRANDED cDNA**

PART A Production of Blunt Ends on the Double-Stranded cDNA (Fig. 4-1, Step 3)

To generate perfectly blunt ends on the ds cDNA, *Pfu* polymerase will be used to fill in or degrade any ragged ends using its 5'-3' polymerase activity and its 3'-5' exonuclease activity.

Step 1   Remove your cDNA synthesis tube (tube 1) from storage at –20°C, and allow it to defrost on ice.

Step 2   Add the following *ice-cold* components to the tube containing your ds cDNA.

PROCEDURAL NOTE

This reaction is designed to blunt the ragged ends of the cDNA, but if allowed to proceed too rapidly, it will cause unwanted degradation of your cDNA. Therefore, it is necessary to carefully control the temperature and time at which this reaction proceeds. Keep all of your tubes on ice, and do not hold these tubes for long periods in your 37°C hands!

    23.0 µl Blunting dNTP Mix

    2.0 µl *Pfu* polymerase (2 U/µl)

Step 3   Quickly mix the reaction components with your pipette tip, and spin the tube for 5 seconds in your nanofuge. Place the tube in a 72°C temp block for *exactly* 30 minutes.

PROCEDURAL NOTE

It is imperative not to exceed 30 minutes! Immediately place your tube on ice following the 30-minute incubation.

»»»

LAB PERIOD IV.1.3 CONT.

PART B Purification of the Double-Stranded cDNA (Fig. 4-1, Steps 4 and 5)

At this point, you need to get rid of the various buffer constituents that have accumulated from the previous reactions as well as any remaining enzymes and prepare your cDNA for resuspension in fresh buffer for the next series of reactions. To accomplish this, you will perform an organic extraction using phenol and chisam (24:1 chloroform:isoamyl alcohol), followed by precipitation in ethanol. To facilitate recovery of only the aqueous phase (containing the DNA) after the organic extraction, you will use a "phase lock gel" tube (PLG Light™, Eppendorf) as you may have done in Project II (Lab Period II.1.3). After centrifugation, the gel material in this tube will form an impermeable barrier between the organic phase (on the bottom of the tube) and the aqueous phase (at the top of the tube). This will enable you to easily recover the aqueous phase containing the cDNA without any contaminating organic material. You will then precipitate your cDNA in ethanol so that you can discard the old buffer constituents and resuspend your cDNA in fresh buffer.

Step 1 Your instructor will provide you with a green 1.5-ml phase lock gel tube containing the gel material. Write your number on this tube, and spin this tube for 30 seconds at maximum speed in a microcentrifuge (see figure in Lab Period II.1.3A, Step 1). This brings all of the gel material to the bottom of the tube.

Step 2 Spin the tube containing your ds cDNA (tube 1) briefly in a nanofuge to get all of the liquid to the bottom. Transfer the entire 225 µl ds cDNA reaction to the prespun green phase lock tube.

PROCEDURAL NOTES

Phenol is buffer saturated and is the lower phase in the tube or container. The upper phase is buffer and is clear. When pipetting phenol, you must go through the upper buffer phase with the pipette tip to reach the lower phenol phase. The phenol phase is yellow. When you pipette the phenol, if it is not yellow, you have probably pipetted only buffer.

Discard all organic waste (i.e., tubes and tips that contain or contained phenol) into the appropriate waste containers in the fume hood.

CAUTION!

The following steps require the use of phenol and chisam (24:1 chloroform:isoamyl alcohol vol/vol). Please note the following precautions.

Phenol and chloroform (in chisam) are both toxic and produce noxious fumes. Work in a fume hood! Because phenol and chisam are solvents that are caustic and will burn skin and clothes, be sure to always wear a lab coat and gloves. Keep tubes closed at all times. Open tubes containing phenol and/or chisam for brief periods only.

The tubes containing phenol and chisam are kept in the refrigerator and can be slippery due to condensation. Handle these tubes carefully to avoid dropping them.

Work in a fume hood!

Step 3 Add 200 µl buffered phenol to your green phase lock tube. Close the cap. *Hold the cap securely,* and mix by inverting the tube back and forth for 30 seconds.

»»

LAB PERIOD IV.1.3 CONT.

Step 4 Add 200 µl chisam to the green Phase Lock Tube. Close the cap. *Hold the cap securely*, and mix by inverting the tube back and forth for an additional 30 seconds.

Step 5 Spin this tube (and a balance tube) for 2 minutes in a microcentrifuge at maximum speed.

Step 6 Pipette the upper aqueous phase to a new 1.5-ml microcentrifuge tube labeled with your number and "ds cDNA ppt." Dispose of the organic material, including the phase lock tube, into the appropriate waste containers in a fume hood.

Step 7 To your "ds cDNA ppt" tube, add 20 µl 3 M sodium acetate (pH 5.2). Close the tube, and mix the contents by inverting several times.

Step 8 Add 400 µl ice-cold 100% ethanol. Close the tube, and mix by inverting several times.

Step 9 Place this tube at –20°C to allow the cDNA to precipitate (a minimum of 1 hour, but it can be stored indefinitely in this form).

LAB PERIOD IV.1.4.   **RECOVERY OF DOUBLE-STRANDED cDNA AND LIGATION OF EcoRI ADAPTORS**

PART A Recovery of Double-Stranded cDNA (Fig. 4-1, Step 5)

Step 1 Retrieve your tube containing your precipitated cDNA from storage at –20°C, and place it into a refrigerated microcentrifuge with the hinge oriented away from the center of the rotor. Centrifuge the tubes for 60 minutes at top speed.

Step 2 After your tube has been centrifuged, you may be able to see a pellet at the bottom of the tube on the side underneath the hinge. Use a flame-drawn Pasteur pipette to remove the ethanol — avoid disturbing the pellet! Save the flame-drawn Pasteur pipette in your rack at your bench for use in Step 5 below.

> **CAUTION!**
> Flame-drawn pipettes are very sharp! Handle with care!

Step 3 Wash the pellet by adding 500 µl 70% ethanol (add carefully to the side of the tube away from your pellet). Rock the tube *gently* several times. You do not need to dislodge the pellet. *Do not vortex!*

Step 4 Place the tube, with its hinge oriented away from the center of the rotor, back into the microcentrifuge. Centrifuge for 5 minutes at top speed.

Step 5 Remove the ethanol with the same flame-drawn Pasteur pipette used in Step 2. Do not dislodge the pellet! Save the pipette for use in Step 6 below.

Step 6 Spin the tube in your nanofuge for 30 seconds, and then remove any residual ethanol with the same flame-drawn Pasteur pipette use in Step 5. Place the open tube on ice for 10 minutes to let the pellet air dry.

PART B Ligation of EcoRI Adaptors (Fig. 4-1, Step 6)

You now have converted your single-stranded mRNA molecules into double-stranded cDNA molecules. Next, you must prepare these molecules for ligation into the lambda vector. Right now, each ds cDNA molecule has a similar sequence at the 3'-end but a unique sequence at the 5'-end. To add

»»

LAB PERIOD IV.1.4 CONT.

an EcoRI "sticky end" to the 5'-end of each molecule (and temporarily to the 3'-end of each molecule as well), you will now perform a "blunt-end ligation" procedure to ligate EcoRI adaptors onto the ends of your ds cDNA molecules. EcoRI adaptors are provided as part of the cDNA synthesis kit. They are short, uneven, pre-annealed double-stranded oligonucleotides with the following sequence:

5'-AATTCGGCACGAGG-3'
   3'-GCCGTGCTCC-5'

The key feature is the four-base AATT, 5'-overhang that represents an EcoRI sticky end that will be compatible with a complementary EcoRI sticky end on one of the lambda vector arms. Importantly, in order to ensure that one adaptor will become ligated to each end of your cDNA molecules (which is what you want to happen) and to avoid ligation of the adaptors to each other (which you do not want to happen), the adaptors have no phosphate group at the 5' overhang end but do have a phosphate group at the blunt 5' end. Also, to maximize the likelihood that you will ligate one adaptor to each end of every cDNA, you will add an excess of adaptor molecules.

PROCEDURAL NOTE

Later you will perform a size-selection procedure to get rid of the excess adaptor molecules that fail to ligate to cDNA molecules.

Step 1 Pipette 9.0 µl EcoRI adaptors into the tube containing the dry cDNA pellet. *Do not* pipette up and down. Let the tube stand on ice for 15 minutes to "soften" the pellet.

Step 2 After the 15-minute softening period, resuspend the pellet by pipetting up and down using a clean yellow tip on your P20 pipette.

PROCEDURAL NOTE

Watch your pellet as you pipette! Be careful not to lose the pellet on the inside of your pipette tip!

Step 3 Place the tube with your cDNA (resuspended in the EcoRI adaptors) in your ice bucket.

PROCEDURAL NOTE

Keep your cDNA tube and all reagent tubes on ice as much as possible.

Add the following components:

      1.0 µl 10x ligase buffer

      1.0 µl ATP (10 mM) (This is *not* dATP!)

      1.0 µl T4 DNA ligase (4 units/µl)

Step 4 Mix the reaction components with your pipette tip. Spin for a few seconds in your nanofuge, and incubate the tube at 8°C overnight. The EcoRI adaptors will be ligated to the blunt ends of your ds cDNA during this time.

PROCEDURAL NOTE

The ligation reaction must be stopped the next morning by removing the reaction tube from 8°C, heat inactivating the reaction at 70°C for 30 minutes, and then storing the reaction at −20°C. If it is not possible for you to do this, your instructor will do it for you.

LAB PERIOD IV.1.5.   KINASE THE ENDS OF THE cDNA AND DIGEST WITH XhoI TO GENERATE UNIQUE ENDS

PART A Kinase the Ends of the cDNA (Fig. 4-1, Step 7)

Now that you have added a dephosphorylated EcoRI sticky end onto each end of your cDNA molecules, you now need to add a phosphate group to the 5' ends of each of these molecules so that the ds cDNA can be ligated to the dephosphorylated lambda vector arms. To accomplish this, the ends of the cDNA must be treated with the enzyme polynucleotide kinase.

Step 1 Retrieve your tube containing your cDNA from storage at –20°C. Spin the tube for 30 seconds in your nanofuge to bring all of the contents to the bottom of the tube, and then place the tube on ice.

Step 2 Add the following reaction components to your cDNA tube:

> 1.0 µl 10x ligase buffer (This name is really correct! Ligase buffer works for the kinase reaction as well as the ligase reaction.)
>
> 2.0 µl ATP (10 mM) (This is *not* dATP!)
>
> 5.0 µl ddH$_2$O
>
> 2.0 µl T4 polynucleotide kinase (5 units/µl)

Step 3 Mix the reaction components with your pipette tip. Then spin the tube for 5 seconds in a nanofuge, and place the tube in a 37°C water bath for 30 minutes. A phosphate group will be added to the 5' ends of your double-stranded cDNA during this time.

Step 4 When the 30-minute incubation is complete, stop the reaction by placing the tube at 70°C for 30 minutes to heat-inactivate the kinase enzyme.

Step 6 After heat inactivating the kinase, spin your tube for 5 seconds in your nanofuge.

Step 7 Place the tube on your bench for 5 minutes to return the contents to room temperature.

PART B Digest with XhoI to Generate Unique Ends (Fig. 4-1, Step 8)

To generate a "unidirectional" library that will optimally facilitate the expression of proteins from the cloned cDNAs, the insert cDNA must be ligated into the vector in the correct orientation with respect to the promoter contained in the "upstream" vector arm. To ensure that *all* of the insert cDNAs will be ligated into the vector in the correct orientation (as opposed to half being ligated in the correct orientation and half in the incorrect reverse orientation), it is preferable to carry out a "unidirectional" ligation. We can accomplish this by manipulating the cDNA insert molecules so that they will have unique upstream and downstream sticky ends that will be compatible with complementary unique sticky ends on each arm of the lambda vector. The "upstream" lambda vector arm (the left arm) comes with an EcoRI overhang (sticky end), whereas the "downstream" lambda vector arm (the right arm) comes with an XhoI overhang (sticky end) (see Fig. 4-2). At this point, your cDNA molecules have EcoRI sticky ends on both ends of each molecule. Thus, they are ready to be ligated to the upstream lambda vector arm, but not to the downstream arm. However, when you synthesized first-strand cDNA from your mRNA, you primed that reaction with a "linker-primer" (Fig. 4-1, Step 1). As a result, you have an intact, double-stranded XhoI site near the downstream end of each of your ds cDNA molecules (Fig. 4-1). Thus, to generate an XhoI sticky end on the downstream end of each of these cDNA molecules, you can now simply digest with the restriction enzyme XhoI.

»»»

LAB PERIOD IV.1.5 CONT.

PROCEDURAL NOTE

You may be concerned that some of your cDNA molecules might also contain naturally occurring XhoI sites in addition to the special XhoI site you added by priming with the linker-primer. Digestion with XhoI would cleave these cDNA molecules into cDNA fragments; however, this will not occur because when the first strand of cDNA was synthesized, *methylated* cytosines were incorporated in the synthesis. The methylated cytosines incorporated into these XhoI sites renders them refractory to cleavage. Only the XhoI site added by the linker-primer contains unmethylated cytosines, and thus, only that site will now be cleaved by digestion with XhoI.

Step 1 Add 28 µl XhoI buffer supplement to your cDNA tube and mix by pipetting up and down.

Step 2 Add 3.0 µl XhoI enzyme. Mix briefly with your pipette tip, and spin in the nanofuge to return all of the liquid to the bottom of the tube.

Step 3 Incubate your tube in a waterbath at 37°C for 1.5 hours.

Step 4 When the 1.5-hour incubation is complete, remove the tube from the water bath. Spin it for 5 seconds in your nanofuge, and store the reaction at −20°C. If necessary, your instructor will do this.

LAB PERIOD IV.1.6.   SIZE SELECTION, EXTRACTION, AND DROP DIALYSIS OF cDNA

PART A Size Selection of Insert cDNA Molecules (Fig. 4-1, Step 9)

Your ultimate goal is to clone full-length cDNA molecules representing full-length or near full-length copies of the original mRNA molecules into the lambda vector with maximum efficiency. At this point, your cDNA synthesis reaction contains such desirable full-length cDNA molecules, *plus* some undesirable, smaller, partial cDNA molecules. These short cDNA molecules are generated by synthesis of cDNA from degraded mRNA templates and from partial cDNA molecules formed because of the reverse transcriptase falling off the mRNA template before completing synthesis. In addition, there are excess EcoRI adaptor molecules that were not ligated to the cDNAs. To selectively recover the full-length cDNA molecules, you will pass this heterogeneous mixture over a size-selection column that will exclude molecules less than 400 bp in length (partial cDNAs and unligated EcoR1 adaptors). You will be provided with one Chromaspin-400 (Clontech) sizing column for this purpose. This will exclude small cDNAs made from degraded mRNA templates, unligated EcoRI adaptors, small pieces of cDNA generated by the XhoI digestion, and any other low molecular weight contaminants.

Step 1 Retrieve your XhoI digest from storage at −20°C, and defrost it on ice.

Step 2 To prepare your Chromaspin-400 column, start with the caps still on both ends, and vortex the column several times to distribute the column material in the buffer (the buffer is already in the column from the manufacturer, Clontech).

Step 3 Remove the cap from the top end of the column and break off the bottom tab (**Figure 4-3**). Place the narrow end of the column in the special screw-top microcentrifuge tube that the manufacturer provides.

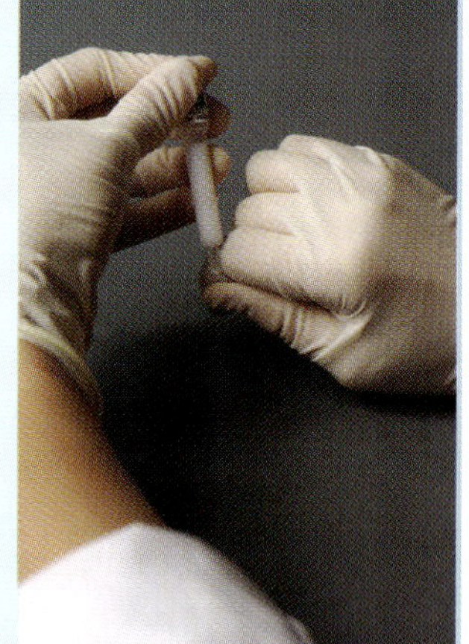

Fig 4-3 Preparation of the Chromaspin-400 column.

LAB PERIOD IV.1.6 CONT.

Step 4 Place the column/microfuge tube assembly into a 15-ml conical centrifuge tube. Spin the tube at 700x g for 3 minutes in a tabletop centrifuge.

Step 5 Remove the column from the microfuge tube, and discard the tube into a waste container. Place the column into a second new screw-top microfuge tube (provided).

Step 6 Add 5 µl 0.1% bromphenol blue dye to the tube containing your ds cDNA, and mix by tapping the tube with your finger.

PROCEDURAL NOTE

This is *not* the same as 5x Blue Juice loading dye (5x BJ) that is used for loading agarose gels!

Step 7 Spin the cDNA tube for 5 seconds in your nanofuge to get all of the liquid to the bottom of the tube.

Step 8 Pipette all of your cDNA plus bromphenol blue dye onto the Chromaspin column that you prepared in Steps 1–4. Try to pipette the cDNA directly onto the middle of the resin bed and not near the sides of the column.

Step 9 Put the column assembly into a 15-ml plastic conical centrifuge tube, and spin the tube at 700x g for 5 minutes. You should recover about 65 µl of solution.

This is your cDNA! Save this tube! You can now discard the column into a waste container.

Step 10 Add about 35 µl ddH$_2$O (or whatever amount is needed) to bring the total volume of your cDNA solution to 100 µl.

PART B Extraction of Size-Selected cDNA (Fig. 4-1, Step 10)

You have now produced and recovered a population of full-length cDNA molecules. To prepare these molecules for ligation to the lambda vector arms, you will first carry out another organic extraction to remove any residual enzymes and other contaminants. As in Lab Period IV.1.3, Part B, you will use Phase Lock Light gel tubes (see figure in Lab Period II1.3A, Step 1) for the extraction.

PROCEDURAL NOTE

The following steps require the use of phenol and chisam (24:1 chloroform/isoamyl alcohol vol/vol). Please note the following precautions.

CAUTION!

Phenol and chloroform (in chisam) are both toxic and produce noxious fumes. Because phenol and chisam are solvents that are caustic and will burn skin and clothes, be sure to always wear a lab coat and gloves. Keep tubes closed at all times. Open tubes containing phenol and/or chisam for brief periods only.

The tubes containing phenol and chisam are kept in the refrigerator and can be slippery because of condensation. Handle these tubes carefully to avoid dropping them.

LAB PERIOD IV.1.6 CONT.

PROCEDURAL NOTES

Work in a fume hood!

Phenol is buffer saturated and is the lower phase in the phenol container. The upper phase is buffer. When pipetting phenol, you must go through the upper buffer phase with the pipette tip to reach the lower phenol phase. The phenol phase is yellow. When you pipette the phenol, if it is not yellow, you have probably just pipetted buffer.

Discard all organic waste (i.e., tubes and tips that contained phenol) into the appropriate waste containers in the fume hood.

Step 1 Centrifuge two green 1.5-ml Phase Lock Light gel tubes for 1 minute.

Step 2 Pipette your 100 µl cDNA solution into one prespun green phase lock gel tube.

Step 3 Add 50 µl buffered phenol to the same green phase lock tube. Close the cap. *Hold the cap securely*, and mix by inverting for 30 seconds.

Step 4 Add 50 µl chisam to the same green phase lock tube. Close the cap. *Hold the cap securely*, and mix by inverting for an additional 30 seconds.

Step 5 Spin this tube and a balance tube for 2 minutes at maximum speed in a microcentrifuge.

Step 6 Transfer the upper aqueous phase to the second prespun green phase lock tube.

Step 7 Add 100 µl chisam to the green phase lock tube. Close the cap. *Hold the cap securely*, and mix by inverting for 30 seconds.

Step 8 Spin this tube and a balance tube for 2 minutes at maximum speed in a microcentrifuge.

Step 9 Remove the upper aqueous phase and place on a dialysis membrane as described below.

PART C Drop Dialysis of Size-Selected Insert cDNA (Fig. 4-1, Step 11)

We have found that for ligation reactions to proceed with optimal efficiency, it is preferable to purify the ligation components (in this case the insert cDNA) by drop dialysis before the ligation reaction. You will, therefore, perform drop dialysis, as described in Lab Period I.1.3, to purify your cDNA before ligating it to the lambda vector arms.

PROCEDURAL NOTE

Be sure to set up the dialysis in a protected area of your bench!

Step 1 Pour 200 ml sterile ddH$_2$O into a 250-ml breaker.

Step 2 Using two pairs of forceps, float a 45-mm diameter Millipore membrane (catalog #VSWP 04500, pore size: 0.025 mm, type VS) filter *shiny side up* on the water surface in the beaker (see figures in Lab Period I.1.3, Step 2). Allow the filter to wet for about 5 minutes, but *do not submerge* the filter.

Step 3 *Carefully* pipette all of the aqueous phase from Step 9 above (100 µl or less) onto the center of the dialysis filter.

Step 4 *Carefully* cover the beaker with parafilm, and *do not disturb* for 3 hours to overnight.

»»

LAB PERIOD IV.1.6 CONT.

PROCEDURAL NOTES

The minimum dialysis time is 3 hours, but 4 to 6 hours is preferable.

Samples must then be removed from the dialysis filter and stored at 4°C.

Step 5 *Carefully* retrieve as much of the cDNA as possible with a P200 pipette and transfer to a new, yellow microfuge tube (label it "Dial. cDNA") (see figure in Lab Period I.1.3, Step 6).

PROCEDURAL NOTE

To prevent submerging the membrane, be sure to hold the pipette at a 90° angle to the filter when retrieving the cDNA liquid. Store the sample at –20°C.

LAB PERIOD IV.1.7.   **CONCENTRATION OF cDNA AND UNIDIRECTIONAL LIGATION OF cDNA INSERTS INTO THE LAMBDA PHAGE VECTOR**

PART A Concentration of cDNA before Ligation to the Lambda Phage Vector (Fig. 4-1, Step 12)

To maximize the efficiency of the ligation reaction, it will be performed in a very small volume (5 µl total). Because you want to maximize the amount of double-stranded cDNA in the ligation reaction to produce the largest possible library, you must first reduce the volume of the solution containing your cDNA to 5 µl or less. This will be achieved using a vacuum concentrator device ("Speed-Vac").

Step 1 Centrifuge your dialyzed sample in a speed-vacuum concentrator until it is evaporated to less than or equal to 5 µl. The time required for this varies greatly and depends on the characteristics of the particular speed-vacuum concentrator and pump you are using, as well as the number of samples that are being evaporated simultaneously. Therefore, you or your instructor will need to carefully monitor the progress of the evaporation of your sample. To a new 1.5-ml microfuge tube, add 5 µl water. Periodically remove your cDNA sample tube from the concentrator, and estimate the volume in your sample tube by holding it beside the control tube containing exactly 5 µl water. When the volume in your sample tube has been reduced to 5 µl or less, you have completed the concentration step. If you have reduced the volume of your sample to less than 5 µl, use a pipette to carefully add back ddH_2O to bring the volume to exactly 5 µl.

PROCEDURAL NOTE

Avoid evaporating your sample to complete dryness, as it may be difficult to get all of your cDNA back into solution when you add back ddH_2O. If possible, try to stop the evaporation while your sample is at a volume of 1 to 5 µl. If you do concentrate your sample to dryness, add back 5 µl ddH_2O, and pipette your sample up and down several times to completely redissolve the cDNA.

If possible, you should now proceed directly to the ligation reaction (Part B below). However, if necessary, you may store the sample at 4°C for several days prior to the ligation step.

PART B Unidirectional Ligation of cDNA Inserts into the Lambda Phage Vector (Fig. 4-1, Part 2)

The lambda Uni-ZAP XR vector you will use was obtained from Stratagene along with the cDNA synthesis kit and comes precut into two cloning arms that are both dephosphorylated. One of the arms (the left arm that has the promoter) was cut with EcoRI, whereas the other arm (the right

»»

arm) was cut with XhoI (flowchart, blue area). Because the insert cDNA you have synthesized now has one EcoRI end and one XhoI end, there is only one orientation in which it can combine with the two lambda arms to generate a complete, recombinant bacteriophage DNA (flowchart, blue area). That orientation will ensure that the template strand will be the one that gets transcribed when the cDNA is expressed in a host bacterium (if expression of the cloned cDNA is desired). Lambda Uni-ZAP XR vector is the same vector as lambda ZAP II from Stratagene (Uni-ZAP XR is simply lambda ZAP II digested with EcoRI and XhoI and dephosphorylated).

Because the lambda vector arms are dephosphorylated, it is not possible for the left arm and right arm to ligate to one another with no intervening insert. The phosphorylated ends on each of the double-stranded cDNA molecules you have synthesized will provide the necessary phosphate group to generate the phosphodiester bonds between a cDNA insert and two vector arms. This will yield the desired "recombinant lambda genome." It is also possible, however, for two cDNA molecules (which are phosphorylated on each end) to ligate to one another. This is an undesirable result because it will lead either to a single lambda clone with two insert cDNAs or to concatenated cDNA molecules that may be too large to be packaged into an infectious lambda bacteriophage when combined with a left and right lambda arm. To optimize the likelihood that each individual cDNA molecule will become ligated between two lambda vector arms, we will strive for a ratio of approximately 1:1 between cDNA molecules and lambda vector arms. One microgram of lambda vector is provided with the cDNA synthesis kit for each ligation reaction. The typical yield of double-stranded cDNA from 5 µg of poly A+ mRNA is sufficient for two ligation reactions, each with 1 µg of lambda vector. Therefore, you will ligate half of your cDNA to lambda vector arms and store the remaining half at –20°C.

PROCEDURAL NOTE

If desired, this remaining cDNA could be used for a second ligation reaction identical to the one described below, but this will not be included in this project.

After adjusting the total volume of your cDNA to 5 µl as described in Part A, proceed with the ligation reaction described below.

Step 1   Add half of your cDNA (2.5 µl) to a new, labeled 0.5-ml microfuge tube. Label this tube "Lig cDNA." Place the tube on ice. Store the remainder of your ds cDNA (2.5 µl) at –20°C.

Step 2   Add the following components to your tube labeled "Lig cDNA" containing the 2.5 µl cDNA:

    0.5 µl 10x ligase buffer

    0.5 µl ATP (10 mM) (This is *not* dATP!)

    1.0 µl Uni-ZAP XR lambda vector (already cut with EcoRI and XhoI and dephosphorylated: 1 µg/µl)

    0.5 µl T4 ligase

    5.0 µl Total volume

Step 3   Mix the reaction components with your pipette tip. Spin the tube for 5 seconds in a nanofuge, and then incubate at 12°C overnight.

PROCEDURAL NOTE

It is necessary to stop the ligation the next morning by placing the ligation tube at –20°C. The instructor can do this if necessary. The ligation reaction can be stored at –20°C until the next lab period.

LAB PERIOD IV.1.8. **PACKAGE THE RECOMBINANT LAMBDA PHAGE GENOMES INTO INFECTIOUS PHAGE PARTICLES AND PLATE AN ALIQUOT OF THE LIBRARY TO ASSESS THE TITER**

PART A Package the Recombinant Lambda Genomes into Infectious Phage Particles

At this point, you have produced recombinant lambda genomes, the majority of which include one cDNA insert ligated between one upstream (left) arm and one downstream (right) arm. Next you will package these recombinant genomes into functional bacteriophage particles. This is done by incubating the recombinant phage genomes with a protein ("packaging") extract that Stratagene provides. The packaging extract contains all of the lambda proteins necessary for the spontaneous assembly of complete lambda phage particles. However, this packaging reaction will only package a lambda DNA molecule of the appropriate size (lambda ZAPII with an insert between 0 and 10 kb in size). Thus, when you add the packaging extract to your recombinant lambda genomes, functional bacteriophage particles will be formed, each carrying a different cDNA insert as part of the phage genome and each capable of infecting host (*E. coli*) bacterial cells.

You will do *one* packaging reaction.

Step 1 Remove your ligation reaction from storage at –20°C, and place it at room temperature at your bench.

Step 2 Remove a tube of packaging extract (Gigapack Gold, Stratagene) from the –80°C freezer, and place it on dry ice.

PROCEDURAL NOTES

Your instructor will provide the tube containing the packaging extract on dry ice.

Keep your packaging extract tube on dry ice until immediately before use. Do not thaw until you are ready!

Step 3 When you are ready to proceed, quickly warm the packaging extract tube between your fingers until it *just* begins to thaw (about 10 seconds).

Step 4 *Immediately* add all 5 µl of your cDNA–lambda ZAP ligation reaction from the tube labeled "Lig cDNA" to the extract tube. Stir gently with your pipette tip. You may pipette up and down gently, but *do not* introduce air bubbles.

PROCEDURAL NOTE

No matter how careful you are, you will likely introduce a few air bubbles. This is not a problem, but you do not want to pipette up and down or vortex and make foam. This can reduce the packaging efficiency.

Step 5 Spin the extract tube 5 seconds in your microcentrifuge.

Step 6 Label this tube "cDNA LIB," and place it in your rack at room temperature. Incubate the reaction at room temperature for exactly *2 hours*. Discard the empty "Lig cDNA" tube.

PROCEDURAL NOTE

Allow the packaging reaction to proceed for *exactly* 2 hours. *Do not exceed 2 hours!* Two hours is the optimal time for this reaction. If the reaction proceeds for more than 2 hours, the number of functional, infectious phage particles will be reduced.

Step 7 After the 2-hour incubation, add 500 µl SM buffer (phage storage buffer) and 20 µl chloroform (*not chisam!*) to the tube labeled "Lig cDNA," which now contains your packaged cDNA library.

»»»

LAB PERIOD IV.1.8 CONT.

Step 8 Spin the "cDNA LIB" tube briefly in your nanofuge to sediment the debris. You have now produced a cDNA library — congratulations!

PROCEDURAL NOTES

Always store the tube containing your cDNA library at 4°C. *Never* store this tube at –20°C! If you place this tube at –20°C, you will inactivate your phage by shearing the phage tails from the phage heads, and the phage will no longer be able to infect host *E. coli* cells!

Save your "cDNA LIB" tube! Do not discard it!

Do not add chisam to your "cDNA LIB" tube or else you will inactivate your phage!

PART B Plating an Aliquot of the cDNA Library to Assess the Titer

As noted above, you have now produced your cDNA library. To characterize the quality of your library, however, various tests are typically performed. The most common assessment is to determine the "titer" of the cDNA library. The titer refers to the number of infectious phage particles per µl of the library or in the entire library. This is assessed by infecting host *E. coli* cells with a small aliquot of the lambda phage library into host *E. coli* cells and then plating on the appropriate growth medium to determine how many infectious phage are present in that aliquot of your library. Each infectious phage will cause the formation of a "plaque" (a relatively clear, small circular area) in a lawn of host bacterial cells. Thus, each infectious phage particle is referred to as a "plaque-forming unit" (PFU).

A second common assessment is the fraction of recombinant phage in the library. In theory, only recombinant phage genomes were formed during the ligation reaction and subsequently packaged into functional, infectious phage particles. However, some uncut lambda vector is inevitably included in the preparation of vector arms provided by the manufacturer, and these uncut vectors can also be packaged to form infectious phage particles. To distinguish the phage particles containing these nonrecombinant phage genomes from those containing the desired recombinant phage genomes (i.e., those that include cDNA inserts), you will employ a colorimetric assay. The nonrecombinant phage genomes carry an intact gene segment (*lacZ*) encoding beta-galactosidase (β-gal), whereas incorporation of a cDNA insert into the recombinant phage genome will result in "insertional inactivation" of the *lacZ* gene so that no functional β-gal enzyme will be produced. The colorimetric assay that you will use results in plaques derived from nonrecombinant phages (that produce β-gal) to develop a blue color; recombinant phages that do not produce β-gal will produce clear plaques. By determining the number of clear plaques as a function of the total number of plaques, you can determine the fraction of recombinant phage in your library.

PROCEDURAL NOTE

To determine the titer of the cDNA library, an aliquot of the phage is used to infect *E. coli* cells. This process is referred to as "transfection." To have *E. coli* cells ready for transfection, it is necessary to begin cultures *the night before* the transfection step. A special strain of *E. coli* called XL-1 Blue MRF' will be used for this process. The manufacturer (Stratagene) has engineered these cells to be compatible with the Uni-ZAP XR lambda vector in which you constructed your cDNA library. This strain is supplied with the cDNA synthesis kit. Preparing a culture of these cells is done as follows: The XL-1 Blue MRF' cells come from Stratagene as a glycerol stock. The cells are best revived by scraping crystals with a sterile loop onto an LB+ tetracycline plate, streaking them out (see Appendix IV), and growing overnight at 37°C. A single colony is then picked to start the LBMM culture in this step.

》》

a. Innoculate 50 ml LB mediium with a single colony from a fresh plate of an appropriate strain of host *E. coli* designed to foster optimal growth of the recombinant lambda bacteriophage. For lambda ZAP Express, the optimized host bacterial strain is XL-1 Blue MRF' (Stratagene). These cells are grown overnight at 30°C with active rotation in a bacterial shaker/incubator. Growth at 30°C insures that the cells will not overgrow; 50 ml will typically provide sufficient cells for titering at least 40 lambda libraries.

b. The next morning, pellet the bacteria by centrifuging them at 1000x *g* for 10 minutes in a tabletop centrifuge.

c. Gently resuspend the pelleted bacteria in 10 ml of freshly prepared 10 mM $MgSO_4$. Just before using the cells, dilute them to an OD_{600} of 0.5 (based on spectrophotometric measurements using a 1.0-cm path length cuvette). The cells can then be kept on ice for up to 12 hours before use.

Step 1 You will prepare two samples of your packaged recombinant lambda bacteriophage to be titered:

a. Undiluted: to a new 1.5-ml microfuge tube containing 200 µl of *E. coli* plating cells (XL-1 Blue MRF'), prepared as described above, add 1 µl of packaged bacteriophage from your "cDNA LIB" tube. Flick the tube to mix. Do *not* centrifuge or you will pellet your cells! Label this tube with your number and "Undil."

b. Diluted: To another new 1.5-ml microfuge tube, mix 1 µl of packaged bacteriophage from your "cDNA LIB" tube with 49 µl of SM buffer. Label this tube "Dil LIB." Mix by tapping the tube, and then spin the tube for 5 seconds in your nanofuge. To yet another new 1.5-ml microfuge tube containing 200 µl plating cells (prepared as described above), add 1 µl from this "Dil LIB" tube. Flick the tube to mix. Do *not* centrifuge or else you will pellet your cells! Label this tube with your number and "1:50 Dil." You can then discard the "Dil LIB" tube with its remaining contents.

Step 2 Incubate both your "Undil" and "1:50 Dil" tubes at 37°C for 15 minutes. The lambda bacteriophage will adhere (adsorb) to the host bacterial cells during this time. After this incubation, keep the tubes at room temperature until you plate out the contents.

Step 3 Store the remainder of your "cDNA LIB" tube at 4°C.

PROCEDURAL NOTES

Save the "cDNA LIB" tube! Do not discard this tube! This is your cDNA library! Do not put the "cDNA LIB" tube at −20°C! Storing your library at −20°C will inactivate your phage by shearing off the tails!

For Steps 4 to 8, prepare one plate with your *undiluted* XL-1 Blue MRF' mixture and one plate with your 1:50 *diluted* XL-1 Blue MRF' mixture.

Step 4 Add 15 µl 0.5 M IPTG to a disposable 15-ml tube containing 3 ml of molten NZY top agarose in a water bath at 55°C, as shown in Lab Period II.1.6, Step 6.

Step 5 To one tube, add 50 µl of 250 mg/ml X-gal (made up in dimethyl formamide). Vortex briefly, and place the tube back into the water bath. Next, flick the tube of cells + phage to mix and add the entire "undil" mixture (about 200 µl) to the tube of top agarose (see Lab Period II.1.6, Step 7).

Step 6 Vortex this mixture, and pour *immediately* onto your warm NZY agar plate labeled "undil." Swirl the plate to evenly distribute the liquid agarose on the top of the agar on the plate (refer to Lab Period II.1.6, Step 7).

»»

LAB PERIOD IV.1.8 CONT.

Step 7   Repeat Steps 4 to 6 for your tube containing the 1:50 diluted cells ("dil" tube).

PROCEDURAL NOTE

The top agarose is 0.6% agarose in NZY medium. It is autoclaved and placed in a 55°C water bath. This agarose will quickly solidify upon removal from the water bath. To avoid problems with solidification, take your plates to the water bath, and plate your libraries next to the water bath. In each case, *immediately* after adding the cell/phage mixture to the top agarose and vortexing, pour the contents of this tube onto the agar plate, and gently swirl to distribute the top agarose evenly over the bottom agar.

Step 8   Let the plates sit upright for at least 5 to 10 minutes to solidify.

Step 9   Incubate the plates *inverted* at 37°C for 8 to 24 hours.

During this time, the bacteria will grow and those infected by phage will be lysed to form plaques. Bacteriophage with the intact *lacZ* gene fragment (encoding part of the protein ß-gal) do *not* carry a cDNA insert and will form blue plaques. Bacteriophage with a disrupted *lacZ* gene fragment will not produce functional ß-gal (because they *have* a cDNA insert) and will form clear plaques. After the overnight incubation, you will be thrilled to see all of the clear plaques on your titer plates! Examples of plaques are shown in Figure 2-3.

PROCEDURAL NOTES

After the incubation period, the plates should be removed from the incubator and stored at 4°C until the next lab period.

One important characteristic of any cDNA library is the total number of cDNA insert-containing (clear) PFUs in the original library. The more recombinant PFUs, the better the chance that even those mRNAs expressed at very low levels in the tissue from which the mRNA was isolated will be represented in your library.

Another important characteristic of a cDNA library is the average size of the inserts (see Module IV.2). To assess the average insert size in your cDNA library, you will pick several individual plaques and determine the insert size for each as described in Module IV.2.

Each plaque on your titer plates represents a single original PFU (infectious phage particle). The phage in each plaque will continue to grow until all the host *E. coli* cells are lysed. Therefore, after the plaques have become large enough to be clearly visible, it is necessary to stop further growth by removing the titer plates from the incubator and storing them at 4°C. This will enable you to distinguish each individual plaque. In addition, phage particles can diffuse through the agar plates so that they may contaminate a neighboring plaque. Therefore, it is also necessary to pick representative clear plaques for size analysis (as described in Module IV-2) no later than the day after the plating of the library to obtain phage representing only the original infectious phage from which each individual plaque was derived.

The procedures for calculating the titer of your library and for picking individual plaques for insert size analysis are described below in Lab Periods IV.1.9 and IV.2.1.

LAB PERIOD IV.1.9.   **CALCULATE OF THE TITER AND PERCENTAGE OF RECOMBINANT PHAGE IN THE LAMBDA cDNA LIBRARY**

THE TITER of your library is the number of PFUs per microliter in your *original* stock library. To calculate this titer, you will count the plaques on one of your titer plates, take into account any appropriate dilution factor, and derive the number of PFUs per microliter of your original library. You prepared two different titer plates with two different dilutions of your original library so that one of the two plates is likely to have a density of plaques that is sufficiently low to allow you to count individual plaques but sufficiently high so that your count will be representative of the entire library. Typically, a total of 50 to 200 plaques on an individual plate is ideal for counting, but more densely populated plates can still yield accurate counts if necessary.

Step 1   Choose one of your two plates which has well-separated plaques that can be easily counted. Count all of the plaques (clear and blue) on that plate. If there are too many plaques to count easily on your least densely populated plate, use a marker to divide the plate into quarters, and count the plaques in one quarter (and then multiply by 4). Also, count the number of blue plaques and the number of clear plaques among this total. Clear and blue plaques should be easily distinguishable, as shown in Lab Period II.1.7, Step 3.

Step 2   Calculate the titer of your library (clear + blue plaques). This is the number of plaques per microliter. If you counted your "undil" plate, the total number of plaques on this plate is your titer per microliter. If you counted your "dil" plate, you need to multiply the number of plaques on that plate by 50 to determine your titer per microliter because that plate represents a 1:50 dilution of 1 µl of your stock library.

Step 3   Determine the total titer of your library by multiplying the titer per microliter by the total volume of your library. The total volume of your library should be approximately 530 µl. Therefore, your total titer = PFU/µl x 530 µl.

Step 4   Determine the percentage of recombinant phage in your library by dividing the number of clear plaques by the total number of plaques (clear + blue) on the titer plate. Multiply by 100 to derive the percentage of recombinant phage.

Step 5   Record your results (total titer and percentage of recombinant phage). How did your results compare with those of the other students in the class?

MODULE IV.2

Characterization and Screening of a cDNA Library Using the Polymerase Chain Reaction

MODULE SUMMARY

Use of PCR to determine the average insert size in a lambda cDNA library and use of PCR to detect gene-specific sequences in a lambda cDNA library.

MODULE BACKGROUND

The polymerase chain reaction (PCR) can be used to perform a variety of analyses or manipulations on a cDNA library. You will perform two of these: (1) determination of the average insert size of cDNAs cloned in your library and (2) detection of a specific cDNA within the library. To analyze the average insert size in your cDNA library, you will pick individual plaques at random and amplify the insert from each using PCR primers that are complementary to sequences in the lambda vector arms that immediately flank each cloned cDNA insert. Because the primers are complementary to vector sequences common to all clones in your cDNA library, you do not require knowledge of the identity or sequence of the particular cDNA in each clone — the same primers will work on every cDNA clone no matter what cDNA is inserted. After you have amplified the inserts from many cDNA clones, you can calculate an "average insert size."

PCR also provides a very quick and easy method to detect a particular cDNA in a library. Before the advent of PCR, detection of a specific cDNA in a library required laborious plaque lifts, hybridization, and plaque purification procedures followed by DNA sequencing. However, with sufficient knowledge of the sequence of an individual cDNA, specific primers can be designed to directly detect the cDNA sequence by PCR amplification from an aliquot of the library. The detection of a PCR product of the expected size from such gene-specific primers indicates the presence of that particular cDNA in the library.

    ISOLATION OF PHAGE FROM INDIVIDUAL PLAQUES FOR INSERT SIZE ANALYSIS

YOU WILL choose four clear plaques at random and prepare these for a PCR-based analysis of the size of the insert in each lambda clone.

PROCEDURAL NOTE

An average insert size of >1.0 kb is generally indicative of full-length cDNAs derived from mammals.

You will recover a sample of the phage from each individual plaque and determine the size of the insert cDNA in the recombinant lambda from which each plaque was derived. You will then determine an "average insert size" by calculating the average of the sizes of the four individual inserts you analyze.

PROCEDURAL NOTE

In your own laboratory, you would want to perform this procedure on many more plaques (50–100) to get a more accurate estimate of the true average insert size in the library. However, in this laboratory exercise, you will examine only four plaques to give you a rough estimate of the average insert size in your library and to allow you to learn the procedure.

»»

LAB PERIOD IV.2.1 CONT.

Step 1 Pipette 50 µl of SM buffer into each of four new 1.5-ml microcentrifuge tubes, each labeled with your number and either "cDNA-1," "cDNA-2," "cDNA-3," or "cDNA-4."

Step 2 Use a P1000 pipette tip to "pick" an *isolated* clear plaque from one of your titer plates and transfer it into the tube labeled "cDNA-1." Your instructor will demonstrate this procedure (see figures in Lab Period II.2.6, Step 2).

PROCEDURAL NOTE

When you pick each plaque, it is critical that you ensure that the agarose plug, containing the phage particles, is actually transferred to each corresponding microfuge tube. The agarose plug should be easily visible in your pipette tip during the transfer process. Look carefully at your pipette tip during each transfer to ensure that you have recovered an agarose plug from each plaque. Then look at your pipette tip again after you have transferred the plug into the tube of SM buffer to ensure that you have actually transferred it into the buffer.

Step 3 Repeat this procedure using a new P1000 pipette tip for each plaque to transfer three other individual clear plaques into the tubes labeled "cDNA-2, "cDNA-3," and "cDNA-4," respectively.

Step 4 Vortex each tube for 1 minute to help release the phage particles from the agarose plugs into the SM buffer.

Step 5 Briefly spin the tubes in your nanofuge to pellet the residual agarose.

Step 6 Store the tubes at 4°C until the next lab period. During this time, phage particles will diffuse out of the agarose plug and into the SM buffer.

LAB PERIOD IV.2.2. **USE OF PCR TO DETERMINE THE AVERAGE INSERT SIZE IN THE cDNA LIBRARY AND TO DETECT A SPECIFIC cDNA CLONE IN THE LIBRARY**

PART A PCR to Determine the Average Insert Size in the cDNA Library

You will use PCR primers complementary to sites flanking the cloning site in the Uni-ZAP lambda vector (flowchart, blue area) to amplify the insert from each of the four different plaques you previously picked into SM buffer. You will then electrophorese the amplified products on an agarose gel and determine their sizes by comparison with molecular size standards. By calculating the average among the four fragment sizes you detect in this experiment, you will derive a rough estimate of the average insert size in your cDNA library.

Step 1 Remove your "cDNA-1" "cDNA-2," "cDNA-3," and "cDNA-4" tubes from storage at 4°C, and incubate them at 37°C for 1 hour. The higher temperature helps to release the phage particles from the agarose plug.

Step 2 Remove your four tubes from the 37°C incubator, and vortex each for 1 minute.

Step 3 Spin all four tubes for 2 minutes in your nanofuge (this sediments unwanted agarose particles).

Step 4 Label the sides of four 0.5-ml PCR tubes with your number and 1, 2, 3, or 4. Label a new 1.5-ml microfuge tube with "PCR Master Mix" and your number. Then add the following components to the "PCR Master Mix" tube.

》》

LAB PERIOD IV.2.2 CONT.

PROCEDURAL NOTE

Except for the primers and ddH$_2$O, the following reagents are from the AmpliTaq Gold PCR Kit (Applied Biosystems).

> 126 µl ddH$_2$O
>
> 32.0 µl dNTP mix
>
> 20.0 µl 10x PCR reaction buffer
>
> 12.0 µl MgCl$_2$ (25 mM)
>
> 4.0 µl T3 primer (forward) 10 pmol/µl
>
> 4.0 µl T7 primer (reverse) 10 pmol/µl
>
> 2.0 µl AmpliTaq Gold DNA polymerase
>
> 200 µl Total volume

PROCEDURAL NOTE

T3 primer: 5' AATTAACCCTCACTAAAGGG 3'
T7 primer: 5' GTAATCAGACTCACTATAGGGC 3'

Step 5 Mix by tapping the tube, and then spin briefly in your nanofuge to get all of the liquid to the bottom.

Step 6 Aliquot 46 µl of the "PCR Master Mix" to each of the four 0.5-ml PCR tubes labeled 1, 2, 3, and 4.

Step 7 Add 4.0 µl of stock phage solution from your tube labeled "cDNA-1" to the 0.5-ml PCR tube labeled 1. Do the same for the other three tubes: that is, add 4.0 µl of stock phage template from your tubes labeled "cDNA-2," "cDNA-3," and "cDNA-4" to the 0.5 ml tubes labeled 2, 3, and 4, respectively.

Step 8 Place all four tubes into the thermal cycler.

PROCEDURAL NOTE

The PCR program has been set as follows:

> a. An initial denaturing step at 94°C for 12 minutes is performed to insure that the lambda phage are lysed and that all of the double-stranded DNA has been converted to single-stranded DNA (and to activate the AmpliTaq Gold DNA polymerase).
>
> b. An annealing step at 55°C for 5 minutes to enable initial binding of the oligonucleotide primers to the template.

The following three steps (c–e) will then be cycled 35 times:

> c. 72°C for 90 seconds for DNA synthesis to extend from the primers (this is the temperature at which *Taq* polymerase is most active).
>
> d. 94°C for 45 seconds to denature the double helix.
>
> e. 55°C for 45 seconds to allow primers to anneal to the template DNA.

>>>

After the 35 cycles, the instrument will perform the following steps:

    f. A final extension step at 72°C for 10 minutes after the 35th cycle to insure that the extensions are complete and strands are annealed.

    g. The program will then maintain the tubes at 4°C.

Step 9 When the PCR reaction is complete, the samples can be stored at –20°C. If necessary, your instructor can do this. Electrophoretic analysis of the products of these PCR reactions will be conducted during the next lab period.

PART B PCR to Detect a Specific cDNA Clone in the cDNA Library

Typically, the most important characteristic of a cDNA library is whether you can use it to isolate a clone of your gene of interest. With the cDNA library you have constructed, you could use any of several different methods to screen for a specific cDNA. You could hybridize a specific nucleic acid probe to plaque lifts of this library as done in Project II to detect the reverse transcriptase repeat sequence in your genomic DNA library. Alternatively, because you constructed this cDNA library in an expression vector, you could allow the host bacteria to produce proteins from the cloned cDNAs and then probe plaque lifts with an antibody to the protein you are interested in studying. This is analogous to a Western blot and would allow you to directly isolate a cDNA clone of a gene that encodes the protein recognized by the antibody.

The advent of PCR has provided another means for screening a gene library, and this is the method you will use to screen the cDNA library in this experiment. To directly screen a library by PCR, you must have access to sufficient sequence information about the desired gene or cDNA to allow you to design sequence-specific primers. In this experiment, you will use the same set of *Ttr*-specific primers as were used for DNA PCR and RNA PCR in Projects I and III, respectively. A typical cDNA library contains a total of 100,000 or more PFU representing about 10,000 different genes. However, PCR enables the detection of a particular gene of interest even if it is represented at very low frequency in the library.

Step 1 Retrieve your cDNA library tube containing your packaged cDNA library from storage at 4°C (it is labeled "cDNA LIB"). This tube was prepared and stored in Lab Period IV.1.8, Part B.

Step 2 Transfer 5 µl of your packaged cDNA library into a new 0.5-ml microfuge tube.

Step 3 Add 26.5 µl ddH$_2$O to the 5 µl of packaged library.

Step 4 Vortex the tube briefly and spin for 5 seconds in your nanofuge. This is your "PCR template."

PROCEDURAL NOTE

Except for the primers and ddH$_2$O, the following reagents are from the AmpliTaq Gold PCR Kit (Applied Biosystems).

Step 5 Prepare a PCR mix by adding the following components to a new 0.5-ml PCR microfuge tube.

»»

LAB PERIOD IV.2.2 CONT.

8.0 µl dNTP mix

5.0 µl 10x PCR reaction buffer

3.0 µl $MgCl_2$ (25 mM)

1.0 µl *Ttr* Forward Primer (10 pmol/µl)

1.0 µl *Ttr* Reverse Primer (10 pmol/µl)

0.5 µl AmpliTaq Gold DNA polymerase

18.5 µl Total volume

PROCEDURAL NOTE

Ttr Forward Primer: 5' ACACAGATCCACAAGCTCCTGACAG 3'
Ttr Reverse Primer: 5' CGGACAGCATCCAGGACTTTGAC 3'

Step 6 Pipette the 31.5 µl of "PCR template" (from Step 4 above) into the 0.5-ml "PCR Mix" tube and mix by pipetting up and down. This will result in a final volume of 50 µl. Label the side of the tube with your number and "*Ttr* cDNA PCR."

Step 7 Place the 0.5-ml tube containing your PCR reaction into a well in the thermal cycler.

Step 8 The PCR program will be set as follows:

a. An initial denaturing step at 94°C for 12 minutes is performed to insure that the lambda phage are lysed and that all of the double-stranded DNA has been converted to single stranded. The 12 minutes at 94°C is also required to activate the AmpliTaq Gold DNA polymerase.

The following three steps (b–d) will be cycled 35 times:

b. 94°C for 1 minute to denature the DNA.

c. 58°C for 1 minute to allow the primers to anneal to the template DNA.

d. 72°C for 3 minutes for DNA synthesis to extend from the primers (this is the temperature at which AmpliTaq DNA polymerase is most active).

After the 35 cycles, the instrument will perform the following temperature step:

e. A final extension step at 72°C for 7 minutes is added after the 35th cycle to insure that the extensions are complete.

The program will then maintain the tubes at 4°C until the tubes are removed and the instrument is turned off.

Step 9 The tubes can then be removed from the PCR instrument and stored at –20°C. If necessary this can be done by your instructor. Electrophoretic analysis of the products of this PCR reaction will be conducted during the next lab period.

LAB PERIOD IV.2.3.   ELECTROPHORETIC ANALYSIS OF PCR PRODUCTS TO CALCULATE THE AVERAGE LIBRARY INSERT SIZE AND SPECIFICALLY DETECT THE *Ttr* GENE PRODUCT

YOU CAN now use agarose gel electrophoresis to examine the products of the PCR reactions you performed on four randomly selected cDNA clones with vector-specific primers, as well as the product of the PCR reaction you performed with gene-specific primers to detect the *Ttr* cDNA. In each case, you will be testing for the appearance of a band indicative of either the presence of an insert in the case of the reactions using vector-specific primers or the specific presence of the *Ttr* cDNA in the case of the reaction using the gene-specific primers. In addition, you will determine the size of each band by comparison with appropriate size markers. This will allow you to calculate the average insert size in your library, as indicated by the average size of the amplification products generated by the vector-specific primers. It will also allow you to confirm the identity of the *Ttr*-specific cDNA amplification product, as the predicted size of that fragment is known (148 bp).

Step 1 Pour a 1.2% agarose gel (refer to Lab Period I.1.4, Part B, and associated figures).

Step 2 Retrieve the tubes containing the PCR products from the amplification of the four randomly selected plaques (tubes labeled "cDNA 1–4," respectively).

Step 3 Retrieve the tube containing the PCR product from the gene-specific amplification designed to detect the *Ttr* cDNA (tube labeled "*Ttr* cDNA PCR").

Step 4 Remove 12 µl from each of the tubes from Steps 2 and 3 above, and transfer each to new correspondingly labeled ("PCR 1–PCR 4" or "Ttr PCR") 1.5-ml microfuge tubes.

Step 5 Add 3 µl of 5x BJ to each tube and mix.

Step 6 Retrieve your tube of 100-bp ladder in 5x BJ (50 ng/µl) (stored at 4°C). Remove 10 µl, and place it in a new microcentrifuge tube labeled "100 bp." Return the marker stock solution to storage at 4°C.

PROCEDURAL NOTE

The 100-bp ladder is a set of DNA fragments of known size and concentration, supplied by New England Biolabs. Refer to Appendix III for a complete description.

Step 7 Retrieve your tube of lambda BstEII DNA marker in 5x BJ (150 ng) (stored at 4°C). Transfer 6 µl to a new microcentrifuge tube labeled "lambda Bst." Return the marker stock solution to storage at 4°C.

PROCEDURAL NOTE

The lambda BstEII DNA marker is a set of DNA fragments of known size and concentration provided by New England Biolabs. It was created by digesting lambda DNA with the restriction enzyme BstEII. Refer to Appendix III for a complete description.

Step 8 Heat the "Lambda Bst" marker at 65°C for 3 minutes and place on ice.

Step 9 Load the PCR products and markers onto the 1.2% agarose gel in the following order:

Lane 1: Empty

Lane 2: 15 µl sample containing the PCR product from random cDNA clone 1 (= "PCR 1")

Lane 3: 15 µl sample containing the PCR product from random cDNA clone 2 (= "PCR 2")

Lane 4: 15 µl sample containing the PCR product from random cDNA clone 3 (= "PCR #3")

»»

LAB PERIOD IV.2.3 CONT.

Lane 5: 15 µl sample containing the PCR product from random cDNA clone 4 (= "PCR 4")

Lane 6: 10 µl 100 bp ladder in 5x BJ (50 ng/µl)

Lane 7: 6 µl lambda BstEII DNA in 5x BJ (150 ng) (heated to 65°C for 3 minutes and placed on ice before loading)

Lane 8: 15 µl sample containing the PCR product to detect the *Ttr* sequence ("*Ttr* cDNA PCR")

Lanes 9 and 10: Empty

Step 10 Run the gel at 60 V for 1.5 hours.

Step 11 After electrophoresis, turn off the power supply, and unplug the leads from the power supply.

Step 12 Carefully stain the gel for 8 to 10 minutes in an ethidium bromide bath using a spatula to transfer the gel from the electrophoresis box to the gel stain container (the instructor will demonstrate this). Using the spatula, transfer the gel to a water bath (to destain the gel) for 15 to 30 minutes.

CAUTION!

Wear gloves, lab coats and safety glasses. Handle these gels with extreme care! They are slippery. Support the entire gel with the spatula to keep it from breaking.

Ethidium bromide is a known mutagen and carcinogen. Wear gloves, lab coats, and safety glasses! Handle gels carefully! Do not splash ethidium bromide! Always rinse the spatula in the water destain bath after it has been in contact with ethidium bromide! Always rinse the gel tray after sliding gels into the ethidium bromide bath. You may unknowingly contact the ethidium bromide! Be very careful not to drip the ethidium bromide anywhere!

PROCEDURAL NOTE

Appendix IV describes an alternative DNA stain, SYBR Safe (Molecular Probes).

Step 13 Place your gel on an ultraviolet (UV) transilluminator, and photograph the gel (or take a digital image with a gel documentation system). Your instructor will demonstrate this. After obtaining a high-quality photograph, discard the gel.

Step 14 An example of the expected pattern of the DNA fragments in this gel is shown in **Figure 4-4.**

Step 15 Determine the size of each random cDNA clone insert (1–4) by comparison with the molecular size standards. What is the average size of the four inserts?

CAUTION!

Handle these gels with extreme care! Support the entire gel with a spatula to keep it from breaking.

Be careful not to cause unwanted contact with the ethidium bromide DNA gel stain by splashing or spreading this solution unnecessarily — always rinse the spatula and gel tray with water after each use.

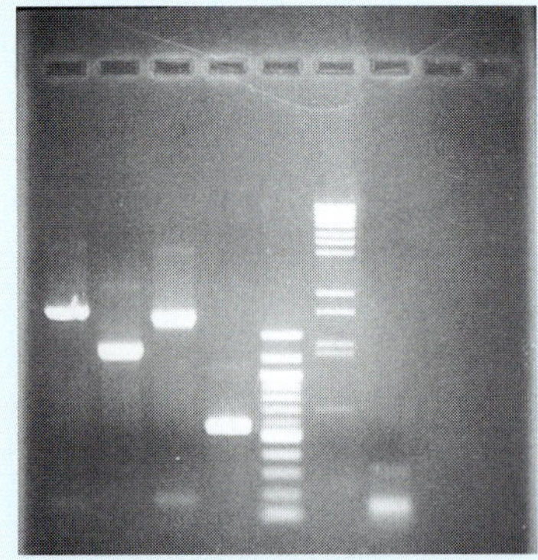

Fig 4-4 Expected gel pattern of the DNA fragments.

LAB PERIOD IV.2.3 CONT.

PROCEDURAL NOTE

Remember that to get an accurate average insert size you would need to amplify the inserts from 50 to 100 plaques.

Step 16 Examine the *Ttr*-specific PCR reaction product (lane 8) to see whether there is a band. If there is a band, determine its size by comparison with the molecular size standards. Does the size of this band correspond to that expected for the *Ttr* sequence (see Figure 1-8 for a diagram of the *Ttr* gene and the primer sequence locations)? Recall that in a cDNA library, the phage contain only transcribed exon sequences and are devoid of intron sequences. The PCR product from a cDNA clone will therefore be smaller than the same PCR product derived from genomic DNA (if the primers used for PCR cross an exon/intron boundary). Further confirmation of the identity of this PCR product (other than the fact that it is the right size) can be obtained by Southern blotting this agarose gel (as in Project I) and probing the blot with a specific oligonucleotide complementary to a sequence in the middle of the expected *Ttr* PCR fragment. Direct confirmation of the identity of the sequence of this band could also be obtained by purifying this PCR product from the agarose gel and subjecting it to DNA sequence analysis (as in Project II). Because you have already experienced those procedures in the other projects, however, we will rely simply on observing a band of the expected length. If you obtained a band of 148 bp, you can conclude that the cDNA library likely contains the *Ttr* gene.

MODULE MATERIALS AND REAGENTS (PER PAIR OF STUDENTS)

MODULE IV.1

ZAP cDNA Synthesis Kit (Stratagene)

Gigapack Gold™ packaging extract (Stratagene)

Poly A+ mRNA (cDNA RNA from Project III.1.2, Step 20 or supplied) (637 µl) (American Bioanalytical)

Flame-drawn Pasteur pipette (2)

Phenol (200 µl). See Appendix II.

Chiasm (300 µl). See Appendix II.

DEPC-treated ddH2O (32.5 µl). See Appendix II.

Phase Lock Light™ tubes (4) (Eppendorf)

3 M NaOAC, pH 5.2 (20 µl). See Appendix II.

100% Ethanol (400 µl) (American Bioanalytical)

70% ethanol (500 µl). See Appendix II.

1% Bromophenol Blue (*not* 5x BJ) (5 µl). See Appendix II.

XL-1 Blue MRF' plating cells (400 µl). See Appendix II.

Chromospin-400 spin column (1) (Clontech)

45-mm diameter Millipore membrane (1) NZY agar plates (2). See Appendix II.

LBMM medium (50 ml). See Appendix II.

10 mM $MgSO_4$ (10 ml). See Appendix II.

0.5% IPTG (30 µl). See Appendix II.

NZY top agarose (6 ml). See Appendix II.

250 mg/ml X-gal (in dimethyl formamide) (150 µl). See Appendix II.

SM buffer (1.1 ml). See Appendix II.

Chloroform (20 µl) (American Bioanalytical)

LB tetracycline plates (1). See Appendix II.

Tetracycline (5 mg/ml) (75 µl). See Appendix II.

XL-1 Blue MRF' (Stratagene)

Nanofuge

Microfuge

Heat block

MODULE IV-2

AmpliTaq Gold™ PCR kit (5 reactions) (Applied Biosystems)

T3 Primer (10 pmol/µl, 4 µl) (American Bioanalytical)

Ethidium bromide staining solution. See Appendix II.

Stain and destain plastic trays

T7 Primer (10 pmol/µl, 4 µl) (American Bioanalytical)

Ttr Forward Primer (10 pmol/µl, 1 µl) (American Bioanalytical)

Ttr Reverse Primer (10 pmol/µl, 1µl) (American Bioanalytical)

1x TAE (600 ml). See Appendix II.

5x Blue Juice loading dye (5x BJ) (15 µl). See Appendix II.

100-bp ladder in BJ (50 µg/µl, 10 µl). See Appendix II.

Lambda BstEII DNA in BJ (25 ng/µl, 6 µl). See Appendix II.

Agarose (1.2 g) (American Bioanalytical)

Electrophoresis power supply, electrophoresis apparatus

Power supply

UV transilluminator

UV photography device

Spatula

Heat block

Thermal cycler

Vortex mixer

Incubator

Nanofuge

Study Questions for Project IV

Module IV.1

1. In Lab Period IV.1.1, why is the dCTP in the dNTP mix methylated?

2. In Lab Period IV.1.1, why is there a CTCGAG sequence in the linker-primer?

3. In Lab Period IV.1.2, why is the dNTP mix not methylated?

4. In Lab Period IV.1.2, why do you think the RNase H and DNA polymerase I are added together rather than sequentially?

5. In Lab Period IV.1.3, Part A, why is it necessary to blunt the ends of the ds cDNA?

6. In Lab Period IV.1.3, Part B, how do the phase lock gel tubes make organic extractions easier and more efficient?

7. In Lab Period IV.1.4, Part B, why are the EcoRI adaptors blunt at one end and sticky on the other? Why is the 5' sticky end not phosphorylated?

8. In Lab Period IV.1.4, Part B, what is the purpose of the ATP (not dATP!) in Step 3?

9. In Lab Period IV.1.5, Part A, what is the purpose of the polynucleotide kinase step?

10. In Lab Period IV.1.5, Part B, what is the purpose of cutting your ds cDNA with the restriction endonuclease XhoI?

11. In Lab Period IV.1.6, Part A, what is the purpose of spinning your cDNA through the Chromaspin column? What does the addition of the bromphenol blue show?

12. In Lab Period IV.1.6, Part B, what is the purpose of the organic extraction (i.e., what are you trying to remove from your cDNA sample)?

13. In Lab Period IV.1.6, Part C, what is the purpose of the drop dialysis step (i.e., what are you trying to remove from your cDNA sample with this step)?

14. In Lab Period, IV.1.7, Part A, why is it usually necessary to concentrate your cDNA sample prior to ligation to the vector?

15. In Lab Period IV. 1.7, Part B, describe how this ligation is a "unidirectional" ligation. Draw a diagram to illustrate your answer.

16. In Lab Period IV.1.8, Part A, what is the purpose of adding the chloroform to your packaged phage?

17. In Lab Period IV.1.8, Part B, what is the purpose of adding the IPTG in Step 4?

18. In Lab Period IV.1.8, Part B, what is the purpose of adding the X-gal in Step 5?

19. In Lab Period IV.1.9, how many clear plaques did you have (on the plate that is most reasonable to count)? How many blue plaques?

20. In Lab Period IV.1.9, what is the titer of your library per microliter? What is the total titer of your library?

21. In Lab Period IV.1.9, what percentage of recombinant plaques did you get?

22. Do you believe you were successful in preparing a cDNA library from mouse liver RNA?

23. On the basis of your titer and the percentage of recombinant data, do you feel that your cDNA library includes most or all of the mRNAs present in the original poly A+ RNA sample?

24. Why is it advantageous to begin with poly A+ RNA when constructing a cDNA library?

25. Why is it advantageous to construct a unidirectional cDNA library in an expression vector?

Module IV.2

1. In Lab Period IV.2.1, what is the purpose of selecting four random plaques from which to obtain phage?

2. In Lab Period IV.2.2, Part A, why are the T3 and T7 primers the appropriate ones to use for this polymerase chain reaction?

3. In Lab Period IV.2.2, Part A, Step 8A, what three important things are accomplished by heating the reaction to 94°C for 12 minutes?

4. In Lab Period IV.2.2, Part B, why are the *Ttr* forward and reverse primers the appropriate ones to use for this polymerase chain reaction?

5. In Lab Period IV.2.3, why is the 100-bp ladder an excellent marker to use for this gel?

6. What was the average insert size in your cDNA library?

7. What were the largest and smallest inserts you detected in your cDNA library?

8. On the basis of the data that you have collected, do you feel your cDNA library includes representation of predominantly full-length mRNAs?

9. Could you specifically detect the *Ttr* sequence in your cDNA library?

10. What can you conclude about expression of the *Ttr* gene in mouse liver on the basis of this result?

11. Explain why genomic libraries made from different tissues from the same organism will be essentially identical, whereas cDNA libraries made from different tissues from the same organism will be different from one another.

AVAILABLE KITS AND COMPONENTS

I

Reagent kits may be ordered from American Bioanalytical (http://americanbio.com). Some materials will be shipped directly from the manufacturers. Certain materials will be shipped independently, such as alcohols, phenol, or chisam or items that are to be shipped and stored at alternate temperatures (e.g., −20°C or 4°C).

Module I.1. Mouse Genomic DNA Isolation and Analysis

Module I.2. Southern Blot Experiment to Detect Gene-Specific Sequences in Genomic DNA

Module I.3. PCR Amplify a Specific Gene from Mouse Genomic DNA

Module II.1. Prepare Mouse Genomic DNA for Cloning, Production, and Titering of a Lambda Phage Genomic Library

Module II.2. Screen a Lambda Phage Library to Detect Clones of a Specific Gene

Module II.3. PCR Subclone into a Plasmid Vector, Propagate in Bacterial Cells, and Prepare Purified Recombinant Plasmid DNA

Module II.4. Dideoxy Thermal Cycle DNA Sequencing of a Cloned Gene

Module II.5. Bioinformatic Analysis of the DNA Sequence Data (no kit necessary)

Module III.1. Isolation and Analysis of Total RNA from Mouse Liver

Module III.2. Northern Blot Experiment to Detect a Specific Gene Transcript

Module III.3. Reverse Transcriptase-Polymerase Chain Reaction Amplification of a Specific Gene Transcript

Module IV.1. Construction of a cDNA Library and cDNA Library Analysis

Module IV.2. Characterization and Screening of a cDNA Library Using the Polymerase Chain Reaction (PCR)

GENERAL MOLECULAR BIOLOGY LABORATORY SETUP

We recommend that students work in pairs for most of the procedures described in this manual. In our experience, this consistently increases student accuracy ("two heads are better than one") and reinforces the procedures for each of them. We strongly suggest that the members of the pair do each procedure together so that each student gains experience with the hands-on protocol.

General Laboratory Equipment

PROCEDURAL NOTE

These items are required for most modules!

Distilled water supply (Milli-Q or similar or another source of molecular biology grade water)

Fume hood

Autoclave

Darkroom access (developing and fixer tanks, water) with red or orange safelight

Thermocycler instrument

Ice buckets

Micropipette sets (2 µl, 20 µl, 200 µl, 1000 µl)

Hybridization (rocking) oven or shaking water baths

Ultraviolet cross-linker (optional)

Water bath "floats" for holding tubes

Hybridization bags (Southerns, Northerns, plaque lifts)

Storage racks or boxes for –20°C and 4°C storage

0.22-µm filter sterilization devices

50-ml syringes and rubber stoppers with holes

Heating plates/stirrers

Heat blocks

Vortex mixers

Spin vacuum concentrator

Microcentrifuge

Benchtop "nanofuge"

Tabletop centrifuge

Bag heat sealer

Agarose gel electrophoresis apparatus

Agarose gel staining and destaining trays

Agarose gel photography setup

Tissue homogenizer (Polytron, Qiagen, or equivalent)

Hazardous waste disposal container

Hazardous waste disposal container for a fume hood

Glass or disposable pipettes (5 ml, 10 ml, 25 ml)

Spatulas

Scissors

Pencils and marking pens

Microcentrifuge tube "lid locks"

Dialysis membranes (Millipore type VSWP 04500, type VS)

Phase Lock Light™ tubes (1.5 ml and 15 ml) (Eppendorf)

Microcon™ 100 concentrator tubes (Amicon)

Gloves

Forceps

Safety goggles

Lab coats

Petri plates

Paper towels

BenchCoat™ or bench covering paper

Kimwipes™

Whatman 3MM™ paper

Plasticware (micropipette tips, tubes [0.5 ml, 0.2 ml, 1.7 ml] 15- and 50-ml plastic conical tubes [polypropylene])

General glassware

A general molecular biology laboratory setup is presented in **Figure A2-1**. Photos of some laboratory equipment are shown in **Figures A2-2** to **A2-16** (note some of the equipment shown in these figures are optional depending upon the modules being done).

A2-1. An example of a molecular biology lab setup.

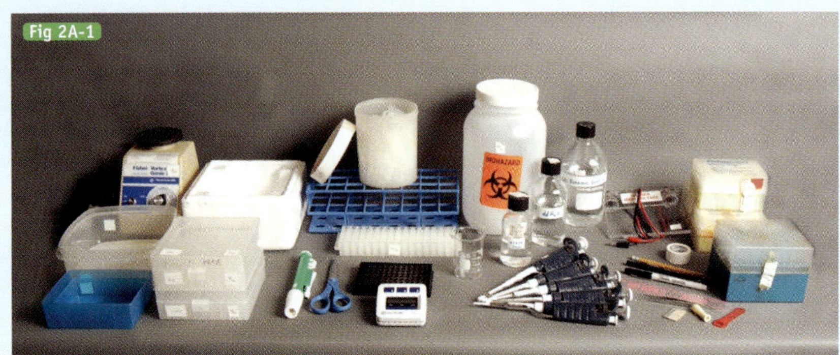

A2-2. Spin vacuum concentrator (Savant Speed Vac; Thermo Savant).

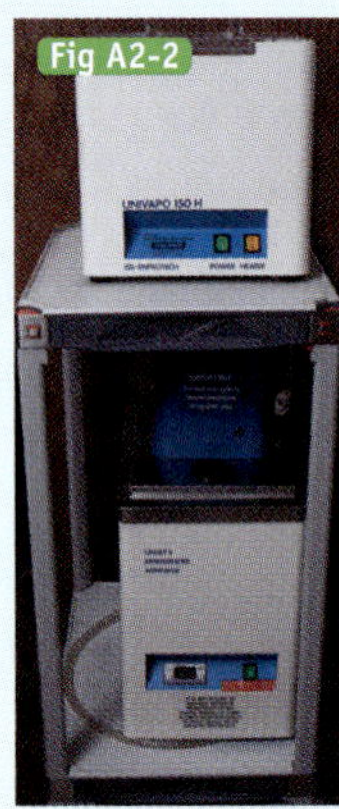

A2-3. Spin vacuum concentrator (inside).

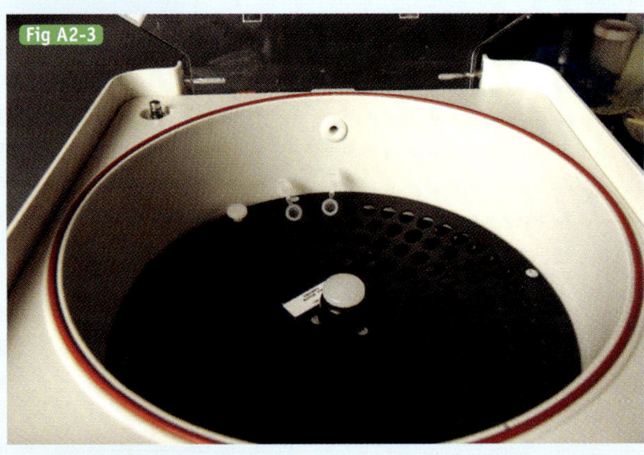

A2-4. Water bath chiller/circulator (Forma Scientific).

A2-5. Waterbath with microcentrifuge tube holder (Precision Scientific).

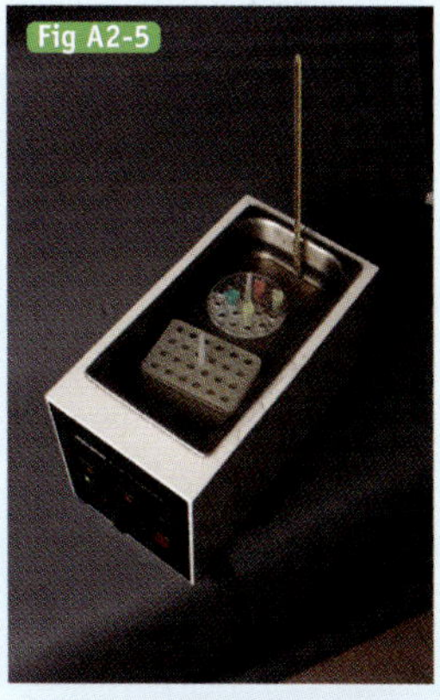

A2-6. Heat sealer (Midwest Pacific).

A2-7. Heat block/lid locks (VUSR or Fischer Scientific).

A2-8. Rocking platform (Hoefer Scientific Instruments).

A2-9. Cross-linker (Bio-Rod).

A2-10. Nanofuge (Fischer Scientific).

A2-11. Cell homogenizer (Polytron).

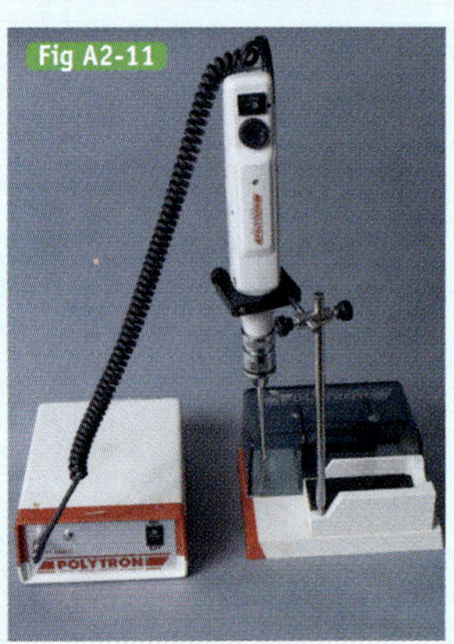

A2-12. Microcentrifuge (refrigerated version is also available) (Eppendorf).

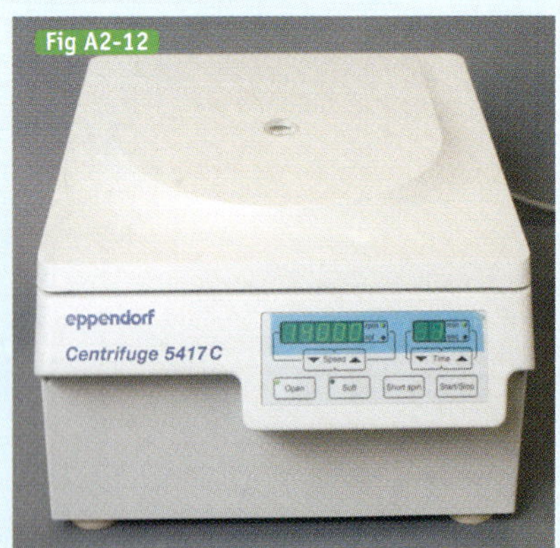

A2-13. Capillary automated DNA sequencer (Applied Biosystems).

A2-14. Thermal cycler (MJ/BioRad).

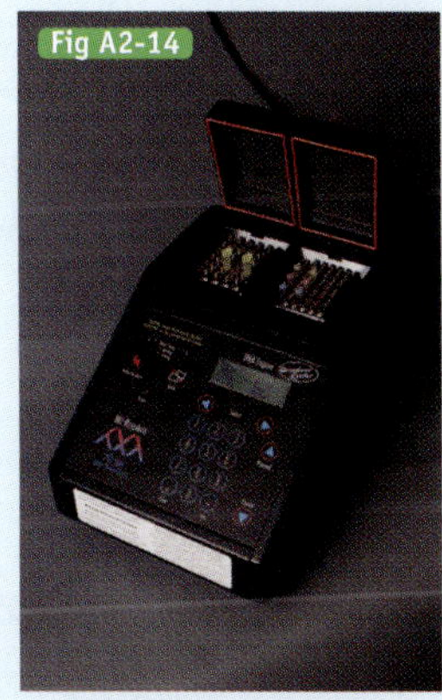

A2-15. Thermal cycler (Applied Biosystems).

A2-16. Q-PCR thermal cycler machine (Applied Biosystems).

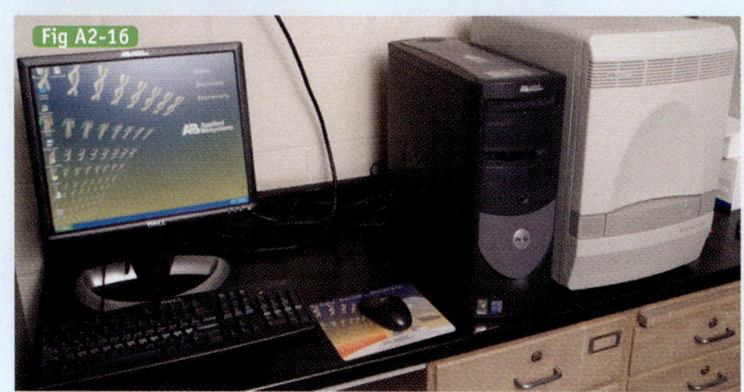

Reagents and Recipes Used in This Manual

All of the general reagent solutions listed below are available from American Bioanalytical (see Appendix I). These can be ordered by specific module. However, recipes for each solution are also provided below for those who prefer to prepare their own reagents. All solutions should be made with double-distilled (Milli-Q or molecular biology grade) water (ddH$_2$O) and sterilized by either autoclaving or 0.2-μm-filter sterilization, except as noted.

General Reagents

Restriction endonucleases and general molecular cloning enzymes and buffers are available from New England Biolabs. These include, restriction endonucleases such as EcoRI, HindIII, AluI, SfiI, MspI, etc., as well as DNA polymerases, ligases, β-agarase, and their buffers and supplements. Further details on their use can be found in the New England Biolabs catalog or at their website (www.neb.com). Also available are agarose gel electrophoresis DNA markers (control markers): HindIII marker DNA, BstEII marker DNA, and 100-bp ladder marker DNA (see Appendix III for a description of these markers).

General chemicals and reagents are available from suppliers such as American Bioanalytical or Fisher Scientific. When specific suppliers are recommended, they are listed with the specific reagent:

Molecular biology grade water (produced by a Milli-Q™ or similar filtration device or purchased commercially)

Agarose, molecular biology grade (American Bioanalytical)

Low-melt agarose, molecular biology grade (American Bioanalytical)

DNAse-free RNAse (Applied Biosystems or New England Biolabs)

Darkroom developer and fixer reagents (Fisher Scientific [Kodak GBX Fixer and GBX Developer])

PCR reagents (PCR and RT-PCR reagents and enzymes are available from New England Biolabs, Applied Biosystems, and/or Roche Diagnostics)

Oligonucleotide primers (all "standard" vector-based DNA primers/probes are available from New England Biolabs, or from the manufacturer of the specific vector; primer sequences are listed in each specific application and are available from American Bioanalytical)

cDNA synthesis/cloning and bacteriophage packaging extract (all reagents for these protocols are available in the Stratagene cDNA synthesis kits; supplementary material for these kits includes competent cells and *E. coli* strains, which are also available from Stratagene).

Lambda phage genomic cloning and bacteriophage packaging (all reagents for these protocols are available in the Stratagene Lambda Zap Express synthesis and cloning kits)

Preparation of plasmid DNA and purification of PCR products (all reagents are in the Qiagen Mini-Prep Kit or QiaQuick DNA Purification Kits, and all recipes are provided in the instruction booklets that accompany the kits)

Purification of RNA from tissue (Qiagen RNEasy kits)

Purification of poly A+ RNA (Ambion Micro Poly(A) Purist™ kit)

DNA sequencing reagents (all reagents for the nonradioactive fluorescence-based cycle sequencing are from Applied Biosystems; the kit is the Applied Biosystems Big Dye Terminator Cycle Sequencing Ready Reaction Mix version 3.1)

Fluorescein-PCR labeling (the nonradioactive labeling kit is available from Roche Diagnostics)

The ECL™ chemiluminescent system (GE Healthcare)

The Phototope™ chemiluminsescent system (New England Biolabs)

Random primer biotin labeling (the kit [NEBlot Phototope™] is available from New England Biolabs)

Stock Solutions

PROCEDURAL NOTES

ddH$_2$O: double-distilled water (18-megohm). The water should be purified through a Milli-Q™ (or similar system) and then sterilized by either autoclaving or 0.2-μm-filter sterilization, or obtained commercially (American Bioanalytical).

Concentrations in parentheses are the final concentrations of the components in the final solution.

5 M NaCl
To make 1 L: dissolve 292 g NaCl in 800 ml ddH$_2$O; fill to 1 L with ddH$_2$O; sterilize.

2 M Tris, pH 8.0
To make 1 L: dissolve 242 g Tris in 800 ml ddH$_2$O; adjust pH to 8.0; fill to 1 L with ddH$_2$O; sterilize.

0.5 M EDTA, disodium dihydrate
To make 1 L: dissolve 186.12 g EDTA disodium dihydrate in 800 ml ddH$_2$O; adjust pH to 8.0; fill to1 L with ddH$_2$O; sterilize.

100x TE Buffer (1 M Tris-HCl + 0.1 M EDTA; pH 8.0)
To make 100 ml: 50 ml 2 M Tris (pH 8.0) + 20 ml 0.5 M EDTA (pH 8.0); fill to 100 ml with ddH$_2$O; sterilize.

1x TE Buffer (0.01 M Tris-HCl + 0.001 M EDTA; pH 8.0)
Dilute 100x TE Buffer 1:100 in sterile ddH$_2$O; sterilize.

TNE buffer (10 mM Tris [pH 7.5] + 10 mM NaCl + 1 mM EDTA [pH 8.0])
To make 500 ml: 1 ml 0.5 M EDTA (pH 8.0) + 2.5 ml 2 M Tris (pH 7.5) + 1 ml 5 M NaCl; fill to 500 ml; sterilize.

70% EtOH
To make 100 ml: 70 ml 100% EtOH + 30 ml sterile ddH$_2$O; do not autoclave.

4 M LiCl
To make 100 ml: dissolve 17.0 g LiCl in 80 ml sterile ddH$_2$O; fill to 100 ml with sterile ddH$_2$O; 0.2-μm-filter sterilize.

1 M MgCl$_2$
To make 100 ml: dissolve 20.3 g MgCl$_2$•6H$_2$O in 80 ml sterile ddH$_2$O; fill to 100 ml with sterile ddH$_2$O; 0.2-μm-filter sterilize.

1 M MgSO$_4$
To make 100 ml: dissolve 12.0 g MgSO$_4$ in 80 ml sterile ddH$_2$O; fill to 100 ml with sterile ddH$_2$O; 0.2-μm-filter sterilize.

20% (w/v) maltose
To make 100 ml: dissolve 20.0 g maltose in 80 ml sterile ddH$_2$O; fill to 100 ml with sterile ddH$_2$O; 0.2-μm-filter sterilize.

20x SSC (3 M NaCl + 0.3 M sodium citrate; pH 7.0)
To make 1 L: 175.3 g NaCl + 88.2 g sodium citrate + 800 ml H_2O; adjust to pH 7.0 with a few drops of 1 M HCl; fill to 1 L with ddH_2O; sterilize.

2x SSC (0.3 M NaCl + 0.03 M sodium citrate; pH 7.0)
To make 1 L: mix 100 ml 20x SSC with 900 ml sterile water.

10x SSC (1.5 M NaCl + 0.15 M sodium citrate; pH 7.0)
To make 1 L: mix 500 ml 20x SSC with 500 ml sterile ddH_2O.

6x SSC (0.9 M NaCl + 0.09 M sodium citrate; pH 7.0)
To make 800 ml: mix 240 ml 20x SSC with 560 ml sterile ddH_2O.

SM (phage storage buffer) (0.1 M NaCl + 0.05 M Tris-HCl [pH 7.5] + 10 mM $MgSO_4$)
To make 500 ml: 10 ml 5 M NaCl + 12.5 ml 2 M Tris (pH 7.5) + 5 ml 1 M $MgSO_4$; fill to 500 ml with ddH_2O; sterilize.

3 M NaOAc (sodium acetate), pH 5.2
To make 100 ml: dissolve 24.61 g NaOAc in 80 ml ddH_2O; adjust pH to 5.2 with glacial acetic acid; fill to 100 ml with ddH_2O; sterilize.

3 M NaOAc (sodium acetate), pH 7.0
To make 100 ml: dissolve 24.61 g NaOAc in 80 ml ddH_2O; adjust pH to 7.0 with glacial acetic acid; fill to 100 ml with ddH_2O; sterilize.

7.5 M NH_4OAc (ammonium acetate)
To make 100 ml: dissolve 57.83 g NH_4OAc in 80 ml ddH_2O; fill to 100 ml with ddH_2O; sterilize.

20% (w/v) SDS (Sodium Dodecyl Sulfate)
To make 1 L: dissolve 200 g SDS in 800 ml sterile ddH_2O; fill to 1 L with sterile ddH_2O; do not autoclave.

CAUTION!

Wear mask and gloves; avoid eye contact.

Agarose Gel Electrophoresis Reagents

50x TAE buffer (2 M Tris + 0.05 M EDTA; pH 8.3)
To make 1 L: 242 g Tris + 900 ml ddH_2O; adjust pH to 8.3 with 57.1 ml glacial acetic acid; add 18.6 g EDTA; fill to 1 L with ddH_2O. This solution can be sterilized but is not required.

1x TAE buffer (40 mM Tris + 1 mM EDTA; pH 8.3)
To make 500 ml: 10 ml 50x TAE + 490 ml ddH_2O. This solution can be sterilized but is not required.

Ethidium bromide stock solution (10 mg/ml)
To make 10 ml: dissolve 100 mg ethidium bromide in 10 ml sterile ddH_2O; store at 4°C as a 10 ml aliquot. Do not sterilize.
Use 300 µl of the stock solution per 500 ml of ddH_2O to prepare a bath for staining agarose gels.

CAUTION!

Ethidium bromide is known as mutagen carcinogen.

5x Blue Juice agarose gel loading dye (5x BJ) (20% [w/v] Ficoll + 0.01 M Tris-HCl [pH 8.0] + 0.01 M EDTA [pH 8.0] + 0.09% [w/v] xylene cyanol FF + 0.09% [w/v] bromphenol blue)
To make 80 ml: 16 g Ficoll (Pharmacia or Sigma) + 400 µl 2 M Tris (pH 8.0) + 1.6 ml 0.5 M EDTA (pH 8.0) + 72 mg xylene cyanol FF + 72 mg bromphenol blue; fill to 80 ml with sterile ddH$_2$O. Do not sterilize.

Plaque Lift Solutions

PROCEDURAL NOTE

These solutions can be sterilized except as noted, but sterilization is not required.

Plaque lift denaturing solution (0.5 M NaOH + 1.5 M NaCl)
To make 1 L: 20 g NaOH + 87.66 g NaCl ; fill to 1 L with ddH$_2$O.

Plaque lift neutralizing solution (0.5 M Tris [pH 7.5] + 1.5 M NaCl)
To make 1 L: 250 ml 2 M Tris (pH 7.5) + 87.66 g NaCl; fill to 1 L with ddH$_2$O.

Plaque wash buffer (50 mM Tris [pH 8.0] + 1 M NaCl + 10 mM EDTA [pH 8.0] + 0.1% [w/v] SDS)
To make 1 L: 50 ml 1.0 M Tris (pH 8.0) + 200 ml 5 M NaCl + 20 ml 0.5 M EDTA (pH 8.0) + 5 ml 20% (w/v) SDS; fill to 1 L with ddH$_2$O. Store at room temperature, but use at 50°C.

Probe-Amp hybridization solution (5x SSC + 0.5% [w/v] ECL Blocking Agent + 0.1% [w/v] SDS + 5% [w/v] Dextran sulfate + 0.1 mg/ml salmon sperm DNA)
To make 1 L: 250 ml 20x SSC + 5 g ECL blocking agent (nonfat dry milk) + 5 ml 20% (w/v) SDS + 500 ml ddH$_2$O; heat to 60°C with stirring until the solution is cloudy and there is no precipitate; remove from heat, and add 50 g dextran sulfate; stir until dissolved; add 100 mg or 10 ml of 10 mg/ml denatured salmon sperm DNA (American Bioanalytical); stir until the solution is cloudy and there is no precipitate; bring to a final volume of 1 L with sterile ddH$_2$O while stirring the solution; aliquot the solution into 50-ml orange cap tubes, and store at −20°C. Do not autoclave.

PROCEDURAL NOTE

10 mg/ml denatured salmon sperm DNA is provided by American Bioanalytical. Or, to make 10 ml: resuspend 100 mg salmon sperm DNA in 10 ml sterile ddH$_2$O, boil 10 minutes and store at −20°C.

Plaque lift first wash (post-hybridization) (1x SSC + 0.1% [w/v] SDS)
To make 1 L: 50 ml 20x SSC + 5 ml 20% SDS; fill to 1 L with ddH$_2$O.

Plaque lift second wash (post-hybridization) (0.5x SSC + 0.1% [w/v] SDS)
To make 1 L: 25 ml 20x SSC + 5 ml 20% SDS; fill to 1 L with ddH$_2$O.

Southern Blot Solutions

PROCEDURAL NOTE

These solutions can be sterilized except as noted, but sterilization is not required.

Southern depurination dolution (0.25 M HCl)
To make 1 L: 20.65 ml concentrated HCl; fill to 1 L with ddH$_2$O). Do not autoclave.

Southern denaturation solution (0.5 M NaOH + 1.5 M NaCl)
To make 1 L: 50 ml 10 M NaOH + 300 ml 5 M NaCl; fill to 1 L with ddH$_2$O.

Southern neutralization solution (1 M Tris-HCl [pH 7.5] + 1.5 M NaCl)
To make 1 L: 500 ml 2 M Tris-HCl (pH 7.5) + 300 ml 5 M NaCl; fill to 1 L with ddH_2O.

Southern Blot Wash and Detection Solutions ("Genomic Southern," Module I.2)

PROCEDURAL NOTE

These solutions can be sterilized except as noted, but sterilization is not required.

Probe-Amp hybridization solution (5x SSC + 0.5% [w/v] ECL Blocking Agent + 0.1% [w/v] SDS + 5% [w/v] Dextran sulfate + 0.1 mg/ml salmon sperm DNA)
To make 1 L: 250 ml 20x SSC + 5 g ECL blocking agent (nonfat dry milk) + 5 ml 20% (w/v) SDS + 500 ml of ddH_2O; heat to 60°C with stirring until the solution is cloudy and there is no precipitate; remove from heat, and add 50 g dextran sulfate; stir until dissolved; add 100 mg (or 10 ml of 10 mg/ml) denatured salmon sperm DNA, and stir until the solution is cloudy and there is no precipitate; bring to a final volume of 1 L with sterile ddH_2O while stirring the solution; aliquot the solution into 50 ml orange cap tubes, and store at −20°C. Do not autoclave.

PROCEDURAL NOTE

10 mg/ml denatured salmon sperm DNA is provided by American Bioanalytical. Or, to make 10 ml: resuspend 100 mg salmon sperm DNA in 10 ml sterile ddH_2O, boil 10 minutes and store at −20°C.

Genomic Southern Wash I (1x SSC + 0.1% [w/v] SDS)
To make 1 L: 50 ml 20x SSC + 5 ml 20% (w/v) SDS; fill to 1 L with ddH_2O.

Genomic Southern Wash II (0.5x SSC + 0.1% [w/v] SDS)
To make 1 L: 25 ml 20x SSC + 5 ml 20% (w/v) SDS; fill to 1 L with ddH_2O.

ECL Buffer 1 (0.15 M NaCl + 0.1 M Tris; pH 7.5)
To make 1 L: 8.77 g NaCl (or 30 ml 5 M NaCl) + 12.1 g Tris (or 50 ml 2 M Tris, pH 7.5); adjust pH to 7.5; fill to 1 L with ddH_2O.

0.1 % (v/v) Tween in ECL Buffer 1 (0.15 M NaCl + 0.1 M Tris + 0.1% [v/v] Tween 20; pH 7.5)
To make 1 L: 8.77 g NaCl (or 30 ml 5 M NaCl) + 12.1 g Tris (or 50 ml 2M Tris, pH 7.5) + 800 ml ddH_2O; stir until dissolved, and add 1 ml Tween 20; stir; adjust pH to 7.5; fill to 1 L with ddH_2O.

Antibody blocking solution (0.5% [w/v] nonfat dry milk in ECL Buffer 1: 0.15 M NaCl + 0.1M Tris; pH 7.5)
To make 1 L: dissolve 5 g instant nonfat dry milk in 1 L ECL Buffer 1. Do not autoclave.

Antibody solution (GE Healthcare) (1:1000 dilution of FL-HRP antibody in a solution of 0.5% [w/v] BSA in ECL Buffer 1).
To make 100 ml: 0.5 g BSA (Bovine Serium Albumin) + 100 ml ECL Buffer 1; stir until dissolved, and add 100 μl FL-HRP antibody; stir; make fresh before use. Do not autoclave.

ECL detection solution (GE Healthcare)
Mix an equal volume of ECL detection solutions 1 and 2 to give sufficient volume to cover the filters. Do not sterilize.

Southern Blot Wash and Detection Solutions ("Ttr PCR Southern," Module I.3)

PROCEDURAL NOTE

These solutions can be sterilized except as noted, but sterilization is not required.

0.5 M Disodium phosphate stock solution (0.5 M Na_2HPO_4 + 0.34% [w/v] H_3PO_4)
To make 1 L: 134 g $Na_2HPO_4 \bullet$ 7 H_2O + 3.4 g H_3PO_4; fill to 1 L with ddH_2O; 0.2-μm-filter sterilize.

Ttr hybridization solution (0.25 M disodium phosphate + 1 mM EDTA + 7% [w/v] SDS; pH 7.2)
To make 1 L: 500 ml 0.5 M disodium phosphate stock solution (see above) + 2 ml 0.5 M EDTA + 350 ml 20% (w/v) SDS; adjust pH to 7.2; fill to 1 L with ddH_2O.

Oligo Southern Wash I (2x SSC + 0.1% [w/v] SDS; pH 7.0)
To make 1 L: 100 ml 20x SSC + 5 ml 20% (w/v) SDS; adjust pH to 7.0; fill to 1 L with ddH_2O.

Oligo Southern Wash II (1x SSC + 0.1% [w/v] SDS; pH 7.0)
To make 1 L: 50 ml 20x SSC + 5 ml 20% (w/v) SDS; adjust pH to 7.0; fill to 1 L with ddH_2O.

Phototope™ blocking solution (125 mM NaCl + 17 mM Na_2HPO_4 + 8 mM NaH_2PO_4 + 5% [w/v] SDS)
To make 1 L: 7.3 g NaCl + 2.41 g Na_2HPO_4 + 0.96 g NaH_2PO_4 + 50 g SDS + 800 ml ddH_2O; adjust pH to 7.2; fill to 1 L with ddH_2O.

Streptavidin solution (a concentrated Streptavidin solution is provided in the New England Biolabs Phototope kit). When used, this solution is diluted 1:1000 in Phototope blocking solution.
To make 10 ml: 10 μl concentrated Streptavidin solution in 10 ml Phototope blocking solution. Do not autoclave.

Biotinylated alkaline phosphatase solution (a concentrated alkaline phosphatase solution is provided in the New England Biolabs Phototope kit). When used, this solution is diluted 1:1000 in Phototope blocking solution.
To make 10 ml: 10 μl concentrated alkaline phosphatase solution in 10 ml Phototope blocking solution. Do not autoclave.

Phototope™ detection wash solution I (this is a 1:10 dilution of Phototope blocking solution in ddH_2O)
(12.5 mM NaCl + 1.7 mM Na_2HPO_4 + 0.8 mM NaH_2PO_4 + 0.5% [w/v] SDS)
To make 1 L: 100 ml Phototope blocking solution + 900 ml ddH_2O; or 0.73 g NaCl + 0.241 g Na_2HPO_4 + 0.096 g NaH_2PO_4 + 5 g SDS + 800 ml ddH_2O; adjust pH to 7.2; fill to 1 L with ddH_2O.

Phototope™ detection wash solution II (10 mM Tris + 10 mM NaCl + 1 mM $MgCl_2$; pH 9.5)
To make 1 L: 10 ml 1 M Tris + 2 ml 5 M NaCl + 1 ml 1 M $MgCl_2$ + 800 ml ddH_2O; adjust pH to 9.5; fill to 1 L with ddH_2O.

PROCEDURAL NOTES

The pH of this solution is critical!

The stock solution can be made as 10x (100 mM Tris + 100 mM NaCl + 10 mM $MgCl_2$; pH 9.5); dilute 1:10 with ddH_2O before use.

Phototope™ Detection Mix (CDP-Star™) (New England Biolabs) — Phototope detection reagent should be diluted just before use in 1x Phototope™ diluent (both provided in the Phototope™ Detection Kit). The diluent is provided as a 25x stock solution that is diluted to 1x in ddH_2O before use.

To make 4 ml of 1:1000 diluted Phototope Detection Mix: 160 μl 25x Phototope™ diluent + 3.84 ml sterile ddH_2O + 4 μl CDP-Star™ detection reagent. Do not sterilize.

RNA Project (Project III) and Northern Blot Reagents

RNA Solutions

Use caution and wear gloves to eliminate RNAse contamination. Refer to Appendix IV for further descriptions.

PROCEDURAL NOTE

All solutions should be made with DEPC-treated double distilled water (DEPC-ddH$_2$O). Tris solutions cannot be DEPC-treated. Solutions should be DEPC treated and autoclaved before adding Tris. After addition of Tris, the solution should be autoclaved again.

DEPC — Diethylpyrocarbonate (Sigma). Stock solution is 10% (v/v) DEPC in ethanol. Store at 4°C.

CAUTION!

DEPC should be handled with caution. Toxic!!

DEPC-ddH$_2$O — Add DEPC stock solution to ddH$_2$O to a final concentration of 0.1% (v/v) in distilled water. Stir the solution slowly for 1 hour at room temperature, and then incubate overnight at 37°C. The next day, autoclave the solution for 15 minutes on liquid cycle.

Northern Blot Solutions

10x MOPS/EDTA (American Bioanalytical) (0.2 M MOPS [3-(N-morpholino) propanesulfonic acid] + 10 mM EDTA; pH 7.0)
To make 1 L: 41.86 g MOPS + 20 ml 0.5 M EDTA + 800 ml DEPC-ddH$_2$O; adjust to pH 7.0; fill to 1 L with DEPC-ddH$_2$O; store at 4°C.

1x MOPS/EDTA (American Bioanalytical) (0.02 M MOPS + 1 mM EDTA; pH 7.0)
To make 1 L: 100 ml 10x MOPS/EDTA + 900 ml DEPC-ddH$_2$O; store at 4°C.

Northern Loading Dye
To make 1.5 ml: 720 µl deionized formamide + 160 µl 10x MOPS (pH 7.0) + 260 µl formaldehyde + 220 µl DEPC-ddH$_2$O + 100 µl 80% glycerol + 40 µl saturated bromphenol blue; do not autoclave.

3 M NaOAc, pH 7.0 (for RNA work)
To make 100 ml: 24.6 g NaOAc + 80 ml DEPC- ddH$_2$O; adjust pH to 7.0 with glacial acetic acid; fill to 100 ml with DEPC- ddH$_2$O.

70% ethanol for RNA work
To make 100 ml: 70 ml 100% ethanol + 30 ml DEPC-ddH$_2$O; store in a DEPC-treated bottle.

Northern prehybridization solution (4x SSPE + 1% [w/v] SDS + 5x Denhardt's solution + 50% [v/v] formamide + 1 mg/ml denatured salmon sperm DNA)
To make 1 L: 200 ml 20x SSPE (see below) + 50 ml 20% (w/v) SDS + 100 ml 50x Denhardt's solution (see below) + 500 ml formamide + 1000 mg or 100 ml 10 mg/ml denatured salmon sperm DNA (American Bioanalytical); fill to 1 L with DEPC-ddH$_2$O.

PROCEDURAL NOTE

10 mg/ml denatured salmon sperm DNA is provided by American Bioanalytical. Or, to make 100 ml: resuspend 1000 mg salmon sperm DNA in 10 ml sterile ddH$_2$O, boil 10 minutes and store at −20˚C.

Northern Hybridization Solution (4x SSPE + 1% [w/v] SDS + 5x Denhardt's Solution + 50% [v/v] formamide + 5% [w/v] dextran sulfate)
To make 1 L: 200 ml 20x SSPE (see below) + 50 ml 20% (w/v) SDS + 100 ml 50x Denhardt's Solution (see below) + 500 ml formamide + 50 g dextran sulfate; fill to 1 L with DEPC-ddH$_2$O.

20x SSPE (3 M NaCl + 0.2 M NaH$_2$PO$_4$ + 20 mM EDTA; pH 7.4)
To make 1 L: 175.32 g NaCl + 27.6 g NaH$_2$PO$_4$•H$_2$O + 7.45 g EDTA disodium dihydrate + 800 ml DEPC- ddH$_2$O; adjust pH to 7.4; fill to 1 L with DEPC-ddH$_2$O.

50x Denhardt's Solution
To make 250 ml: 2.5 g Ficoll + 2.5 g polyvinyl-pyrrolidone + 2.5 g bovine serum albumin (fraction V); fill to 250 ml with DEPC-ddH$_2$O; 0.2-μm-filter sterilize; store at –20°C.

Northern Wash Solution I (2x SSPE + 0.1% [w/v] SDS)
To make 1 L: 100 ml 20x SSPE + 5 ml 20% SDS; fill to 1 L with ddH$_2$O.

Northern Wash Solution II (1x SSPE + 0.1% [w/v] SDS)
To make 1 L: 50 ml 20x SSPE + 5 ml 20% SDS; fill to 1 L with ddH$_2$O.

Guanadinium thiocyanate solution
To make 1 L: 60 g guanadinium thiocynate + 5 ml 1 M Na citrate + 0.5 ml β-mercaptoethanol (BME) + 80 ml DEPC-ddH$_2$O; adjust pH to 7.0; fill to 100 ml with DEPC-ddH$_2$O; 0.2-μm-filter sterilize; then add 0.5 g SDS.

5.7 M CsCl
To make 70 ml: 67.2 g CsCl + 1.16 g disodium EDTA; fill to 70 ml with DEPC-ddH$_2$O.

> **CAUTION!**
>
> BME is hazardous!
> ___
> When handling with SDS, wear mask and gloves; avoid eye contact.

Phototope™ Detection Reagents

PROCEDURAL NOTE

These solutions can be sterilized except as noted, but sterilization is not required.

Phototope™ blocking solution (125 mM NaCl + 17 mM Na$_2$HPO$_4$ + 8 mM NaH$_2$PO$_4$ + 5% [w/v] SDS)
To make 1 L: 7.3 g NaCl + 2.41 g Na$_2$HPO$_4$ + 0.96 g NaH$_2$PO$_4$ + 50 g SDS + 800 ml ddH$_2$O; adjust pH to 7.2; add ddH$_2$O to 1 L.

Streptavidin solution — a concentrated streptavidin solution is provided in the New England Biolabs Phototope kit. When used, this solution is diluted 1:1000 in Phototope blocking solution.
To make 10 ml: 10 μl concentrated streptavidin solution in 10 ml Phototope blocking solution. Do not autoclave.

Biotinylated alkaline phosphatase solution — a concentrated alkaline phosphatase solution is provided in the New England Biolabs Phototope kit. When used, this solution is diluted 1:1000 in Phototope blocking solution.
To make 10 ml: 10 μl concentrated alkaline phosphatase solution in 10 ml Phototope blocking solution. Do not autoclave.

Phototope™ Detection wash solution I (this is a 1:10 dilution of Phototope blocking solution in ddH$_2$O) (12.5 mM NaCl + 1.7 mM Na$_2$HPO$_4$ + 0.8 mM NaH$_2$PO$_4$ + 0.5% [w/v] SDS)
To make 1 L: 100 ml Phototope blocking solution + 900 ml ddH$_2$O; or 0.73 g NaCl + 0.241 g Na$_2$HPO$_4$ + 0.096 g NaH$_2$PO$_4$ + 5 g SDS + 800 ml ddH$_2$O; adjust pH to 7.2; fill to 1 L with ddH$_2$O.

Phototope™ Detection wash solution II (10 mM Tris + 10 mM NaCl + 1 mM MgCl$_2$; pH 9.5)
To make 1 L: 10 ml 1 M Tris + 2 ml 5 M NaCl + 1 ml 1 M MgCl$_2$ + 800 ml ddH$_2$O; adjust pH to 9.5; fill to 1 L with ddH$_2$O.

PROCEDURAL NOTES

The pH of this solution is critical!

The stock solution can be made as 10x (100 mM Tris + 100 mM NaCl + 10 mM $MgCl_2$; pH 9.5); dilute 1:10 with ddH_2O before use.

Phototope™ Detection (CDP-Star™) (New England Biolabs) — Phototope™ detection reagent should be diluted just before use in 1x Phototope™ diluent (both provided in the Phototope™ Detection Kit). The diluent is provided as a 25x stock solution that is diluted to 1x in ddH_2O before use.

To make 4 ml of 1:1000 diluted Phototope Detection Mix: 160 µl 25x Phototope diluent + 3.84 ml sterile ddH_2O + 4 µl CDP-Star™ detection reagent. Do not sterilize.

Bacterial Growth Media

Ampicillin (50 mg/ml)
To make 1 ml: dissolve 50 mg ampicillin in 0.8 ml sterile ddH_2O; fill to 1 ml with sterile ddH_2O; 0.2-µm-filter sterilize (do not autoclave); store at –20°C.

Tetracycline (5 mg/ml)
To make 1 ml: dissolve 5 mg tetracycline in 0.8 ml sterile ddH_2O; fill to 1 ml with sterile ddH_2O; 0.2-µm-filter sterilize (do not autoclave); store at –20°C in an aluminum foil-wrapped tube (light sensitive!).

250 mg/ml X-Gal stock solution (5-bromo-4-chloro-3-indole-β-D-galactopyranoside)
(Do not make this solution in plastic tubes; polypropylene is acceptable)
To make 50 ml: dissolve 12.5 g X-Gal in 40 ml dimethylformamide; fill to 50 ml with dimethylformamide; aliquot in 1.5 ml microcentrifuge tubes; wrap tube with aluminum foil (light sensitive!); store at –20°C; do not autoclave.

CAUTION!

Dimethylformamide is hazardous!

25 mg/ml X-Gal
To make 10 ml: 1 ml 250 mg/ml stock (see above) + 9 ml dimethylformamide; aliquot in 2-ml microcentrifuge tubes; wrap tube with aluminum foil (light sensitive!); store at –20°C.

0.5 M IPTG (isopropyl-β-D-thiogalactopyranoside)
To make 30 ml: dissolve 3.6 g IPTG in 20 ml sterile ddH_2O; fill to 30 ml with sterile ddH_2O; 0.2-µm-filter sterilize (do not autoclave); aliquot in 1.5-ml microcentrifuge tubes; store at –20°C.

NZY Medium
To make 1 L: 5 g NaCl + 2 g $MgSO_4 \bullet 7H_2O$ + 5 g yeast extract + 10 g NZ amine (casein hydrolysate) + 800 ml ddH_2O; adjust pH to 7.5 with NaOH; fill to 1 L with ddH_2O; or dissolve 22 g NZY powder (Qbiogene) in 1 L ddH_2O; autoclave.

NZY top agarose (0.6% agarose in NZY medium)
To make 1 L: add 6 g agarose in 1 L NZY medium; autoclave; store at 55°C.

NZY plates
To make 1 L: mix 15 g agar and components for 1 L NZY medium; fill to 1 L with ddH_2O; autoclave; pour plates (in the hood or other sterile area) and incubate overnight at room temperature (or until dry) before storing at 4°C; 1 L makes about 30 plates.

LB Medium, Miller
To make 1 L: 10 g tryptone + 5 g yeast extract + 10 g NaCl in 800 ml ddH_2O; adjust pH to 7.0; fill to 1 L with ddH_2O; autoclave; or dissolve 25 g LB Broth (Miller) powder (Difco) in 1 L ddH_2O; autoclave; store at room temperature.

LBMM Medium (10 mM $MgSO_4$ + 0.2% [w/v] maltose in LB medium [Miller])
To make 50 ml: 0.5 ml filter-sterilized 1 M $MgSO_4$ + 0.5 ml filter-sterilized 20% (w/v) maltose + 49 ml LB medium (Miller); store at room temperature; make fresh before use.

LB Agar, Miller
To make 1 L: mix 15 g agar with components for 1 L LB medium (Miller); fill to 1 L with ddH_2O; autoclave; pour plates (in the hood or other sterile area), and incubate overnight at room temperature (or until dry) before storing at 4°C; 1 L makes about 30 plates.

LB Top Agarose (0.6% agarose in LB medium) (Miller)
To make 1 L: dissolve 6 g agarose in a total final volume of 1 L LB medium (Miller); autoclave and store at 55°C.

LB + Amp (50 mg/L) plates
To make 1 L: make up LB agar (Miller) as described above; cool to about 60°C, and then add 50 mg ampicillin (1.0 ml of 50 mg/ml solution) to each 1 L before pouring the plates; pour plates (in the hood or other sterile area), and incubate overnight at room temperature (or until dry) before storing at 4°C; 1 L will make about 30 plates.

LB + tetracycline (12.5 mg/L) plates
To make 400 ml: make LB agar (Miller) as described above; after cooling to 65°C, add 5 mg of tetracycline (1 ml of 5 mg/ml solution) to 400 ml LB medium; pour plates; dry at room temperature before storing at 4°C; 400 ml will make about 13 plates.

PROCEDURAL NOTE

Tetracycline is light sensitive. Wrap container in aluminum foil before storing.

LB Agar, with addition of X-gal or IPTG
If adding X-gal and/or IPTG to 30 ml volume plates, spread 50 μl of 25 mg/ml X-gal and/or 50 μl IPTG (0.1 mg/ml) per plate 1 hour before needed, allowing the solutions to soak into plates.

2x YT medium
To make 1 L: 16 g Bacto tryptone + 10 g Bacto Yeast Extract + 5 g NaCl + 900 ml ddH_2O; adjust pH to 7.0 with 5 M NaOH; fill to 1 L with ddH_2O; autoclave.

SOC medium
To make 1 L: 20 g Bacto tryptone + 5 g yeast extract + 0.5 g NaCl + 10 ml 250 mM KCl + 900 ml ddH_2O; adjust pH to 7.0 with 10 M NaOH; bring the total volume to 1 L with ddH_2O, autoclave; when the solution is cool, add 10 ml sterile 1 M $MgCl_2$ (sterilized) and 20 ml sterile 1 M glucose (filter sterilized).

III

MARKER MAPS AND SIZES

Lambda DNA–HindIII Digest

Fragment	Base Pairs	ng loaded on gel[a]					
		100	150	300	400	500	1000
1	23,130	48	71.5	143	191	238	477
2	9416	19	29	58	78	97	194
3	6557	14	20	41	54	68	135
4	4361	9	13.5	27	36	45	90
5	2322	5	7.2	14	19	24	48
6	2027	4	6.3	13	17	21	42
7[b]	564	1	1.7	4	5	6	12
8[b]	125	0.3	0.4	0.8	1	1.5	3

[a]ng of DNA in each fragment.
[b]Gel bands not shown on figure below.

0.5 µg lambda DNA–HindIII digest visualized by ethidium bromide staining on a 1.0% agarose gel.

The total size of lambda DNA is 48,502 bp.

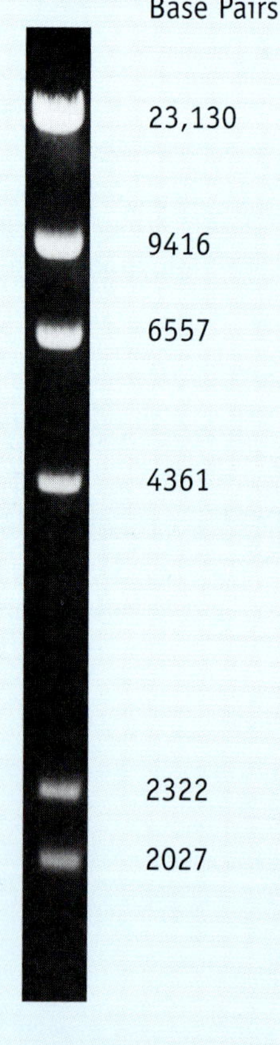

Lambda DNA–BstEII Digest

Fragment	Base Pairs	ng loaded on gel[a]				
		50	150	250	400	1000
1	8,454	8.7	26	44	70	174
2	7,242	7.5	22	37	60	149
3	6,369	6.6	20	33	53	131
4	5,686	5.9	18	29	47	117
5	4,822	5.0	15	25	40	99
6	4,324	4.5	13	22	36	89
7	3,675	3.8	11	19	30	76
8	2,323	2.4	7.2	12	19	48
9	1,929	2.0	6.0	9.9	16	40
10	1,371	1.4	4.2	7.1	11	28
11	1,264	1.3	3.9	6.5	10	26
12	702	0.7	2.2	3.6	5.7	14
13[b]	224	0.2	0.7	1.2	1.8	4.6
14[b]	117	0.1	0.4	0.6	1.0	2.4

[a]ng of DNA in each fragment.
[b]Gel bonds not shown on figure below.

0.5 µg lambda DNA–BstEII digest visualized by ethidium bromide staining on a 1.0% agarose gel.

The total size of lambda DNA is 48,502 bp.

Note: If the lambda DNA-BstEII marker is *not* heated prior to loading, the 8454 bp fragment and the 5686 bp fragment will anneal to give a band at 14140 bp.

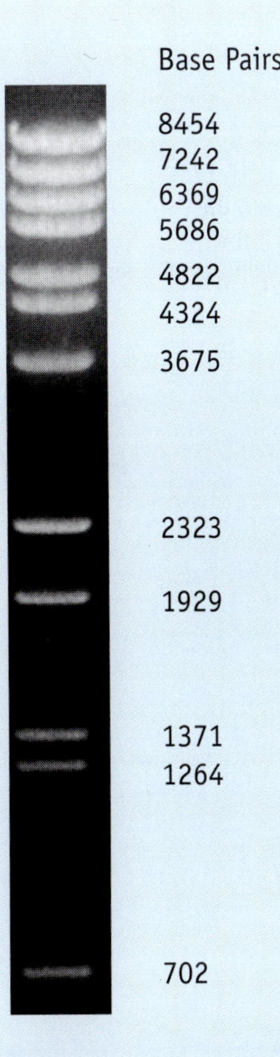

New England Biolabs 100-bp DNA Ladder

Fragment	Base Pairs	ng loaded on gel[a]					
		100	200	300	400	500	1000
1	1517	9	18	27	36	45	90
2	1200	7	14	21	28	35	70
3	1000	19	38	57	76	95	190
4	900	5.4	11	16	21.5	27	54
5	800	5	9.5	14	19	24	48
6	700	4	8.5	12.5	17	21	42
7	600	3.5	7	11	14	18	36
8	500,517	19	39	58	77.5	97	194
9	400	7.5	15	22	30	38	76
10	300	6	11.5	17.4	23	29	58
11	200	5	10	15	20	25	50
12	100	9.5	19	29	38.5	48	96

[a]ng of DNA in each fragment.

0.5 µg 100-bp DNA ladder visualized by ethidium bromide staining on a 1.3% TAE agarose gel.

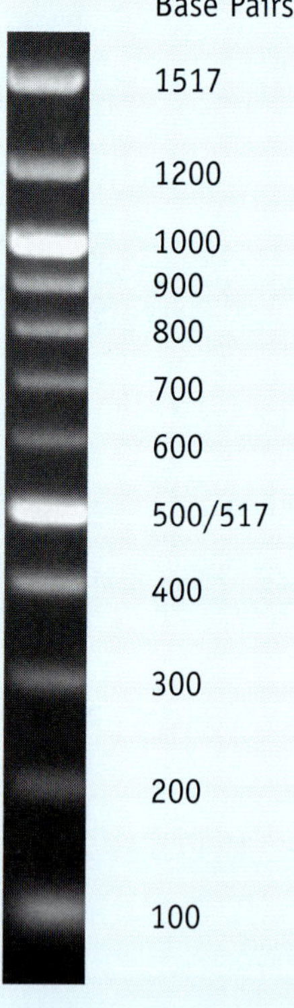

ALTERNATE PROTOCOLS **IV**

> **PROCEDURAL NOTE**
>
> All solutions used in the alternate protocols are listed in Appendix II except for those described within the protocol.

Alternate Protocol AI. Isolation of High Molecular Weight Mouse Genomic DNA (Not Using a "Kit")

> **PROCEDURAL NOTES**
>
> For additional background information on total genomic DNA isolation, refer to the introductory sections to Project I.
>
> Refer to Project I for additional descriptions and for cited figures.

LAB PERIOD AI.1.   MOUSE DNA ISOLATION

EACH PAIR will be given about 0.5 to 1.0 g of frozen mouse liver tissue from which you will isolate genomic DNA. These tissue samples should be prepared by dissecting liver from a freshly euthanized mouse, rinsing the tissue briefly in sterile saline, cutting the tissue into uniform fragments, and snap freezing the fragments in liquid nitrogen. The frozen fragments should be stored at –80°C before use in this experiment. (See figures in Project I, Lab Period 1.1, Steps 1–3 and 6.)

Step 1 Grind the tissue with a mortar and pestle in liquid nitrogen. One partner can add the liquid nitrogen to the mortar while the other grinds the tissue into a fine powder. *Wear safety glasses!* Be sure to work as quickly as possible to prevent the liquid nitrogen from completely evaporating. If it does, the sample will be very difficult to remove from the bottom of the mortar. As the level of the liquid nitrogen becomes low, add small aliquots of liquid nitrogen until the tissue is completely pulverized. Then quickly transfer the sample in liquid nitrogen to a 50-ml conical tube.

> **PROCEDURAL NOTE**
>
> Be sure to use phenol/chisam-resistant plastic (polypropylene) centrifuge tubes through this procedure!

Step 2 As soon as the liquid nitrogen has boiled away and the pellet is dry, add 4 ml digestion buffer (approximately 1.2 ml per 100 mg of tissue), and gently vortex the tube just long enough to dislodge the pellet from the bottom of the tube (approximately 5 seconds).

Step 3 Wrap a piece of parafilm around the cap to prevent leakage. Tape your tube to a rocker platform, and incubate overnight (12–18 hours) at 50°C with gentle rocking.

LAB PERIOD AI.2. CONTINUE MOUSE DNA ISOLATION

PROCEDURAL NOTE

Chisam is 24:1 chloroform:isoamyl alcohol.

CAUTION!

Phenol and chloroform (in chisam) are both toxic and produce noxious fumes. Because phenol and chisam are solvents that are caustic and will burn skin and clothes, be sure to always wear a lab coat and gloves. Keep bottles and tubes closed at all times. Open bottles and tubes containing phenol and/or chisam for brief periods only. Work in a fume hood!

Bottles or tubes containing phenol and chisam are kept in the refrigerator and can be slippery due to condensation. Handle these bottles carefully and with one hand cradling the bottle or tube underneath.

PROCEDURAL NOTES

Phenol is buffer saturated and is the lower phase in brown phenol bottles. The upper phase is buffer. When pipetting phenol, you must go through the upper buffer phase with the pipette to reach the lower phenol phase. The phenol phase is yellow. When you pipette the phenol, if it is not yellow, you have probably just pipetted buffer.

All of the liquid phenol and chisam waste goes into a container in a fume hood designated for liquid organic waste. All of the solid waste (i.e., tubes and tips that contained phenol) goes into a solid waste container in the fume hood.

After 12 hours of incubation, large quantities of high molecular weight DNA will have been released from your mouse cells, resulting in a highly viscous sludge.

Step 1   Add 4 ml of phenol (*do not vortex!*). Wrap the cap of your tube with parafilm, and gently rock the tube for about 4 hours at room temperature (refer to Lab Period I.1.1, Step 6).

Step 2   Centrifuge at 1700x *g* in a swinging bucket rotor in a tabletop centrifuge for 5 minutes at 4°C.

Step 3   Remove the aqueous layer (top) using a 10-ml glass pipette. Do not transfer any of the protein interface. Place the supernatant into a 15-ml conical phase lock tube (Phase Lock Light; Eppendorf) (similar to that in the photo of Lab Period II.1.3A) in your rack. Add 0.1x TE to raise the volume to 4 ml.

Step 4   Add 4 ml of phenol (*do not vortex!*). Wrap the cap with parafilm, and gently rock the tube for about 1 to 2 hours at room temperature.

Step 5   Centrifuge at 1700x *g* in a swinging bucket rotor in a tabletop centrifuge for 2 minutes at 4°C.

Step 6   Remove the aqueous layer (top) using a 10-ml glass pipette. Place the supernatant into a new 15-ml conical phase lock tube (Phase Lock Light; Eppendorf) in your rack. Add 0.1x TE to raise the volume to 4 ml.

Step 7   Add 4 ml chisam. Wrap the cap with parafilm, and gently rock the tube for about 2 minutes by hand.

PROCEDURAL NOTE

Chisam is 24:1 chloroform:isoamyl alcohol and is not yellow.

Step 8   Centrifuge at 1700x g in a swinging bucket rotor in a tabletop centrifuge for 2 minutes at 4°C.

Step 9   Remove the aqueous layer (top) using a 10-ml glass pipette. Place the supernatant in a new 15-ml conical tube in your rack. Add 0.1x TE to raise the volume to 4 ml.

Step 10   Add 0.5 volume (2 ml) 7.5 M ammonium acetate and 2 volumes (8 ml) ice-cold 100% ethanol to precipitate the DNA. Invert the tube several times (be sure to mix the solution completely to get all the DNA precipitated). The DNA should immediately precipitate.

PROCEDURAL NOTE

This is really cool — you are allowed to become excited at this point!

Step 11   Centrifuge at 1700x g in a swinging bucket rotor in a tabletop centrifuge for 5 minutes at 4°C. Pour away the supernatant into a waste beaker.

Step 12   Add 5 ml 70% ethanol to the DNA pellet and invert gently several times. Centrifuge at 1700x g in a swinging bucket rotor in a tabletop centrifuge for 5 minutes at 4°C. Carefully pour away the supernatant into a waste beaker.

PROCEDURAL NOTE

The pellet may not stick as well to the side of the tube after washing with ethanol. Take care not to pour away your pellet.

Step 13   Leave your tube in an inverted position on Kimwipes to air dry for about 1 hour.

Step 14   Add 2 ml TNE buffer to the dry pellet. Be sure to get the pellet off the bottom of the tube before tapping the tube horizontally on a rocker platform.

Step 15   Rock the tube to resuspend the DNA at 4°C (this will take 3 to 4 days).

LAB PERIOD AI.3.   CONTINUE MOUSE DNA ISOLATION

Step 1 Your mouse DNA has been gently rocking horizontally in the cold room for 3 to 4 days and should be completely resuspended in the TNE buffer. Centrifuge your tubes briefly in a tabletop centrifuge at 1700x g to get all of the liquid to the bottom of the tube.

Step 2 Using a P1000 and a cut-off blue tip (use scissors to snip off the bottom 2 to 3 mm of the tip), pipette your solution up and down to help complete the resuspension of the mouse genomic DNA. The larger bore of the tip will help prevent shearing of your DNA (refer to Lab Period I.1.3, Step 5).

Step 3 Remove 4 µl of this solution and transfer to a microfuge tube labeled "mouse DNA + RNA." Store at −20°C.

Step 4 Add 20 µl DNase-free RNase A (10 mg/ml) to your mouse DNA in the 15-ml orange cap tube. Invert several times to mix, and incubate at 37°C for 30 minutes to 1 hour or longer. *Do not vortex!* Place on ice.

Step 5 Transfer the contents of your tube to a new 15-ml Phase Lock Light tube (Eppendorf). Add 2 ml phenol (*do not vortex!*). Wrap the cap of your tube with parafilm, and gently rock the tube at room temperature for about 1 hour.

Step 6 Centrifuge at 1700x g (2800 rpm) for 5 minutes at 4°C in the tabletop centrifuge. Pour the aqueous (top) layer into a new 15-ml conical tube. Add 0.1x TE to raise the volume to 2 ml if necessary.

Step 7 Repeat Steps 5 and 6, except that the tube can be inverted by hand for about 1 minute before centrifugation. If possible, try not to suck up any material from the interface (this is where most of the protein is!). Pour the aqueous solution into a 15-ml phase lock tube instead of a plain 15-ml tube.

Step 8 Add 2 ml chisam (*do not vortex!*). Wrap the cap of your tube with parafilm, and gently invert for 1 minute.

CAUTION!

Work with chisam in the hood! Wear safety glasses and gloves.

Step 9 Centrifuge for 5 minutes at 1700x g in the tabletop centrifuge at 4°C, and pour the aqueous (top) layer into a new (standard orange cap) 15-ml conical tube. Add 0.1x TE to raise the volume to 2 ml if necessary.

Step 10 Add 1/10th volume (200 µl) 3 M sodium acetate (pH 5.2) and 2 volumes (4 ml) ice-cold 100% ethanol. Invert the tube several times to mix. The DNA should immediately precipitate.

Step 11 Centrifuge your tubes for 5 minutes (4°C) at 1700x g in the tabletop centrifuge.

Step 12 Carefully decant the supernatant into a waste beaker. Add 1 ml 70% ethanol to the pellet, and invert gently several times. Centrifuge at 1700x g in the tabletop centrifuge for 5 minutes at 4°C. Carefully pour away the supernatant into a waste beaker, and leave the tube inverted to dry on Kimwipes for 0.5 to 1 hour.

Step 13 Resuspend the DNA in 1 ml TNE buffer. To initiate resuspension, pipette the solution up and down gently with a cut-off blue pipette tip. Cap the tube and wrap with parafilm on the outside of the cap to prevent leaks.

Step 14 Rock the tube horizontally on a rocker platform to resuspend the DNA at 4°C for 24 hours or more.

LAB PERIOD AI.4. **DROP DIALYSIS**

THE DNA solution is now ready for dialysis as described in Project I (see photos in Lab Period I.1.3, Steps 2, 5, and 6). After 12 to 24 hours of dialysis, the DNA can be quantified by spectrophotometry or agarose gel analysis. Starting with this lab period, you can proceed to Genomic DNA Isolation and Analysis: Project I, Lab Period I.1.3.

Alternate Protocol AII: Purification of DNA from Agarose Gels Using β-Agarase and Low-Melt Agarose

A variety of techniques are available for purification of DNA fragments from agarose gels. More recently, these methods include microcentrifugation columns ("spin columns"), which physically shear the agarose during centrifugation or biochemically bind and elute the DNA. Several sources of these kits are provided at the end of this protocol. The procedure outlined below uses β-agarase, which degrades low-melting temperature agarose for purification. After the DNA has been prepared for purification (e.g., by digestion of a plasmid or by polymerase chain reaction [PCR]), the DNA is electrophoresed on a low-melt agarose gel, and the band of DNA of correct size is excised, treated with β-agarase, and purified for subsequent ligation.

LAB PERIOD AII.1.   AGAROSE GEL ELECTROPHORESIS OF DNA FOR GEL PURIFICATION

PROCEDURAL NOTE

Refer to Lab Period I.1.4, Part B, for gel pouring description.

Step 1 To prepare a 1.0% *low-melt* agarose gel, combine the following in a 250-ml flask:

1.0 g low-melt agarose and 100 ml 1x Tris-acetate buffer (TAE).

Melt the agarose is melted in a microwave oven (high power for about 4 minutes) or on a hot plate using a stir bar over the flask and heat until the solution just begins to boil. *Carefully* swirl the solution. Be sure that no solid agarose remains. If you can see unmelted agarose, heat the sample further.

Step 2 Arrange the gel tray in your gel box as described in Project I. Allow the agarose to cool to about 65°C before pouring. Pour approximately 35 ml (depending on your gel rig) of molten agarose into the gel tray. The level of the liquid should not be more than half way up the teeth of the comb. An instructor will mark this position on your comb. Wait at least 20 minutes for the gel to cool and solidify.

PROCEDURAL NOTE

Handle these low-melt gels with extreme care! They are much more fragile and slippery than regular agarose gels.

Step 3 *Carefully* pour gel running buffer (1x TAE) into both side chambers of your gel box. Be sure that the buffer level on both sides is even with the top of your gel and that you have some buffer covering the top of your gel.

Step 4 *Carefully* remove the comb from the gel.

PROCEDURAL NOTE

For these low-melt agarose gels, it is critical that the comb be removed *before* you attempt to move the gel tray (with regular agarose gels, it does not matter whether you remove the comb before or after turning the gel tray).

Step 5 *Carefully* turn the gel 90° in the gel rig. Use your fingers on the ends of the gel tray to hold the gel in place while you are moving it. Get help if you need it.

Step 6 Add more buffer until the wells are full of buffer and the gel is *completely* submerged. The thickest area of the gel is at the wells and should be covered by 1 or 2 mm of buffer.

PROCEDURAL NOTE

Gels may be stored in 1x TAE buffer overnight at 4°C.

Step 7 Add 5x blue juice (BJ) to your DNA sample tube(s). Stir with your pipette tip to mix.

Step 8 Load the samples into the wells underneath the buffer.

Step 9 Electrophorese the gel at 60 V for 1.5 to 2.0 hours.

Step 10 After electrophoresis, turn off the power supply. Unplug the leads from the power supply, and carefully stain the gels for 8 minutes in the ethidium bromide bath (your instructor will demonstrate how to do this safely). Destain your gel in the water bath for at least 15 minutes. Wear gloves, lab coats, and glasses!

»»

LAB PERIOD AII.1. CONT.

PROCEDURAL NOTE

Handle these-low melt gels with extreme care! They are more fragile and slippery than regular agarose gels. Support the entire gel with the spatula to keep it from breaking.

CAUTION!

Ethidium bromide is a known mutagen and carcinogen. Wear gloves, lab coats, and safety glasses! Handle gels carefully! Do not splash ethidium bromide! Always rinse the spatula in the water destain bath after it has been in contact with ethidium bromide! Always rinse the gel tray after sliding gels into the ethidium bromide bath. You may unknowingly contact the ethidium bromide! Be very careful not to drip the ethidium bromide anywhere!

PROCEDURAL NOTE

An alternative procedure for staining involves SYBR Safe and is described later in Appendix IV.

LAB PERIOD AII.2.   SIZE SELECTION AND PURIFICATION OF ELECTROPHORESED DNA

LABEL a microfuge tube with your number before going to the darkroom.

Step 1 *Carefully* place your gel into a plastic container, and take it to the dark room. *Carefully* place your gel onto the ultraviolet (UV) light box. CAUTION! Be sure to wear safety glasses to protect your eyes from the UV light! Each student should carefully cut out the DNA band from his or her lane on the gel using a clean razor blade. Cut the gel as close to the band as possible. Your goal is to make the gel slice as small as you can. If possible, use long-wave UV light to limit damage to your DNA.

Step 2 Place the gel slice into your blue microfuge tube. Dispose of the rest of the gel in the special waste containers in each dark room, and return to your bench.

Step 3 Spin the tube briefly in your nanofuge to bring all of the gel material to the bottom.

Step 4 Estimate the volume of your excised agarose gel slices containing your DNA by weighing the microfuge tube that contains the gel slice (use the empty microfuge tube to tare the balance). The volume of gel in milliliters is approximately equal to the weight of the gel in grams. Generally, cutting out one lane on an agarose minigel yields about 50 to 150 mg of agarose, which equals about 50 to 150 µl volume.

PROCEDURAL NOTE

Use caution and wear gloves while dealing with this sample. Remember that it contains trace amounts of ethidium bromide!

Step 5 Equilibrate the gel slice by soaking it twice with two volumes (about 200 µl) of 1x β-agarase buffer on ice for 30 minutes each. After each soak, remove the buffer with a pipette.

Step 6 Place your microfuge tube containing the gel fragments at 70°C for 10 minutes in order to melt the agarose.

PROCEDURAL NOTE

Low-melt agarose melts at 70°C, whereas normal agarose melts at about 95°C — this is why using low-melt agarose is important; because DNA will denature at 95°C but not at 70°C.

Step 7 Place the microfuge tube at 41°C for 5 minutes to equilibrate the melted agarose to the proper digestion temperature (temperatures above 45°C will heat kill β-agarase enzyme).

PROCEDURAL NOTES

Low-melt agarose is liquid at 41°C, whereas normal agarose solidifies at this temperature.

For the following two steps, try to keep the microfuge tube at 41°C for as much of the time as possible; 41°C is necessary to maintain the gel in its molten state.

Step 8 Add 1 unit of β-agarase enzyme (New England Biolabs) for each 100 µl of solution in the gel slice tube, and mix by tapping the tube. Your total volume should now be about 100 µl. Incubate at 41°C for 60 minutes or longer.

Step 9 Check the completeness of the digestion by placing the microfuge tube on ice for 5 minutes, and then monitor whether solidification of any of the agarose occurs. If any of the agarose solution does solidify, heat the mixture at 70°C for 5 minutes, and return to Step 8.

»»

PROCEDURAL NOTES

The above protocol is not suitable for DNA fragments less than 500 bp because the DNA can diffuse out of the gel slices during the long equilibrations in Step 5. For fragments less than 500 bp, add one-tenth volume of 10x β-agarase buffer in Step 5 instead of the two equilibration steps.

Steps 10 to 16 use a Microcon-100 Microconcentrator (Amicon) to purify your DNA sample without using organic extractions or ethanol precipitations. The molecular weight cut-off of these filters is 125 bp for double-stranded nucleic acids and 300 nucleotides for single-stranded DNA.

Step 10 Place a Microcon-100 sample reservoir into a Microcon vial (these steps using the Microcon-100 Microconcentrators will be demonstrated by your instructors). Label the reservoir with your number.

Step 11 Load the reservoir with your β-agarase digested DNA sample (do *not* load more than 200 µl — check your volume before adding your sample to the reservoir!). Add 300 µl 20% isopropanol to the reservoir.

PROCEDURAL NOTE

The isopropanol helps to remove substances from the low melting point agarose than can inhibit some restriction enzymes and DNA ligase (especially ethidium bromide).

Step 12 Centrifuge the Microcon assembly for 14 minutes in a microfuge at 500x g at room temperature.

Step 13 Remove the filtrate from the Microcon vial using a P1000 pipette and transfer it to a 1.7-ml microfuge tube. Label this tube "Microcon waste."

PROCEDURAL NOTE

The filtrate contains a very small amount of ethidium bromide; wear gloves!

Step 14 Add 450 µl of ddH$_2$O to the reservoir, and centrifuge again at 500x g for 14 minutes. Again, remove the filtrate from the vial and place in the same clear 1.7-ml tube (labeled "Microcon waste").

Step 15 Again, add 450 µl ddH$_2$O to the reservoir, and centrifuge at 500x g for 13 minutes. Again remove the filtrate from the vial and place in the same clear 1.7-ml tube (labeled "Microcon waste"). Dispose the waste tube appropriately.

Step 16 To recover your purified DNA, remove the reservoir from the vial, and invert into a new Micron vial. Centrifuge at 500x g for 2 minutes.

Step 17 Remove the reservoir from the vial, and throw the reservoir away. The vial now contains 5 to 20 µl of your purified DNA.

As an alternative to this method, a large number of companies sell kits for the isolation of DNA from gels. Recommended suppliers of gel purification kits include the following:

NucleoTrap Gel Extraction Kit (BD Biosciences Clontech)

Freeze 'N Squeeze DNA Gel Extraction Spin Columns 25 (Bio-Rad)

Eppendorf Perfectprep Gel Cleanup (Brinkmann)

GFX PCR DNA and Gel Band Purification Kit (GE Healthcare, formerly Amersham Bioscience)

Sephaglas BandPrep Kit (GE Healthcare, formerly Amersham Biosciences)

PureLink Gel Extraction Kit (Invitrogen)

S.N.A.P. Gel Purification Kit (Invitrogen)

S.N.A.P. UV-Free Gel Purification Kit (Invitrogen)

Matrix Gel Extraction System (Marligen Biosciences, Inc.)

Rapid Gel Extraction System (Marligen Biosciences, Inc.)

Montage Gel Extraction Kit (Millipore)

UltraClean GelSpin (S) Kit Sample (Mo Bio Laboratories, Inc.)

UltraClean 15 Kit (Mo Bio Laboratories, Inc.)

UltraClean GelSpin Kit (Mo Bio Laboratories, Inc.)

UltraClean GelSpin Kit (Mo Bio Laboratories, Inc.)

SpinPrep Gel DNA Kit (Novagen)

QIAEX II Suspension (Qiagen)

QIAquick Gel Extraction Kit (Qiagen)

QIAEX II Gel Extraction Kit (Qiagen)

Agarose Gel DNA Extraction Kit (Roche Applied Science)

GenElute Agarose Spin Columns (Sigma-Aldrich)

GenElute Gel Extraction Kit (Sigma-Aldrich)

GenElute Minus EtBr Spin Column (Sigma-Aldrich)

StrataPrep DNA Gel Extraction Kit (Stratagene)

SUPREC 01 (Takara Mirus Bio)

Zymoclean Gel DNA Recovery Kit (Zymo Research)

Alternate Protocol AIII. An Alternative Preparation of Mammalian Total RNA Using an Ultracentrifuge

Tissues are dissected from mice and homogenized in a solution containing a potent RNase inhibitor. The homogenate is then layered onto a cesium chloride cushion in an ultracentrifuge tube in preparation for an overnight run in an ultracentrifuge.

Step 1 Obtain approximately 300 µl (about 0.5 g) of tissue, and place in 2 ml guanidinium thiocyanate solution. Homogenize three times for 5 seconds each with a Polytron (or similar) tissue homogenizer. Cool the sample on ice between each of the three homogenizations.

Step 2 Place 2 ml 5.7 M cesium chloride solution into a 5-ml ultracentrifuge tube.

Step 3 Carefully layer the homogenized sample onto the 2 ml cesium chloride solution in the centrifuge tube.

Step 4 Rinse the empty sample tube with 500 µl guanidinium thiocyanate solution, and add it to the sample in the ultracentrifuge tube. Repeat (total sample layered onto the CsCl = 3 ml).

Step 5 Place the ultracentrifuge tube into a bucket from a Beckman SW 50.1 rotor. Find the opposite bucket (see the bucket numbers on the rotor), and balance the loaded buckets by adding the guanidine isothiocyanate solution dropwise to the tube in the lighter bucket (or prepare another sample in this tube).

Step 6 Screw the tops onto the buckets and hang them on the rotor. Place the loaded rotor into the ultracentrifuge and spin at 35,000 rpm at 18°C (to prevent the guanidinium solution from freezing) for about 18 hours. On the next day, stop the run, and remove the rotor.

Step 7 Carefully unhook the buckets from the rotor.

Step 8 Carefully unscrew the top and use forceps to pull the tube out of each loaded bucket.

Step 9 Invert the tube into a beaker to empty. *Keep the tube inverted*, and place it on a Kimwipe in an empty beaker to drain.

Step 10 Resuspend the "pellet" (you may not see anything) in 200 µl DEPC-H_2O. Transfer to a sterile microfuge tube. Rinse the same tube twice with 100 µl DEPC-H_2O, and transfer each aliquot to the sterile microfuge tube. (The total volume now equals 400 µl.)

Step 11 Add 40 µl 3 M NaOAc (pH 7.0), and precipitate the RNA by adding 1 ml cold 100% ethanol. Invert to mix and place the tube at –70°C for at least 1 hour.

Step 12 Recover the precipitated RNA by spinning the tube in the microfuge for 60 minutes at 4°C. Remove the supernatant, and add 1 ml 70% ethanol (in DEPC-treated water). Invert several times to wash the pellet, and centrifuge 10 minutes at 4°C. Remove the supernatant, and air dry the pellet. (Do not use a speed-vacuum concentrator to aid this drying step.)

Step 13 Resuspend the pellet in 400 µl DEPC-H_2O. Vortex until the pellet is completely resuspended.

Step 14 Dilute a small aliquot of the RNA 1:10 with DEPC-H_2O and read the OD_{260} on a UV spectrophotometer to determine the concentration and total quantity of RNA. Spectrophotometric readings should be taken at wavelengths of 260 and 280 nm to quantitate the amount of RNA and to help determine its purity. An $OD_{260} = 1$ corresponds to approximately 40 µg/ml of single-stranded RNA. The ratio between the readings at 260 and 280 nm gives an estimate of RNA purity. An OD_{260}/OD_{280} of about 2.0 indicates a pure RNA preparation. A continuous scan between OD_{220} to OD_{310} provides an even better estimate of nucleic acid purity (see Project I.1.4, Part A).

V OTHER USEFUL PROTOCOLS AND INFORMATION

AV.1. Pure Culture Techniques*

BACTERIA USUALLY exist in **mixed populations** in soil, water, and various organs of the human body. It is not feasible to study the characteristics of a particular species when it is mixed with other species, and, therefore, a pure culture must be obtained in preparation for further work. As the name implies, a **pure culture** is one that contains only a single species of organism.

In this exercise, the streak plate technique and the pour plate technique will be used to separate the bacteria in a mixed population. The fundamental principle underlying both techniques is that a mixture of bacteria is gradually thinned out in a growth medium so that the individual organisms have room to form separate masses of bacteria called **colonies.** Presumably, each colony arises from the growth and reproduction of a single bacterium, and therefore all the cells in the colony are genetically identical. A sample of the cells may then be selected and further cultivated to form a pure culture.

Bacterial colony characteristics also will be studied in this exercise. These characteristics reflect the genetic composition of a bacterium and serve as markers for different bacterial species. The laboratory microbiologist often uses colony characteristics as a criterion in the identification of an unknown bacterium.

A. STREAK PLATE TECHNIQUE

PROCEDURAL NOTE

To perform the streak plate technique for isolating a single species in pure culture.

The **streak plate technique** is a relatively inexpensive and rapid method for separating bacteria in a mixed population. It requires only a single plate of growth medium and, with practice, it yields excellent distribution of bacterial colonies. For these reasons, the technique is standard practice in many research, industrial, and clinical laboratories.

Special Materials

Mixed population of bacteria

Nutrient agar plates or materials for their preparation

Procedure

Step 1 Select a nutrient agar plate or, at the direction of the instructor, pour a plate of the medium using a Petri dish and melted nutrient agar. With a wax pencil or felt marker, label the bottom side of the plate with your name, the date, and the designation "streak plate." Obtain a mixed population of bacteria with which to work. Before beginning the streak plate technique, read through Steps 2 through 4 and review **Figure A5.1** to familiarize yourself with the technique.

Step 2 Aseptically obtain a loopful of the mixed population, and lightly streak it several times along one area of the plate, as shown in Figure A5.1A. Try to avoid airborne contamination by lifting the lid of the Petri dish only enough to permit entry of the loop.

Step 3 Sterilize the loop to destroy any remaining bacteria. To ensure that it is cool, touch the loop to the center of the plate or between the agar and the edge of the plate. Pass the loop one time across the previous streaks to pick up some bacteria, and continue streaking into a second area of the plate, as illustrated in Figure A5.1B.

*From *Alcamo's Laboratory Fundamentals of Microbiology*, 7th edition, by J.C. Pommerville. Sudbury, MA: Jones and Bartlett, 2005, pp. 43–48.

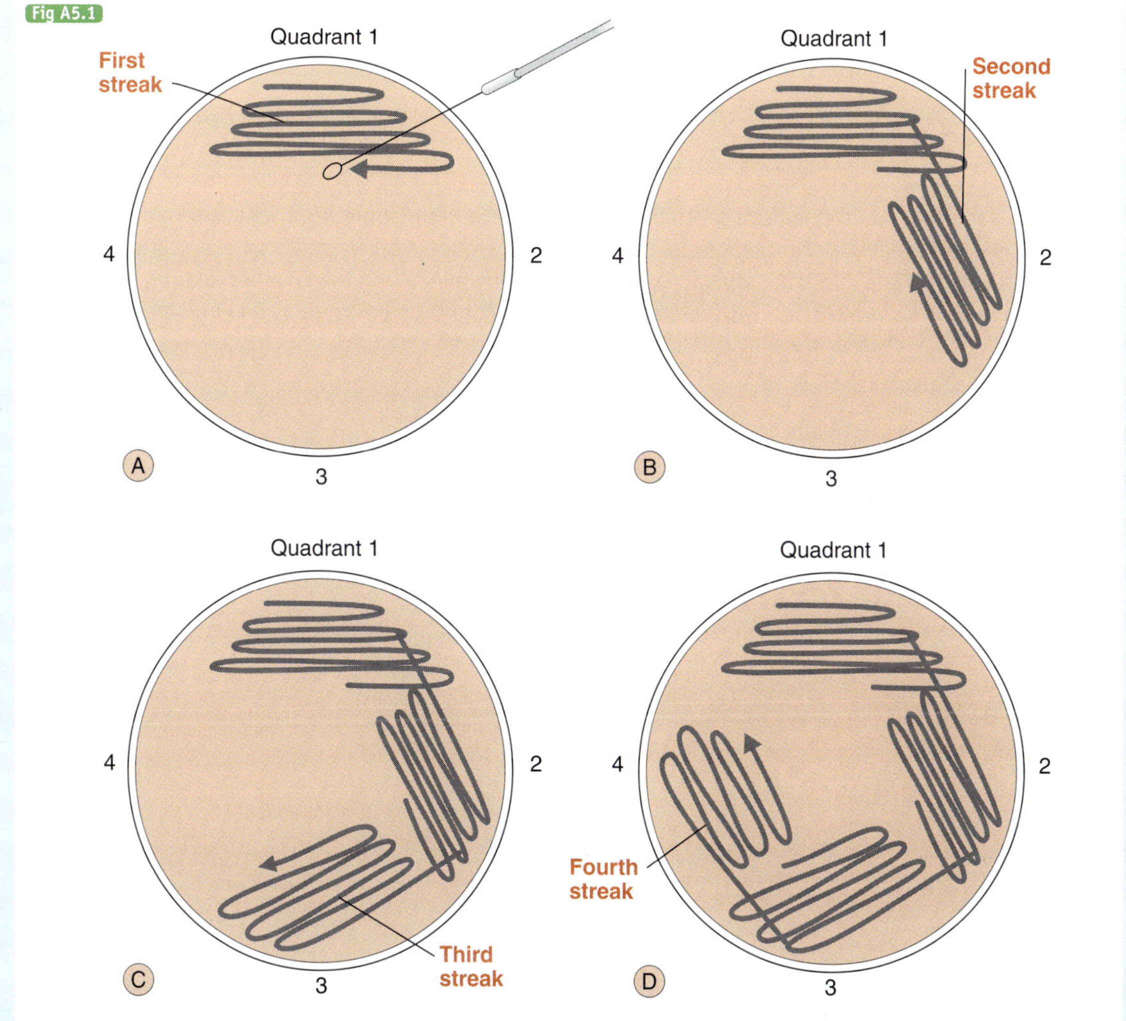

The streak plate isolation technique.

Step 4 Sterilize the loop as before, then be certain that it is cool. To pick up some bacteria, pass the loop one time through the second area, and continue streaking into a third area (Figure A5.1C).

Step 5 Sterilize the loop as before, and streak some bacteria from the third area of the plate into the fourth area (Figure A5.1D). Be sure to use up the remaining space on the plate.

Step 6 To obtain practice, repeat the streak plate technique using additional plates and different mixed populations. The instructor may recommend alternative streaking methods.

Step 7 Additional practice can be obtained by attempting to isolate bacteria on selective media. For example, streptococci can be cultivated on mitis salivarius agar using a swab of material from the gingival (tooth-gum) crevice; staphylococci can be cultivated on mannitol salt agar using material swabbed from the nose cavity; and intestinal bacteria can be cultivated on MacConkey agar and EMB agar using a fecal suspension as a source.

Step 8 Invert all the plates and incubate them for 24 to 48 hours at 37°C. The plates are inverted so that moisture accumulates on the lid rather than on the agar surface, where it may cause colonies to run together. After incubation, refrigerate the plates in the inverted position until the next laboratory session to preserve the bacterial growth and prevent drying of the medium.

CAUTION!

Never touch the bacterial colonies on a plate with your fingers. Each colony contains millions of live organisms.

Step 9 Examine the plates for well-isolated and separated colonies, as shown in **Figure A5.2**, and enter a representation of a good streak plate in the Results section. Add appropriate labels and brief explanations to produce a "talking picture." Describe several isolated colonies by numbering the colonies and noting their size, color, and characteristics, with reference to the standard terminology provided in **Figure A5.3**.

Fig A5.2

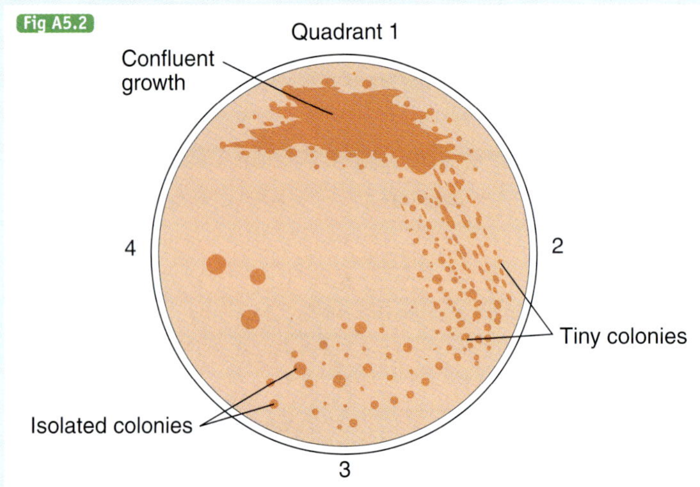

Isolated colonies of bacteria on a well-executed streak plate.

Fig A5.3

Form

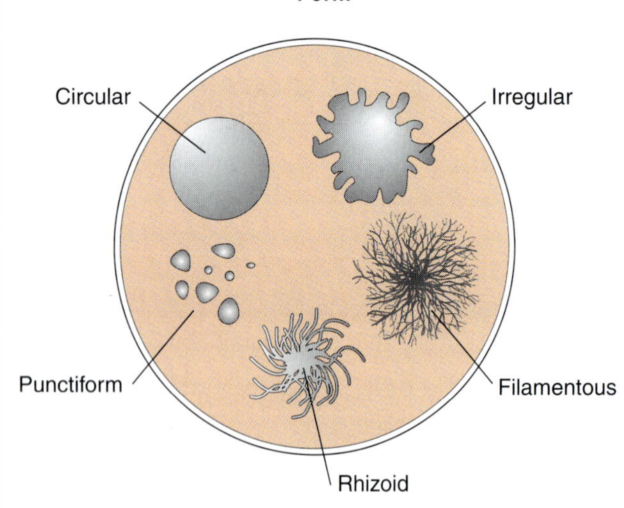

Elevation

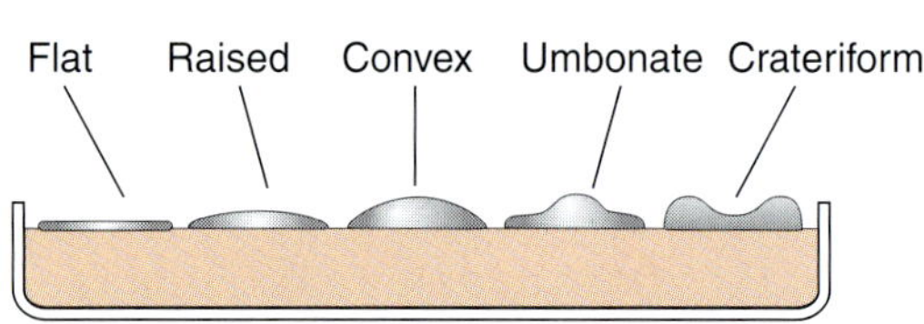

Margin

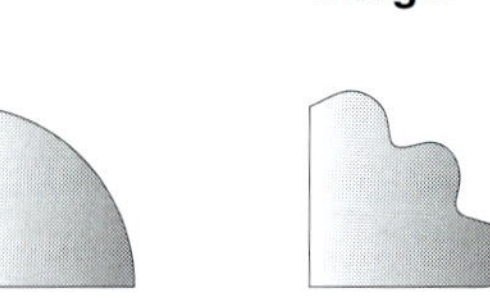
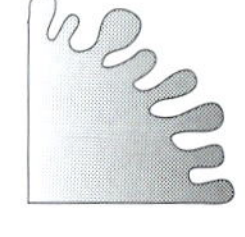

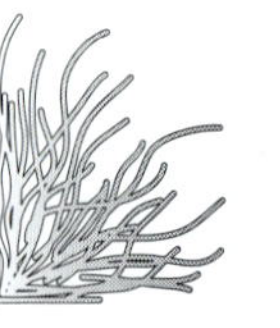
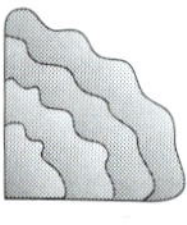

Standard terminology for bacterial colony characteristics.

Size may be determined by measuring the colony diameter in millimeters if rulers are available. Record your observations.

Step 10 At the direction of the instructor, use a bacteriological needle or loop to select samples from various colonies, and inoculate nutrient agar slants to obtain pure cultures. Stained smears may also be made from the colonies to determine the morphological characteristics of the organisms.

B. POUR PLATE TECHNIQUE

PROCEDURAL NOTE

To perform the streak plate technique for isolating a single species in pure culture.

In the **pour plate technique,** a mixed population of bacteria is progressively diluted in three "deep" tubes of liquid nutrient agar. The medium is then poured into Petri dishes for incubation. Colonies of bacteria subsequently appear on the surface as well as within the medium. Once it is started, the technique must be performed without delay because the liquid agar solidifies quickly in the tubes.

Special Materials

Mixed populations of bacteria

Deep tubes of liquid nutrient agar

Sterile Petri dishes

Procedure

Step 1 The pour plate technique must be performed rapidly to prevent premature hardening of the agar medium in the test tubes. Therefore, it is well to review Steps 2 through 6 before beginning the procedure. Refer also to **Figure A5.4** to obtain a thorough understanding of what you will be doing before starting your work.

Fig A5.4

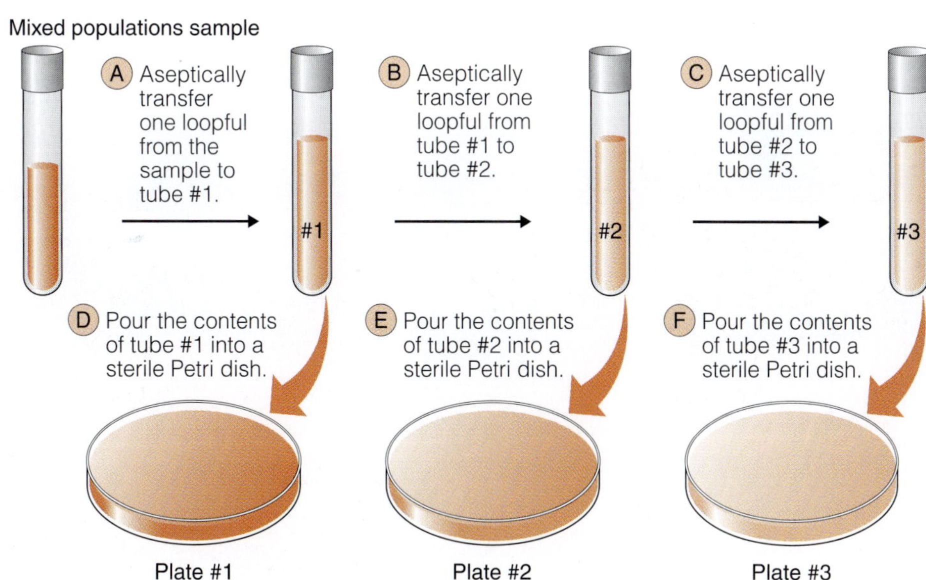

The pour plate isolation technique.

Step 2 Obtain three sterile Petri dishes and label the bottom sides with your name, the date, and the designations "Pour Plate #1," "Pour Plate #2," and "Pour Plate #3." The instructor will explain the preparation of the nutrient agar deep tubes and how they are melted and maintained at approximately 45° to 50° C to prevent solidification. Obtain three tubes for the technique. A mixed population of bacteria should also be available.

Step 3 Aseptically obtain a loopful of the mixed population (Figure A5.4A). Inoculate it into the first melted agar deep tube by placing the loop below the surface and shaking the loop lightly. Resterilize the loop and replace it in its holder. Now mix the contents of the tube by rolling it between the palms of the hands or by vigorously tapping its bottom with the index finger.

Step 4 Sterilize the loop and be certain that it is cool. Obtain one loopful of liquid agar from the first deep tube and inoculate it into the second deep tube (Figure A5.4B). Sterilize the loop, then mix the tube contents well by rolling or tapping the tube.

Step 5 Sterilize the loop, ensure that it is cool, and then obtain one loopful of liquid agar from the second deep tube and inoculate it into the third deep tube (Figure A5.4C). Mix the contents well.

Step 6 Aseptically pour the contents of each tube into the designated Petri dish (Figure A5.4D–F), being sure to flame the neck of the tube after removing the cap. Gently rotate the dishes in a wide arc on the laboratory desk to distribute the medium evenly over the bottom of the plates.

Step 7 Allow the agar to solidify for several minutes, then incubate the plates in the inverted position, as outlined in Part A, Step 8.

Step 8 Examine the plates for well-isolated colonies. Surface as well as subsurface colonies should be visible. Describe, draw, and label the colonies. Compare the number and variety of colonies appearing in this technique with those appearing in the streak plate technique as part of your notes. At the instructor's direction, transfer samples from the plate to nutrient agar slants for pure culture cultivation, and prepare smears for staining. Subsurface colonies may be reached by applying a warm inoculating needle to melt the agar down to the level of the colony.

AV.2. Guidelines for Successful Ligation Reactions

SUCCESSFUL LIGATION of DNA molecules is the cornerstone of any recombinant DNA experiment. In setting up a ligation, the goal is to obtain as many recombinant molecules as possible, while minimizing the frequency of vector/vector and insert/insert ligation products. Both the total concentration of DNA molecules and the molar ratio of insert to vector molecules must be considered. For most ligations, the best ratio of insert molecules to vector molecules is from 3:1 to 1:3. The ideal concentration of DNA molecules for plasmid ligations is in the range of 1 to 10 ng/µl, while for lambda ligations it is in the range of 100 to 200 ng/µl. The goal of combining one insert molecule with one vector molecule also, of course, depends on the insert and vector molecules having compatible ends (i.e., compatible overhangs or blunt ends).

First, the insert to vector ratio must be calculated as a molar ratio (number of molecules or moles of insert to number of molecules or moles of vector). Too often students use the same number of nanograms of insert and vector, which can lead to ligations with ratios that are likely to result in a failed experiment. For example, if a student uses 100 ng of a 44-kb lambda cloning vector and 100 ng of a 440-base-pair insert, the molar ratio will be 100 to 1! The primary product from such a reaction will be concatemers or chains of insert molecules ligated together. This reaction is likely to produce lambda library with a very low titer. Students often repeat the ligation and put in even more insert DNA, reasoning that the experiment didn't work the first time, so more DNA must be needed. In fact, to achieve the desired 1:1 ratio in this case, the investigator must use 100 ng of the vector and only 1 ng of the insert!

What is the easiest way to calculate an appropriate insert to vector ratio? The traditional way of calculating this ratio would be:

$$(\text{moles of insert/molecular weight of the insert}) \, / \, (\text{moles of vector/molecular weight of the vector}) = R$$

where R is the insert to vector molar ratio.

Because most molecular biologists hate to calculate moles and molecular weights, it is more convenient to express the ratio in terms of base pairs and nanograms:

$$(\text{ng of insert/bp of insert}) \, / \, (\text{ng of vector/bp of vector}) = R$$

In this case, R is the insert to vector ratio and bp is the length of the DNA in base pairs. In a typical experiment, the size of the vector and insert are known as is the amount of vector recommended for the ligation. The unknown is the amount of insert to be used. If this value is established as x, then the equation can be rearranged to solve for x as follows:

$$x \text{ ng of insert} = [(\text{ng of vector} \times \text{bp of insert}) \, / \, (\text{bp of vector})] \times R$$

Using this equation, it's very easy to calculate the amount of insert needed for any insert to vector ratio. What specific insert to vector ratio is usually recommended? Although ratios of 3:1 to 1:3 can work well, an insert to vector ratio of 1:1 to 3:1 is most often recommended for plasmid ligations. Some investigators recommend ratios as high as 6:1 for blunt-end ligations. For genomic or cDNA library construction in lambda vectors, a ratio of 1:1 is strongly recommended to limit the number of recombinant vectors with multiple inserts.

The second issue is the total amount of DNA that is needed for a successful ligation reaction. Even if the insert to vector ratio is correct, a ligation can still fail. Imagine, for example, that the above ligation with 100 ng of vector and 1 ng of insert is done in a total volume of 1 liter! In this case, the vector and insert ends will rarely find one another and the ligation will be very inefficient. Clearly, the total volume of the ligation determines the concentration of free ends of the DNA molecules to be ligated. For most ligations, the following can serve as guidelines for successful reactions:

1. For plasmid ligations, the vector concentration should be between 1 ng/µl and 10 ng/µl.
2. For lambda ligations, the vector concentration should be between 100 ng/µl and 200 ng/µl.

If one adheres to the insert to vector ratio guidelines outlined above, the total concentration of DNA will be in the appropriate range.

Other than improper insert to vector ratios or incorrect DNA concentrations, there are several other reasons that ligations can fail or be inefficient. One common reason for failure of ligation reactions is incomplete inactivation of restriction enzymes prior to ligation. Although many restriction enzymes can be heat activated, one must be aware that such heating often does not completely remove all enzyme activity. We prefer organic extractions (see the next paragraph) as a more reliable method of removing all restriction enzyme activity. If the enzyme can be heat inactivated, we recommend heat inactivation followed by organic extractions.

Successful ligations also require very pure DNA. We recommend purifying DNA by organic extractions using the phase lock gel tubes described in the text. Generally, it is best to do two phenol extractions followed by one chisam extraction (chisam is a 24:1 volume:volume ratio of chloroform and isoamyl alcohol). If the starting DNA is already relatively clean, however, one phenol extraction followed by one chisam extraction is often sufficient. We recommend that these organic extractions be followed by drop dialysis. In a typical series of organic extractions, the phenol removes most of the proteins, the chisam extraction removes residual phenol from the aqueous DNA phase, and the dialysis removes trace amounts of both phenol and chisam. Even if you use a column DNA purification system, we recommend doing drop dialysis as the last step in cleaning the DNA before a ligation. Predigested vectors purchased from vendors generally do not need any further purification.

Once a plasmid ligation has been completed, how much DNA should be used to transform the competent cells? In general, 1 to 10 ng of vector per 50 µl of competent cells works well. More DNA gives more total transformed colonies, while less DNA gives fewer transformed colonies.

PROCEDURAL NOTE

You may obtain fewer colonies per plate by using less DNA, but the number of colonies you obtain per ng of vector DNA will be greater.

The same situation is found with the amount of ligated lambda DNA mixed with packaging extract. Again, more ligated DNA per aliquot of packaging extract gives more total plaques per plate. However, less DNA per aliquot of packaging extract gives more plaques per ng of vector DNA.

Test transformations or infections with ligated plasmid or phage DNA can be done to see what amount of DNA gives the best balance of numbers of colonies/plaques per plate and the number of transformants/plaques per ng of vector. For example, using the following ligation of a plasmid to an insert in a 10-µl total volume, one might obtain the following results:

1 µl ligation (1 ng plasmid vector) + 50 µl competent cells = 100 colonies

10 µl ligation (10 ng plasmid vector) + 50 µl competent cells = 800 colonies

The second transformation gives the greatest total number of colonies (800 vs. 100), but the first transformation gives the greatest number of transformed colonies per ng of vector DNA (100 colonies per ng vs. 80 colonies per ng). As you can see, if the goal is to obtain the greatest number of total transformed colonies, the best strategy would be to do 10 transformations with 1 µl of ligation in each reaction. This would result in 1000 total colonies. Whereas, using all 10 µl of ligation in one competent cell transformation would result in only 800 colonies. Keep in mind that the 10 transformations will give more total colonies but will also be more expensive (due to the cost of competent cells) and more time consuming.

To summarize, for successful ligations be sure to do all of the following:

1. Calculate the correct insert to vector ratio.
2. Calculate the correct concentration of DNA.
3. Be sure restriction enzymes have been completely inactivated.
4. Be sure your DNA is very pure.
5. Drop dialyze your DNA before ligation.
6. Be sure not to add too much DNA to your transformations and lambda packaging reactions.

Good luck!

AV.3. Use of DNase to Eliminate DNA Contamination from RT-PCR Reactions

THIS IS the protocol recommended when the gene being analyzed has no introns. In such a situation, the PCR product from contaminating DNA will be the same size as from mRNA, so elimination of contaminating DNA is critical!

Step 1 Pipette 17 μl RT buffer into your "PCR RNA" tube, which already contains 1 μl of your RNA sample. Mix gently by pipetting up and down. This RT buffer contains 100 ng of RNase-free DNase I (Boehringer-Mannheim), 5 mM $MgCl_2$, 50 mM KCl, 10 mM Tris-HCl (pH 8.3), 20 units of RNase inhibitor (Applied Biosystems), and 1 mM of each dNTP. Incubate at 37°C for 30 minutes. This step will degrade any contaminating genomic DNA in your RNA preparation.

Step 2 Heat to 95°C for 5 minutes to heat-kill the DNase I enzyme. Spin the tube briefly in the nanofuge.

Step 3 Pipette 1 μl RT primer (either random hexamers, oligo-dT, or downstream specific primer) into your "PCR RNA" tube. Mix gently by pipetting up and down, and let stand 10 minutes at room temperature. During this time the primers will anneal to the RNA.

Step 4 Add 1 μl 50 U/μl MuLV Reverse Transcriptase (Applied Biosystems), and mix by gently pipetting up and down.

Step 5 Place your "PCR RNA" tube into the thermocycler. Run one cycle to generate cDNA:

a. 42°C for 15 minutes (during which time the RT enzyme will do first-strand cDNA synthesis)

b. 99°C for 5 minutes (to denature the cDNA and RNA)

c. 5°C for 1 minute (to cool the tubes)

Step 6 After the tubes have cooled, remove your tube from the PCR instrument. Spin the tube for 15 seconds in your nanofuge to recover all liquid to the bottom of the tube. Add 80 μl *Taq* mix. *Taq* mix is 1.25 mM $MgCl_2$, 50 mM KCl, 10 mM Tris-HCl (pH 8.3), 2.5 units of AmpliTaq DNA polymerase (Applied Biosystems), and 125 ng of upstream specific primer (125 ng of specific downstream primer was already added for the RT step). Mix by pipetting up and down.

Step 7 Set the thermocycler to run 30 cycles to amplify the cDNA fragment. These cycles will take about 3 hours to complete:

d. An initial denaturing step at 95°C for 2 minutes.

The following two steps (e and f) will be cycled 30 times:

e. 95°C for 1 minute — denaturing step

f. 60°C for 1 minute — combined annealing and elongation step. This temperature will be different for different oligo primers.

g. After the 30 cycles have been completed — 60°C for 7 minutes, final extension step.

h. The program will then maintain 8°C.

AV.4. Alternate Protocol: SYBR Safe DNA Agarose Gel Stain

SYBR Safe DNA gel stain (Molecular Probes) has been specifically developed for reduced mutagenicity, making it safer than ethidium bromide for staining DNA in agarose or acrylamide gels. (Refer to http://probes.invitrogen.com/products/sybrsafe.) SYBR Safe stain comes either as a concentrate or as a ready-to-use solution, which can be used just like an ethidium bromide solution. The detection sensitivity with SYBR Safe stain is comparable to that obtained with ethidium bromide. DNA bands stained with SYBR Safe DNA gel stain can be detected using a standard UV transilluminator, a visible-light transilluminator or a laser-based scanner. The stain is also suitable for staining RNA in gels. Bound to nucleic acids, SYBR Safe stain has fluorescence excitation maxima at 280 and 502 nm and an emission maximum at 530 nm. It should not be stored in a secondary container where it could be unnecessarily exposed to light.

HANDLING AND DISPOSAL

SYBR Safe DNA gel stain showed no or very low mutagenic activity when tested by an independent, licensed testing laboratory, and this stain is not classified as hazardous waste under U.S. Federal regulations. The safety testing included three well-established mammalian cell-based tests, a battery of well-established Ames-test bacterial strains, and extensive tests for environmental safety. Nevertheless, please exercise appropriate care and judgment when using this reagent and dispose the stain in compliance with all pertaining local regulations.

PROCEDURAL NOTE

The agarose gel can be stained postelectrophoresis (Protocol A) or the gel can be cast with the SYBR Safe stain included (Protocol B).

PROTOCOL A. STAINING NUCLEIC ACIDS AFTER ELECTROPHORESIS

Step 1 Soak the gel in SYBR Safe stain. If using SYBR Safe gel stain concentrate, dilute 10,000x in TAE or TBE agarose gel buffer (as appropriate) before use. Place the gel in a plastic container, such as a pipette-tip box lid or a household food-storage container. Do not use a glass container, as the dye in the staining solution may adsorb to the walls of the container, resulting in poor gel staining. Add sufficient SYBR Safe DNA gel stain to cover the gel, and ensure that the entire gel is fully immersed during staining.

Step 2 Incubate for 30 minutes. Protect the gel and staining solution from light by covering it with aluminum foil or by placing it in the dark. Gently and continuously agitate the gel at room temperature (e.g., on an orbital shaker at 50 rpm). No destaining is required.

Step 3 View and photograph the gel, as described in Protocol B, Step 3, below.

PROTOCOL B. PRECASTING SYBR SAFE STAIN IN AGAROSE GELS

Step 1 You can prepare the agarose gel directly in SYBR Safe DNA gel stain. SYBR Safe stain is provided in buffer; simply substitute SYBR Safe stain for the buffer when preparing the molten agarose. If using the concentrate, dilute appropriately before use. The agarose/SYBR Safe stain mixture may be heated in a microwave oven. As with precasting gels with ethidium bromide, the mobility of nucleic acid fragments in the gel may be somewhat slower when run in these gels compared to their mobility in gel without stain.

Step 2 Electrophorese the gel. Use a running buffer appropriate to the SYBR Safe gel stain formulation. No poststaining or destaining is needed.

Step 3 Viewing and photographing the gel. Stained gels can be viewed using a standard 300-nm transilluminator, a 254-nm epi- or transilluminator, or a blue-light transilluminator. DNA stained with SYBR Safe stain can also be visualized and analyzed using imaging systems equipped with an excitation source in the UV range or between 470 to 530 nm. Stained gels can be photographed using Polaroid Type 667 black-and-white print film and a SYBR Safe photographic filter (S37100) (Molecular Probes). A SYPRO photographic filter (S6656) or a Kodak Wratten #9 filter will also work well. Using this film and one of these filters, SYBR Safe DNA gel stain provides the same detection sensitivity as ethidium bromide using a photographic filter appropriate for ethidium bromide. A standard ethidium bromide photographic filter with SYBR Safe DNA gel stain does not provide sensitivity comparable to that provided by ethidium bromide stained gels. Gels stained with SYBR Safe stain can also be imaged using a CCD[LB2] camera or a laser-based scanner. Refer to Table A-1 for suggestions for use of some of these instruments.

Table A-1 Filter Selection Guide for Use with SYBR Safe Stain

Instrument (Manufacturer)	Excitation Source	Emission Filter
AlphaImager (Alpha Innotech)	302 nm	SYB-500
AlphaImager HP (Alpha Innotech)	302 nm	SYB-500
AlphaDigiDoc RT (Alpha Innotech)	UV transilluminator	
Shroud, Camera Stand (Alpha Innotech)	UV transilluminator	SYB-100
DE500 or DE400 light cabinet 2.17" diam. (Alpha Innotech)	UV transilluminator	SYB-500
DE500 or DE400 light cabinet 2" diam. (Alpha Innotech)	UV transilluminator	SYB-400
VersaDoc Imaging Systems (Bio-Rad)	Broadband UV	520LP
Molecular Imager FX Systems (Bio-Rad)	488 nm	530 nm
BP Gel Doc Systems (Bio-Rad)	302 nm	520DF30 (#170-8074)
Typhoon 9400/9410 (GE Healthcare)	488 nm	520 BP 40
Typhoon 9200/9210/8600/8610 (GE Healthcare)	488 nm	526 SP
FluorImager (GE Healthcare)	488 nm	530 DF 30
Storm (GE Healthcare) Blue	(fluorescence mode)	
VDS-CL (GE Healthcare)	Transmission	UV Low
Ultracam/Gel Imager (Ultra-Lum)	UV	Yellow Filter (#990-0804-07)
Omega Systems (Ultra-Lum)	UV	520 nm
Polaroid Camera (Polaroid)	UV	SYBR Safe Photographic Filter (S27100)
FOTO/Analyst Express/Investigator/Plus/Luminary (FOTODYNE)	UV	Fluorescent Green (#60-2034)
FOTO/Analyst Minivisionary (FOTODYNE)	UV	Fluorescent Green (#62-4289) FOTO/Analyst
Apprentice (FOTODYNE)	UV	Fluorescent Green (#62-2535) FOTO/Analyst
Luminary (FOTODYNE)	UV	Fluorescent Green (#60-2056)
FCR-10 (Polaroid)	UV	#3-4218
FUJI FLA-3000 (FUJI Film)	473 nm	520LP
BioDocIt/AC1/EC3/BioSpectrum (UVP)	302 nm	SYBR Green (#38-0219-01) or SYBR Gold (#38-0221-01)
Gel Logic (Kodak)	UV 535 nm	WB50
Syngene Instruments (Syngene)	UV	500–600 nm Shortpass filter

AV.5. Using DEPC for RNA Solutions

PROCEDURAL NOTE

All RNA solutions should be made with DEPC-treated water. Tris solutions cannot be DEPC-treated. Solutions containing Tris should be DEPC-treated and autoclaved *before* adding Tris. After addition of Tris, the solution should be autoclaved again.

DEPC: diethylpyrocarbonate (Sigma). The stock solution from Sigma, Inc., is 10% DEPC in ethanol (vol/vol).

DEPC-treated water: Add DEPC stock solution to ddH_2O to a final concentration of 0.1% (vol/vol) in distilled water. This solution should be stirred slowly for 1 hour at room temperature and then incubated overnight at 37°C. The next day, the solution is autoclaved for 15 minutes on liquid cycle. The autoclaving removes the DEPC and leaves the remaining solution RNase-free.

AV.6. Preparation of Phenol for DNA Extractions

Step 1 Phenol should be redistilled at 160°C. Store the purified phenol in 100 ml aliquots at –20°C until needed. Redistillation is a hazardous procedure. It is much more convenient and safer to purchase molecular biology grade phenol (American Bioanalytical).

Step 2 As needed, remove from freezer and warm to room temperature. Then melt the crystals at 70°C.

Step 3 Stir the melted phenol with an equal volume of ddH_2O for several hours at 4°C. Then allow it to stand at room temperature until the phases separate.

Step 4 Remove the lower phase to a fresh brown glass container (shield the phenol from light). At this stage, the phenol can be stored for several months at 4°C.

Step 5 To equilibrate the phenol, add an equal volume of 1 M Tris-HCl, pH 8.0. Stir for several hours at 4°C. Then let stand at room temperature.

Step 6 After the phases have separated, discard the top (aqueous) layer.

Step 7 Mix the phenol layer with an equal volume of 0.1 M Tris HCl, pH 8.0. Stir for several hours. Store at 4°C. Do not store for longer than 1 month.

PROCEDURAL NOTE

Phenol is dangerous and can cause burns. Always wear gloves and safety glasses, and handle phenol with care. If phenol touches your skin, rinse quickly with large quantities of water, followed by soap and water. *Do not* rinse with ethanol. Phenol solutions in Tris buffers will turn yellow quickly and should be discarded. The antioxidant 8-hydroxyquinoline can be added at 0.1% as a preservative and has a bright yellow color that aids in identification of phenol layers during extractions. *Always work with phenol in a fume hood!*

AV.7. Preparation of Chisam for DNA Extractions

CHISAM IS chloroform:isoamyl alcohol (24:1 vol/vol). Store chisam in a tightly capped bottle protected from light. *Always work with chloroform and chisam in a fume hood!*

AV.8. Preparation of CaCl$_2$ Competent Cells

Step 1 Grow a 10-ml overnight culture of *E. coli* in LB medium.

Step 2 In the morning, make a 1:100 dilution of the overnight culture in 30 ml of 2x YT medium.

Step 3 Grow cells at 37°C in a shaking incubator with good aeration until the OD_{550} = 0.25–0.30 (about 1.5 to 2 hours). It is very important not to let the cells overgrow.

Step 4 While the cells are growing, prepare fresh 100 mM $CaCl_2$ and 50 mM $CaCl_2$/50 mM $MgCl_2$ by diluting 1 M stocks of $CaCl_2$ and $MgCl_2$ and place solutions on ice.

Step 5 When the cells have reached the proper OD_{550}, spin at 1700x g for 10 minutes at 4°C.

Step 6 Pour off the supernatant, and tap the tube against the side of the counter several times to loosen the cell pellet.

PROCEDURAL NOTE

For Steps 7 through 11, chill all tubes and tips at 4°C before use (this helps to keep the cells cold throughout the protocol).

Step 7 Add 15 ml ice-cold 100 mM $CaCl_2$, and swirl the tube to resuspend the cells. Pipette the solution only if necessary to resuspend the cells, as it is important to have minimal contact with the cells. Keep the cells on ice as much as possible. This step should take no longer than 1 minute.

Step 8 Chill on ice for 30 minutes.

Step 9 Spin at 1500x g for 10 minutes at 4°C.

Step 10 Once again, pour off the supernatant and tap the tube against the counter several times. Add 3 ml of ice-cold 50 mM $CaCl_2$/50 mM $MgCl_2$, and swirl to resuspend. Do not pipette the solution at this step because it will greatly decrease your transformation efficiency.

Step 11 Chill on ice for a minimum of 60 minutes before using.

PROCEDURAL NOTE

You can save the cells overnight on ice at 4°C with a slight decrease in transformation efficiency.

AV.9. Salt Concentrations for Nucleic Acid Precipitation

	Stock Solution	Final Concentration before Ethanol
Sodium acetate	2.5 M (pH 5.2)	0.25 M
Sodium chloride	5.0 M	0.2 to 1.0 M
Ammonium acetate	7.5 M	2.5 M
Lithium chloride	8.0 M	0.8 M

AV.10. Notes on Ethanol Precipitation of DNA and RNA

Step 1 Adjust the salt concentration according to the previous chart, and mix the samples thoroughly.

Step 2 Add 2 volumes of ice-cold ethanol and mix again. (RNA requires 2.5 volumes of ethanol.)

PROCEDURAL NOTE

NaCl concentrations higher than 1.0 M will cause NaCl to precipitate with the DNA, resulting in a "salt pellet."

Step 3 The DNA precipitate forms at low temperatures and is recovered by centrifugation. Chilling for 1 hour at –20°C or 20 minutes at –70°C is usually sufficient unless the DNA concentration is very low (< 0.1 µg/ml) or the DNA is smaller than approximately 200 bp. Addition of $MgCl_2$ to 0.01 M will help to precipitate small fragments. Spinning for 15 minutes at 4°C in a microcentrifuge (12,000x g) is sufficient to pellet most DNA preparations, although up to 1 hour can be used for small fragments or low concentrations of DNA.

Step 4 Drain the pellet by inversion or with a flame-drawn pipette. Remove traces of ethanol by air drying for 15 to 30 minutes. Wash the pellet once or twice with 70% ethanol (cold) when it is necessary to be sure that no solutes remain in the pellet (especially salts). To wash, fill the tube about half full. Invert the tube several times and spin again to ensure recovery of the pellet. After the wash, the pellet will not adhere as well to the wall of the tube; thus, take care in removing the supernatant. If salt is a problem, two 70% ethanol washes should be performed.

PROCEDURAL NOTES

Nucleotide triphosphates can be removed from DNA made up to 2 M in ammonium acetate using two sequential ethanol precipitations. This is often done to remove free label after kinasing DNA.

Short oligonucleotides (primers, probes) can be removed from DNA by precipitating using lithium chloride (0.8 M final concentration).

Ethanol versus isopropanol: Isopropanol is an effective alternative to ethanol and has the advantage of precipitating DNA at lower concentrations. Instead of mixing two volumes of ethanol with the DNA-salt solution, addition of 0.7 to 1.0 volumes of isopropanol will suffice.

Precipitation of RNA: the methods used for precipitation of RNA are essentially identical to those described above for DNA. The major concern in any type of RNA work is avoiding ribonuclease contamination. For ethanol precipitation of RNA, researchers generally use 2.5 to 3 volumes of ethanol instead of the 2 volumes commonly used for precipitation of DNA.

AV.11. Separation of DNA and Marker Dyes in Agarose and Acrylamide Gels

Separation of DNA in agarose gels

% Agarose	Optimal Range of Separation (kb)
0.6	1–20
0.7	0.8–10
0.9	0.5–7
1.2	0.4–6
1.5	0.2–4
2.0	0.1–3

PROCEDURAL NOTE

Low-percentage gels (< 0.6%) are very difficult to handle and are best poured and run at 4°C. Separation of fragments more than 20 kb is best done using pulsed-field gel electrophoresis. Separation of fragments smaller than 0.2 kb is best done on acrylamide gels. NuSieve and Metaphor agarose are especially designed to separate small DNA fragments (1000 bp to as small as 8 bp).

Separation of fragments in acrylamide gels

% Acrylamide	Optimal Range of Separation (bp)
3.5	100–1000
5.0	80–500
8.0	60–400
12.0	40–200
20.0	10–100

Migration of marker dyes in polyacrylamide

% Gel	Bromphenol blue (bp)	Xylene cyanol (bp)
3.5	100	460
5.0	65	260
8.0	45	160
12.0	20	70
20.0	12	45

PROCEDURAL NOTE

The fragment lengths indicate the approximate size DNA fragment with which the dye migrates in the given percentage gel.

AV.12. Estimating DNA and RNA Concentrations Using Spectrophotometry and Gel Electrophoresis

TWO METHODS can be used to assess the quality and quantity of purified genomic DNA: gel electrophoresis and UV spectrophotometry. Each of these methods provides different, but overlapping information about the quality and quantity of genomic DNA. Gel electrophoresis can be used to estimate the quantity of DNA and to confirm that the isolated DNA is initially of high molecular weight and free of contaminating RNA. UV spectrophotometry can be used to provide a precise estimate of the quantity of recovered DNA and to confirm that the DNA is free of other chemical contaminants including organic solvents.

The concentration and purity of DNA can be determined by reading the optical density of a sample at 260 and 280 nm. The reading at 260 nm allows calculation of the concentration of nucleic acid in the sample. For pure double-stranded DNA, 1 OD_{260} = 50 mg/μl (for a 1.0-cm path length cuvette). The ratio between the readings at 260 and 280 nm (OD_{260}/OD_{280}) provides an estimate of the purity of the nucleic acid. Pure preparations of DNA have an OD_{260}/OD_{280} value of 1.6–1.9. If there is contamination with significant amounts of protein, lipid or phenol, the OD_{260}/OD_{280} will be less than 1.6. Values greater than 1.9 indicate possible contamination with RNA.

The most informative estimate of purity of a nucleic acid preparation is obtained from a continuous scan of UV absorbance from 220 to 320 nm. This is because phenol will give a scan that is quite different in shape from that seen with pure DNA (see Figure 1-14). Phenol is particularly problematic because it increases the reading at both 260 and 280 nm, and this can result in an OD_{260}/OD_{280} ratio that is still in the 1.6 to 1.9 range. However, the phenol will cause the 260 reading to be much higher than it should be, and this will result in a gross overestimate of the amount of DNA in the sample. In addition, the phenol in the sample will be very effective at preventing enzyme function in downstream reactions. In other words, OD_{260}/OD_{280} ratios alone are often not sufficient to detect phenol contamination. Protein contamination will cause an increase at 280 relative to 260 nm and thus will decrease the OD_{260}/OD_{280} ratio of the sample. In addition, the readings are much higher at 220 nm when there is significant protein contamination (see Figure 1-14). Contamination with lipid or RNA results in a scan that looks very similar to that of pure DNA; however, the OD_{260}/OD_{280} ratio will be higher with RNA contamination (often over 1.9) and lower with lipid contamination. However, unless there is a lot of contamination, the ratios are still likely to be within the 1.6 to 1.9 range. Remember that RNA contamination can also be seen by running an agarose gel. All of these contaminants as well as free nucleotides cause an increase in the readings at 260 nm and thus an overestimate of the amount of DNA. An excellent solution to contamination with any of these molecules is to extract once with phenol and once with chisam using the Phase Lock Light gel tubes (Eppendorf) described in the text. After two extractions, dialyze the sample overnight against 0.1x TE as described in the text for mouse DNA in Project I. For RNA contamination, samples should first be digested with RNase A, then extracted with phenol and chisam, and then drop dialyzed.

AV.13. Other Useful Molecular Biology Information*

THE GENETIC CODE

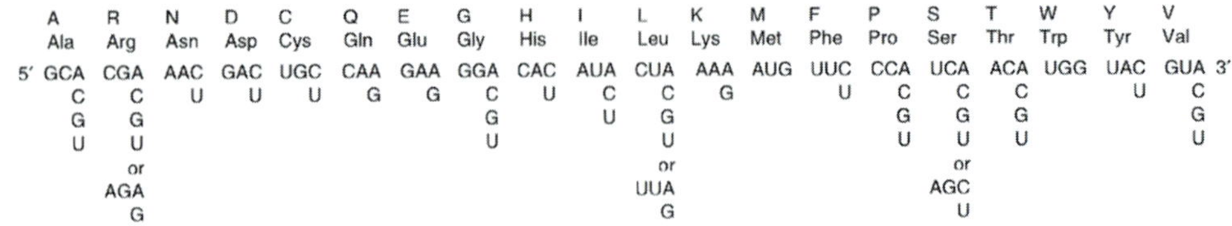

A Ala	R Arg	N Asn	D Asp	C Cys	Q Gln	E Glu	G Gly	H His	I Ile	L Leu	K Lys	M Met	F Phe	P Pro	S Ser	T Thr	W Trp	Y Tyr	V Val	
5′ GCA C G U	CGA C G U or AGA G	AAC U	GAC U	UGC U	CAA G	GAA G	GGA G C G U	CAC U	AUA C U	CUA C G U or UUA G	AAA G	AUG	UUC U	CCA C G U	UCA C G U or AGC U	ACA C G U	UGG	UAC U	GUA C G U 3′	

Second Position

First Position (5′ end)		U	C	A	G		Third Position (5′ end)
	U	UUU ⎤ UUC ⎦ Phe UUA ⎤ UUG ⎦ Leu	UCU ⎤ UCC ⎥ UCA ⎥ Ser UCG ⎦	UAU ⎤ UAC ⎦ Tyr UAA Stop UAG Stop	UGU ⎤ UGC ⎦ Cys UGA Stop UGG Trp	U C A G	
	C	CUU ⎤ CUC ⎥ CUA ⎥ Leu CUG ⎦	CCU ⎤ CCC ⎥ CCA ⎥ Leu CCG ⎦	CAU ⎤ CAC ⎦ His CAA ⎤ CAG ⎦ Gln	CGU ⎤ CGC ⎥ CGA ⎥ Arg CGG ⎦	U C A G	
	A	AUU ⎤ AUC ⎥ Ile AUA ⎦ AUG Met	ACU ⎤ ACC ⎥ ACA ⎥ Thr ACG ⎦	AAU ⎤ AAC ⎦ Asn AAA ⎤ AAG ⎦ Lys	AGU ⎤ AGC ⎦ Ser AGA ⎤ AGG ⎦ Arg	U C A G	
	G	GUU ⎤ GUC ⎥ GUA ⎥ Val GUG ⎦	GCU ⎤ GCC ⎥ GCA ⎥ Ala GCG ⎦	GAU ⎤ GAC ⎦ Asp GAA ⎤ GAG ⎦ Glu	GGU ⎤ GGC ⎥ GGA ⎥ Gly GGG ⎦	U C A G	

Termination Signals

UAA (Ochre)
UAG (Amber)
UGA (Opal)

Single Letter Code

A = adenosine B = C or G or T
C = cylidine D = A or G or T
G = guanosine H = A or C or T
T = thymidine K = G or T
U = uridine M = A or C
N = A or C or G or T
R = A or G
S = C or G
V = A or C or G
W = A or T
Y = C or T

AMINO ACID STRUCTURES

Each amino acid is accompanied by its three- and one-letter code, residue molecular weight (actual molecular weight minus water) and side-chain pK_a where appropriate.

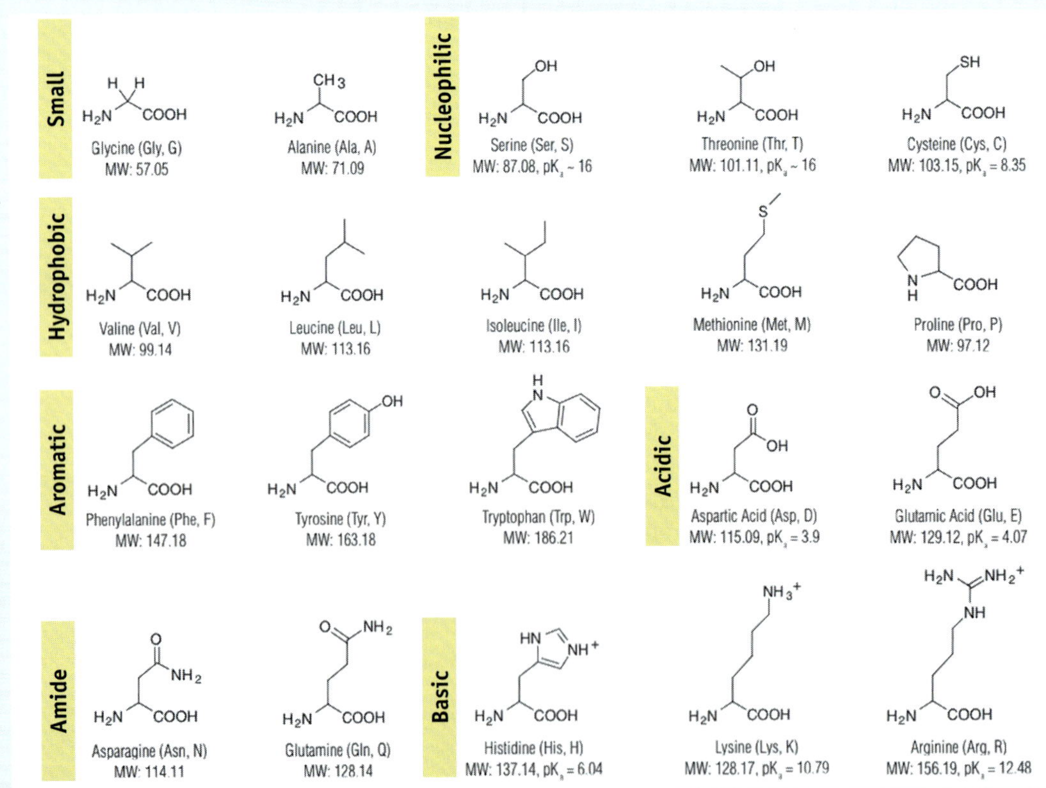

*Source: All tables, chart, and figures in this section copyright © New England Biolabs. Reprinted with permission.

DNA BASE PAIRS

The structures of the adenosine:thymidine and guanosine:cytidine base pairs are shown in the context of the ribose phosphodiester backbones. The numbering schemes of the ribose and nucleotide moieties are indicated. Arrows indicate the polarity of each strand from 5' to 3'.

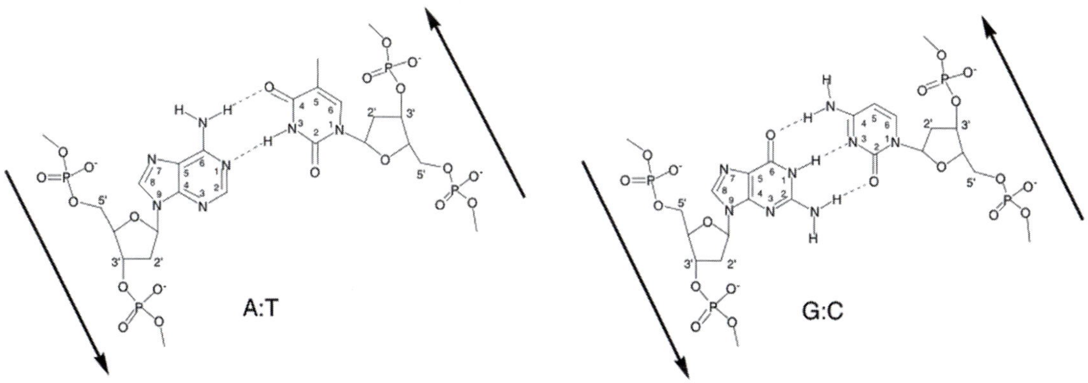

NUCLEIC ACID DATA

Average weight of a DNA basepair (sodium salt) = 650 daltons

1.0 A_{260} unit ds DNA = 50 µg/ml = 0.15 mM (in nucleotides)

1.0 A_{260} unit ss DNA = 33 µg/ml = 0.10 mM (in nucleotides)

1.0 A_{260} unit ss RNA = 40 µg/ml = 0.11 mM (in nucleotides)

MW of a double-stranded DNA molecule = (# of base pairs) X (650 daltons/base pair)

Moles of ends of a double-stranded DNA molecule = 2 X (grams of DNA) / (MW in daltons)

Moles of ends generated by restriction endonuclease cleavage:

   (a) circular DNA molecule: 2 X (moles of DNA) X (number of sites)

   (b) linear DNA molecule: 2 X (moles of DNA) X (number of sites) + 2 X (moles of DNA)

1 µg of 1000 bp DNA = 1.52 pmol = 9.1×10^{11} molecules

1 µg of pUC18/19 DNA (2686 bp) = 0.57 pmol = 3.4×10^{11} molecules

1 µg of pBR322 DNA (4361 bp) = 0.35 pmol = 2.1×10^{11} molecules

1 µg of M13mp18/19 DNA (7249 bp) = 0.21 pmol = 1.3×10^{11} molecules

1 µg of λ DNA (48502 bp) = 0.03 pmol = 1.8×10^{10} molecules

1 pmol of 1000 bp DNA = 0.66 µg

1 pmol of pUC18/19 DNA (2686 bp) = 1.77 µg

1 pmol of pBR322 DNA (4361 bp) = 2.88 µg

1 pmol of M13mp18/19 DNA (7249 bp) = 4.78 µg

1 pmol of λ DNA (48502 bp) = 32.01 µg

1.0 kb DNA = coding capacity for 333 amino acids ≈ 37,000 dalton protein

10,000 dalton protein ≈ 270 bp DNA

50,000 dalton protein ≈ 1.35 kb DNA

Isotope Data

Isotope	Particle Emitted	Half Life			
^{14}C	β	5,730 years	1 Ci	=	1,000 mCi
^{3}H	β	12.3 years	1 mCi	=	1,000 µCi
^{125}I	λ	60 days	1 µCi	=	2.2×10^6 disintegrations/minute
^{32}P	β	14.3 days	1 Becquerel	=	1 disintegration/second
^{33}P	β	25 days	1 µCi	=	3.7×10^4 Becquerels
^{35}S	β	87.4 days	1 Becquerel	=	2.7×10^{-5} µCi

Acids and Bases

Compound	Formula	Molecular Weight	Specific Gravity	% by Weight	Conc Reagent Molarity
Acetic acid, glacial	CH_3COOH	60.0	1.05	99.5	17.4
Formic acid	$HCOOH$	46.0	1.20	90	23.4
Hydrochloric acid	HCl	36.5	1.18	36	11.6
Nitric acid	HNO_3	63.0	1.42	71	16.0
Perchloric acid	$HClO_4$	100.5	1.67	70	11.6
Phosphoric acid	H_3PO_4	98.0	1.70	85	18.1
Sulfuric acid	H_2SO_4	98.1	1.84	96	18.0
Ammonium hydroxide	NH_4OH	35.0	0.90	28	14.8
Potassium hydroxide	KOH	56.1	1.52	50	13.5
Sodium hydroxide	$NaOH$	40.0	1.53	50	19.1
β-mercaptoethanol	$HSCH_2CH_2OH$	78.1	1.11	100	14.3

Protein Data

Bacterial Cells: *E. coli* or *Salmonella typhimurium*

Cell Data	per cell	per liter at 10^9 cells per ml	
Wet Weight	9.5×10^{-13} g	0.95 g	Theoretical maximum yield for a 1 liter culture
Dry Weight	2.8×10^{-13} g	0.28 g	(10^9 cells /ml) if protein of interest is:
Total Protein	1.55×10^{-13} g	0.15 g	0.1% of total protein: 150 µg/liter
Volume	1.15 µm^3 = 1 femtoliter		2.0% of total protein: 3 mg/liter
Protein Conc. in the cell:	135 mg/ml		50.0% of total protein: 75 mg/liter

Common Plasmid Gene Products

Gene	Gene Product # of Residues	Molecular Weight (daltons)
tet (pBR322)	401	43,267
amp (pBR322)	286	31,515
kan (pACYC177)	264	29,047
cam (pACYC184)	219	25,663
lacZ α (pUC19)	107	12,232
lacZ	1,023	116,351

Nucleotide Physical Properties

Compound	Molecular Weight	λ Max (pH 7.0)	Absorbance at λ Max 1 M Solution (pH 7.0)
ATP	507.2	259	15,400
CTP	483.2	271	9,000
GTP	523.2	253	13,700
UTP	484.2	262	10,000
dATP	491.2	259	15,200
dCTP	467.2	271	9,300
dGTP	507.2	253	13,700
dTTP	482.2	267	9,600

pH vs. Temperature for Tris Buffer

pH of Tris Buffer (0.05 M)		
5°C	25°C	37°C
7.76	7.20	6.91
7.89	7.30	7.02
7.97	7.40	7.12
8.07	7.50	7.22
8.18	7.60	7.30
8.26	7.70	7.40
8.37	7.80	7.52
8.48	7.90	7.62
8.58	8.00	7.71
8.68	8.10	7.80
8.78	8.20	7.91
8.88	8.30	8.01
8.98	8.40	8.10
9.09	8.50	8.22
9.18	8.60	8.31
9.28	8.70	8.42

Agarose Gel Resolution

% Gel	Optimum Resolution for Linear DNA (kb)
0.5	30 to 1.0
0.7	12 to 0.8
1.0	10 to 0.5
1.2	7 to 0.4
1.5	3 to 0.2

AVOIDING RIBONUCLEASE CONTAMINATION

RNA is more susceptible to degradation than DNA, due to the ability of the 2′ hydroxyl groups adjacent to the phosphodiester linkages in RNA to act as intramolecular nucleophiles in both base- and enzyme-catalyzed hydrolysis. Whereas deoxyribonucleases (DNases) require metal ions for activity and can therefore be inactivated with chelating agents (e.g., EDTA), many ribonucleases (RNases) bypass the need for metal ions by taking advantage of the 2′ hydroxyl group as a reactive species.

Many microbial ribonucleases, including *E. coli* RNase I and RNase M (1), apparently share many properties with bovine pancreatic RNase A. RNase A is a single-strand specific endoribonuclease that is resistant to metal chelating agents and can survive prolonged boiling or autoclaving. RNase A-type enzymes rely on active site histidine residues for catalytic activity (2) and can be inactivated by the histidine-specific alkylating agent diethyl pyrocarbonate (DEPC).

Sources of RNase contamination

It is important to realize that, even though New England Biolabs' enzymes certified for RNA work have been purified free of ribonucleases, it is possible to reintroduce RNases during the course of experimentation via aqueous solutions or from the environment at large. Autoclaving a solution will kill contaminating bacteria, but RNases liberated from the dead bacteria will still be active. Additionally, ungloved fingers can introduce bacteria into solutions resulting in RNase contamination. RNase contamination can be avoided by using RNase-free solutions and following a few common sense laboratory procedures.

Laboratory precautions (3,4)

Always wear gloves when working with RNA. It is a good idea to maintain a separate area for RNA work that has its own set of pipetters. This is especially important if your work requires the use of RNase A (e.g., plasmid preps). Sterile, disposable plasticware can safely be considered RNase-free and should be used when possible. Metal spatulas can quickly be decontaminated by holding in a burner flame for several seconds. Any piece of glassware that was ever in common use should be viewed with suspicion. Contaminating RNases can be inactivated by baking glassware at 180°C or higher for several hours. Alternatively, glassware can be soaked in freshly prepared 0.1% (v/v) DEPC in water or ethanol for 1 hour, drained, and autoclaved (necessary to destroy any unreacted DEPC that can otherwise react with other proteins and RNA). DEPC will destroy polycarbonate or polystyrene materials (e.g., electrophoresis tanks), which should instead be decontaminated by soaking in 3% hydrogen peroxide for 10 minutes. Remove peroxide by extensively rinsing with RNase-free water (see below) prior to use.

Preparation of solutions (3,4)

DEPC treatment of solutions is accomplished by adding 1 ml DEPC (Sigma) per liter of solution, stirring for 1 hour, and autoclaving for one hour to hydrolyze any remaining DEPC. Compounds with primary amine groups (e.g., Tris) will react with DEPC. Consequently, Tris buffers should be prepared by dissolving Tris base (from a fresh bottle reserved for RNA work) in DEPC-treated and autoclaved water, adjusting the pH (with an electrode reserved for RNA work), and re-autoclaving to sterilize. Solutions of thermolabile materials (e.g., DTT, nucleotides, manganese salts) should be prepared by dissolving the solid (highest available purity) in DEPC-treated and autoclaved water and passing the solution through a 0.2-μm filter to sterilize. As an alternative to DEPC, which can inhibit enzymatic reactions if not completely hydrolyzed, we have found that Milli-Q™ (Millipore) purified water is sufficiently free of RNases for most RNA work.

Inhibitors of ribonucleases

RNA can also be protected from RNase activity by using an RNase inhibitor. Ribonucleoside Vanadyl Complex (NEB #S1402S, page 164) is a transition-state analog inhibitor of RNase A-type enzymes, with $K_i = 1 \times 10^{-5}$ M. Unfortunately, the complex also inhibits many other enzymes used in RNA work. One exception is reverse transcriptase, so this complex has proved useful in protecting cellular mRNA from degradation during purification prior to cDNA synthesis (5). An angiogenin-binding protein isolated from human placenta is a better RNase inhibitor ($K_i = 4 \times 10^{-14}$ M; 6) and is specific for RNases, but it is much more expensive. Consequently, this inhibitor is useful only for small-scale applications (e.g., *in vitro* transcription).

References:

1. Meador III, J. and Kennell, D. (1990) *Gene* 95, 1–7.

2. Fersht, A.R. (1977) *Enzyme Structure and Mechanism*, Freeman, Reading, PA, 325–329.

3. Blumberg, D.D. (1987) *Methods Enzymol.* 152, 20–24.

4. Sambrook, J., Fritsch, E.F. and Maniatis, T. (1989) *Molecular Cloning: A Laboratory Manual*, Cold Spring Harbor Press, Cold Spring Harbor, NY, 7.3–7.5.

5. Berger, S.L. (1987) *Methods Enzymol.* 152, 227–234.

6. Lee, F.S. and Vallee, B.L. (1990) *Proc. Natl. Acad. Sci. USA* 87, 1879–1883.

USEFUL CALCULATIONS   **VI**

---

1. Calculation of Picomoles of DNA or DNA Ends

To determine the number of pmol for a given number of micrograms (µg) of double-stranded DNA, use the following formula:

$$\# \text{ pmol} = \frac{\text{µg DNA x } 10^6}{\text{\# nucleotide pairs x } 660}$$

For single-stranded DNA, use the following formula:

$$\# \text{ pmol} = \frac{\text{\# µg DNA x } 10^6}{\text{\# nucleotides x } 330}$$

PROCEDURAL NOTE

660 is the approximate molecular weight of a DNA nucleotide pair, and 330 is the approximate weight of a nucleotide.

To calculate the number of pmol ends, you have in a sample of DNA, use the following equation:

pmol ends = pmol DNA x 2 (for a single-stranded DNA fragment)

For calculations of the number of pmol ends where the number of DNA fragments is greater than one, estimate the number of fragments of DNA and multiply the pmol end calculation, as listed above, by the estimated number of fragments.

2. Calculation of the Melting Point Temperature (T_m)

Different equations can be used to calculate the T_m of a double-stranded nucleic acid. Which equation you use depends on what kind of duplex you are dealing with, on what reagents are in your solutions, and how accurately you need to know the T_m. The following is a compilation of equations that allow you to calculate the T_m for different kinds of duplexes. Select the equation that matches the type of duplex you are trying to melt or reanneal.

PROCEDURAL NOTE

The following equations are for double-stranded nucleic acids longer than 100 bp.

DOUBLE-STRANDED DNA

$$T_m = 81.5°C + 16.6 \log_{10} [M] + 0.41(\%[G + C]) - 500/n - F$$

T_m = Melting temperature (°C)

[M] = Concentration of monovalent cations in the buffer [molar]

G = Number of G residues in two strands of the dsDNA

C = Number of C residues in two strands of the dsDNA

%(G + C) = Sum of the number of Gs and Cs in the helix divided by the total number of nucleotides (i.e., the G + C content)

n = length of duplex in base pairs

F = 0.63°C $\times$ percentage of formamide in the solution (°C)

DOUBLE-STRANDED RNA

$$T_m = 78.0°C + 16.6 \log_{10} [M] + 0.7(\%[G + C]) - 500/n - F$$

DNA-RNA HYBRIDS

$$T_m - 67.0 + 16.6 \log_{10} [M] + 0.8(\%[G + C]) - 500/n - F$$

OLIGONUCLEOTIDES

Calculation of T_m for double-stranded nucleic acids less than 100 bp is most accurately done using a computer program. Many computer programs are commercially available or are on the Internet. These include PRIMER 3 (http://frodo.wi.mit.edu/cgi-bin/primer3/primer3_www.cgi) and IDT (http://www.idtdna.com/Support/Technical/TechnicalBulletinPDF/Calculation_of_Tm_for_Oligonucleotide_Duplexes.pdf).

PROCEDURAL NOTE

Many of these websites use thermodynamic principles and take into account "nearest neighbor" effects of adjacent nucleotides.

For oligonucleotides less than 100 nucleotides in length, and if the cation concentration is less than 0.5 M and the G + C is 30% to 70%, the following equation can provide an estimate of the T_m:

$$T_m = 81.5°C + 16.6 (\log_{10} [Na^+]) + 0.41 (\%[G + C]) - 675/n$$

PROCEDURAL NOTE

This calculation only provides a rough estimate of the T_m.

THE "WALLACE" EQUATION

PROCEDURAL NOTE

The Wallace equation should, in general, *not* be used, as it is very inaccurate. It is provided here simply to warn you to avoid its use.

<u>Do not use this equation!</u>

$$T_m = 2°C \times [A + T] + 4°C \times [G + C]$$

where (A + T) is the sum of A and T residues in the oligonucleotide and (G + C) is the sum of G and C residues in the oligonucleotide.

3. Choice of Hybridization and Annealing Temperatures Based on T_m

After you have calculated the T_m for your probe/target combination, you are ready to select hybridization or annealing temperatures. Remember that the T_m is the temperature at which half the molecules should be hybridized. Thus, the T_m is not an appropriate temperature for hybridization. For nucleic acid hybridization experiments, it is necessary to use temperatures below the T_m.

For long DNA or RNA (more than 100 bases):

Stringent hybridization temperature: 15°C below T_m

Moderately stringent hybridization temperature: 25°C below T_m

Low stringency hybridization temperature: 35°C below T_m (or lower)

Oligonucleotide annealing temperature: 5°C to 10°C below T_m

4. Molar Ratio Calculations for Ligations

To calculate molar ratios for DNA ligations into cloning vectors, use the following formula. To calculate an insert: vector ratio of 1.0:

(ng of insert/length of the insert in base pairs) / (ng of vector/length of the vector in base pairs) = 1

PROCEDURAL NOTE

An insert to vector ratio of 1.0 is best for most ligations.

The following references should be very helpful to anyone interested in performing laboratory molecular biology. They are divided them into three categories: Laboratory Manuals, Textbooks on Molecular Biology, and Books on the Human Genome Project. Volumes indicated with an asterisk have proved to be especially useful.

Laboratory Manuals

Current Protocols in Molecular Biology, by F.M. Ausubel, R. Brent, R.E. Kingston, D.D. Moore, J.G. Seidman, J.A. Smith, and K. Struhl. New York: John Wiley and Sons, 2006. This is often referred to as the "The Red Book," a unique and huge laboratory manual (four big binders!) in loose-leaf form that is updated (for a price) four times each year. Protocols are gleaned from many molecular biology laboratories around the world. It is an excellent and extremely useful resource and is also available online and on CD-ROM.

Molecular Cloning: A Laboratory Manual, 3rd ed., by J. Sambrook and D.W. Russell. Woodbury, NY: Cold Spring Harbor Laboratory Press, 2001. This is the update of the classic 1982 manual, and it consists of three volumes. It is associated with a website (www.molecularcloning.com) and is evolving into an online laboratory manual. Using a purchase code found within the first volume, you can log in to the site and search protocols and register for e-mail updates to the protocols. Like "The Red Book," new protocols are added on a regular basis. This makes this set a strong competitor with "The Red Book."

Genome Analysis: A Laboratory Manual Series, edited by B. Birren, E.D. Green, S. Klapholz, R.M. Myers, H. Riethman, and J. Roskams. Woodbury, NY: Cold Spring Harbor Laboratory Press, 1997–1999. This is a huge set that includes the following four volumes: *Analyzing DNA*, *Detecting Genes*, *Cloning Systems*, and *Mapping Genomes*. This series is very complete and is highly recommended.

Short Protocols in Molecular Biology, 5th ed., by F.M. Ausubel. New York: John Wiley & Sons, 2002. This is a two-volume distillation of "The Red Book."

PCR Primer: A Laboratory Manual, 2nd ed., edited by C.W. Dieffenbach and G.S. Dveksler. Woodbury, NY: Cold Spring Harbor Laboratory Press, 2003. This is a very useful PCR manual with many protocols and troubleshooting guides.

RNAi: A Guide to Gene Silencing, by G.J. Hannon. Woodbury, NY: Cold Spring Harbor Laboratory Press, 2003. This is a good collection of RNAi protocols and strategies.

DNA Microarrays: A Molecular Cloning Manual, edited by D. Bowtell and J. Sambrook. Woodbury, NY: Cold Spring Harbor Laboratory Press, 2003. This is a collection of protocols, strategies, and information on bioinformatics and available software for microarray analysis.

A–Z of Quantitative PCR, edited by S.A. Bustin. La Jolla, CA: International University Line, 2004.

Real-Time PCR: An Essential Guide, by K. Edwards, J. Logan, and N. Saunders. New York: Garland Science/BIOS, 2004.

Rapid Cycle Real-Time PCR: Methods and Applications, by C. Wittwer, K. Meinhard, and K. Kaul. New York: Springer, 2004.

A Short Course in Bacterial Genetics: A Laboratory Manual and Handbook for Escherichia coli *and Related* Bacteria, by J.H. Miller. Woodbury, NY: Cold Spring Harbor Laboratory Press, 1992. This two-volume work is long, detailed, and complete (despite its name!). It is an invaluable reference for anyone struggling with *E. coli* genetics.

Methods for General and Molecular Bacteriology, edited by P. Gerhardt. Washington, DC: ASM Press, 1993.

The Practical Approach Series. Oxford University Press. This amazing series of soft-cover books has over 200 titles, including several that we have used regularly. Each is edited by experts in various areas of molecular biology. These include DNA Cloning, vol. 1, 2, 3 and 4; PCR 1, 2 and 3; Gene Probes 1 and 2; Essential Molecular Biology 1 and 2; Gene Targeting; In Situ Hybridization; Non-isotopic Methods in Molecular Biology; Plant Molecular Biology 1 and 2; Pulsed Field Gel Electrophoresis; RNA Processing; Transcription Factors; DNA Microarrays; Molecular Plant Biology; Functional Genomics; Eukaryotic DNA Replication; The Internet for Molecular Biologists; Bioinformatics: Sequence, Structure, and Databanks; Differential Display; Genome Mapping; Mouse Genetics and Transgenics; and Phage Display.

The Polymerase Chain Reaction, by K. Mullis, F. Ferre, and R. Gibbs. Boston: Birkhauser Press, 1994. Mullis received the Nobel prize for his invention of the PCR technique.

PCR Protocols, 2nd ed., by J.M.S. Bartlett and D. Stirling. Totowa, NJ: Humana Press, 2003.

Molecular Biology Labfax, Volumes 1 and 2: Recombinant DNA, 2nd ed., by T.A. Brown. San Diego: Academic Press, 1998. This is a compendium of facts that are useful to any molecular biologist.

Lab Ref: A Handbook of Recipes, Reagents, and Other Reference Tools for Use at the Bench, edited by J. Roskams. Woodbury, NY: Cold Spring Harbor Laboratory Press, 2002.

Calculations for Molecular Biology and Biotechnology: A Guide to Mathematics in the Laboratory, by F.H. Stephenson. San Diego: Academic Press, 2003.

Lab Math: A Handbook of Measurements, Calculations, and Other Quantitative Skills for Use at the Bench, by D.S. Adams. Woodbury, NY: Cold Spring Harbor Laboratory Press, 2003.

Pulsed Field Gel Electrophoresis: A Practical Guide, by B. Birren and E. Lai. San Diego: Academic Press, 1993. This is the best guide available that includes both theoretical background and protocols for PFGE.

Textbooks on Molecular Biology

Genes VIII, by B. Lewin. Cambridge, MA: Oxford University Press and Cell Press, 2004. This is the most popular book on molecular genetics at the first-year graduate student level. Lewin is the former editor of the journal *Cell* and has put together a readable, although encyclopedic, survey of the field.

Essential Genes, by B. Lewin. Upper Saddle River, NJ: Pearson Education, Pearson Prentice Hall, 2006. This is a concise version of *Genes VIII* that includes access to an e-book with Flash illustrations. The text is updated online and provides links to primary research articles.

Recombinant DNA, 2nd ed., by J.D. Watson, M. Gilman, J. Witkowski, and M. Zoller. New York: W.H. Freeman, 1992. This is an excellent, although somewhat dated, introduction to molecular cloning.

Principles of Gene Manipulation, 6th ed., by S.B. Primrose, R.M. Twyman, and R.W. Old. Oxford: Blackwell Publishing, 2001. This is another excellent introduction to recombinant DNA technology and is more up-to-date than Watson's book.

Molecular Biology of the Gene, 5th ed., by J.D. Watson T.A. Baker, S.P. Bell, A. Gann, M. Levine, and R. Losick. San Francisco: Benjamin Cummings, Pearson Education, 2004. This is an excellent update of the classic textbook, with especially good chapters on eukaryotic molecular biology.

A Primer of Genome Science, 2nd ed., by G. Gibson and S.V. Muse. Sunderland, MA: Sinauer Associates, 2005. This is an excellent introduction to genomics, transcriptomics, proteomics, integrative genomics, and bioinformatics.

Principles of Genome Analysis and Genomics, 3rd ed., by S.B. Primrose and R.M. Twyman. Oxford: Blackwell Publishing, 2002.

Bioinformatics: Sequence and Genome Analysis, 2nd ed., by D.W. Mount. Cold Spring Harbor, NY: Cold Spring Harbor Laboratory Press, 2004. This is a detailed volume covering all aspects of bioinformatics relating to nucleic acids and proteins. The volume comes with a user code that enables access to a website that contains additional useful information and links to useful molecular biology websites.

Bioinformatics for Dummies, by J.M. Claverie and C. Notredame. New York: Wiley Publishing, 2003.

Molecular Genetics of Bacteria, 2nd ed., by L. Snyder and W. Champness. Washington, DC: ASM Press, 2002. This is an excellent introduction to molecular genetics and molecular biology of bacteria (especially *E. coli*).

Plasmid Biology, edited by B.E. Funnell and G.J. Phillips. Washington, DC: ASM Press, 2004.
Mobile DNA II, by N.L. Craig, R. Craigie, M. Gellert, and A.M. Lambowitz. Washington, DC: ASM Press, 2002.

DNA Replication, 2nd ed., by Arthur Kornberg and Tania A. Baker. New York: W.H. Freeman and Co., 1992. This book has excellent chapters on plasmids, viruses, and DNA repair. It is especially good on prokaryotic replication and molecular biology.

The Bacterial Chromosome, edited by N.P. Higgins. Washington, DC: ASM Press, 2004.
A Genetic Switch: Phage Lambda Revisited, 3rd ed., by M. Ptashne. Woodbury, NY: Cold Spring Harbor Laboratory Press, 2004. This book has everything you wanted to know about phage lambda but were afraid to ask.

Genes and Signals, by M. Ptashne and A. Gann. Woodbury, NY: Cold Spring Harbor Laboratory Press, 2002. This is a review of gene regulation in both prokaryotes and eukaryotes with a link to register for an online version of the book with additional information and lectures.

Molecular Biology of the Cell, 4th ed., by Bruce Alberts, A. Johnson, J. Lewis, M. Raff, K. Roberts, and P. Walter. New York: Garland Publishing, 2002. This is an excellent undergraduate textbook that integrates cell biology, genetics, biochemistry, and molecular biology.

Molecular Cell Biology, 4th ed., by H. Lodish, A. Berk, L. Zipursky, P. Matsudaira, D. Baltimore, and J. Darnell. New York: W. H. Freeman and Co., 2000. This is still another excellent undergraduate textbook covering much the same material as the previous text.

The Eighth Day of Creation: Makers of the Revolution in Biology, by H.F. Judson. Woodbury, NY: Cold Spring Harbor Laboratory Press, 1996. This is not a textbook but a fascinating account of the people, events, and experiments comprising the early history of the molecular biology revolution.

The Code of Codes, by D. Kevles and L. Hood. Cambridge, MA: Harvard University Press, 1992. This is a discussion of the scientific and social issues surrounding advances in recombinant DNA, molecular cloning, and genomics. The book is a little out of date but still makes a good read.

Books on the Human Genome Project

Life Script, by N. Wade. New York: Simon and Schuster, 2001. This is an easy read by an excellent science writer.

Genome, by M. Ridley. New York: Perennial Press, 2000. This has 23 chapters, each on a different chromosome; it is a very interesting, readable book.

The Common Thread, by J. Sulston and J. Ferry. Washington, DC: The Joseph Henry Press, 2002. This is great book that includes the memoirs of a Nobel-prize-winning author (Sulston) and the genome "race."

Understanding the Genome, edited by G. Olshevsky. New York: Bryan Press, 2002. This is a readable series of short articles from the editors of *Scientific American*.

The Human Genome, by C. Dennis and R. Gallagher. Hound Mils, UK: Palgrave McMillan, 2001. This is a mini "coffee table" book with an introduction by Jim Watson and original articles at the end of the book; it has nice descriptions and great illustrations.

The Genomic Revolution, edited by M. Yudell and R. De Sale. Washington, DC: The Joseph Henry Press, 2002. This is a series of short articles by leaders in the field covering techniques, applications, and future prospects.

Transducing the Genome, by G. Zweiger. New York: McGraw-Hill, 2001. This is an easy read and is more directed to the bioinformatics revolution.

Genetically Yours, by H. Lim. Hackensack, NJ: World Scientific, 2002. This is a little cut and dry but is full of information on many topics, such as biopharming, bioinformatics, aquatic biotechnology, and stem cell research.

Your Genes, Your Choice, by C. Baker. Available as a pdf download from the National Academy of Sciences webpage, 1999, http://www.beyonddiscovery.org/content/view.article.asp?a=239.

Additional Resources

Each year *Nucleic Acids Research* publishes a special issue describing a wide variety of databases containing useful compilations of sequence and other information. This year's update includes 548 databases. Dr. Michael Galperin compiled the list. An abbreviated list of databases is shown here. The entire list is available as a searchable, up-to-date, online resource at http://nar.oupjournals.org/content/vol32/suppl_1.

General Molecular Biology Database

Database	URL	Description
Nucleotide Sequence		
GenBank	www.ncbi.nlm.nih.gov	All publicly available nucleotide and protein sequences
EMBL Nucleotide Sequence Database	www.ebi.ac.uk/embl.html	All publicly available nucleotide and protein sequences
DNA Data Bank of Japan (DDBJ)	www.ddbj.nig.ac.jp	All publicly available nucleotide and protein sequences
DNA Sequences: Genes, Motifs, and Regulatory Sites		
TIGR Gene Indexes	www.tigr.org/tdb/tgi.shtml	Organism-specific databases of EST and gene sequences
ExInt	intron.bic.nus.edu.sg/exint/exint.html	Exon-intron structure of eukaryotic genes

TRANSFAC	transfac.gbf.de/TRANSFAC/index.html	Transcription factors and binding sites
RDP	rdp.cme.msu.edu	Ribosomal database project: rRNA sequence data
Gene Expression		
PIR	pir.georgetown.edu	A collection of protein sequence databases
SWISS-PROT	www.expasy.ch/sprot	Curated protein sequence databases
PROSITE	www.expasy.ch/prosite	Biologically significant protein patterns and profiles
Pfam	www.sanger.ac.uk/Software/Pfam	Sequence alignments and profile hidden Markov models
Carbohydrate		
CCSD	bssv01.lancs.ac.uk/gig/pages/gag/carbbank.htm	Complex carbohydrate structure databases (CarbBank)
Protein Structure		
PDB	www.rcsb.org/pdb	All available 3-D structures of proteins and nucleic acids
Genomics		
GO	www.geneontology.org	Gene ontology consortium database
KEGG	www.genome.ad.jp/kegg	Databases of genes, proteins, and metabolic pathways
EcoCyc	ecocyc.org	*E. coli* K-12 genes, metabolic pathways, transporters, regulation
Ensembl	www.ensembl.org	Annotated information on eukaryotic genomes

Source: Copyright © New England Biolabs. Reprinted with permission.

Websites for Molecular Biology Protocols

Here is a list of some of the many websites that feature molecular biology protocols. Some are interactive (you can ask questions and receive answers from other scientists).

Additional sources of suppliers of specific reagents are at via www.biocompare.com

Protocols for recombinant DNA isolation, cloning, and DNA sequencing are at www.genome.ou.edu/protocol_book/protocol_index.html

Cell and molecular biology online protocols are via www.cellbio.com/protocols.html

Protocol Online — your lab's reference book — is at www.protocol-online.org

Molecular biology protocols are at www.protocol-online.org/prot/Molecular_Biology

NWFSC molecular biology protocols are at micro.nwfsc.noaa.gov/protocols

Molecular biology protocols are at www.dartmouth.edu/~ambros/protocols/molbio.html

Laboratory protocols are at wheat.pw.usda.gov/homepage/lazo/methods

Molecular biology protocols are at www.highveld.com/protocols.html

Molecular biology protocols are at www.biotaq.com/Protocols/Protocols.htm

The Original CPC (Comprehensive Protocol Collection) is at www.dartmouth.edu/~ambros/protocols.html

Molecular method and protocol links are at www.uwm.edu/People/zeng/method.htm

123 Genomics is at www.123genomics.com/files/protocols.html

Protocols zbio.net is at molbiol.ru/eng/protocol

Biowww.net is at www.biowww.net